GEOLOGICAL SURVEY.

# SECOND REPORT

OF THE

# GEOLOGICAL SURVEY

IN

# KENTUCKY,

MADE DURING THE YEARS 1856 AND 1857,

BY


DAVID DALE OWEN,

PRINCIPAL GEOLOGIST,

ASSISTED BY

ROBERT PETER, Chemical Assistant;

SIDNEY S. LYON, Topographical Assistant.



FRANKFORT, KENTUCKY.

A. G. HODGES, PUBLIC PRINTER.

1857.

# INTRODUCTORY LETTER.

HIS EXCELLENCY, C. S. MOREHEAD,

GOVERNOR OF KENTUCKY:

SIR:—Since the manuscript of the First Report of the Geological Survey of Kentucky was submitted the Geological Survey has been continued, both in the field and office, with unremitting diligence.

The two Topographical Corps which the law required to be organized have been operating—the one in the east, chiefly in Greenup county; the other in the west, chiefly in Hopkins county. The Western Corps was detailed on the Base Line in August, on which work it is still engaged.

In consequence of the indisposition of the gentleman who was to take the direction of the Eastern Corps, in the early part of last season, in the absence of the Topographical Assistant, when engaged in other parts of the State, the work in Greenup was somewhat retarded, but nevertheless it is hoped that during the present season the Topographical work of that county will not only be completed, but the detailed survey extended across part of Carter into Lawrence.

The Topographical Survey of Hopkins county was nearly completed last August, when the Western Corps had to be detailed on the Base Line. As that line traverses the State nearly centrally, and is intended to afford stations of departure, return, and interconnection, and is the main line to which all the Topographical work is to be referred, it has not only been executed with all the accuracy which our means would afford, but it has been run open—i. e. cut out—so as to give it permanency of location easily recognizable.

My own Corps has been chiefly occupied in continuing the reconnoissance through the central counties of the State, defining the general boundaries of the formations, and unravelling their stratigraphical geology, making collections for the Chemical Department and the State cabinet of ores, coals, rocks, fossils, clays, soils, and sub-soils, and testing the mineral waters, qualitatively, at their fountain head.

The Chemical Assistant has already completed and reported on, since his last report, two hundred and six analyses, as follows:

48 iron ores of the limonite variety.
22 iron ores of the carbonate variety.
43 soils, sub-soils, and marls.
31 limestones.
30 coals.
16 mineral waters and salts.
4 copper and zinc ores, and bitumens.
4 sandstones.
2 pig-iron.
2 shales and slates.

206

For the convenience of the printer, who is desirous of commencing the printing of the report early, in order to be able to have it completed, as the law provides, by the time the Legislature is next in session, all that part of the report which is now completed, both General, Chemical, and Topographical, has been arranged in the First Part of this report, while the reports to be subsequently completed will form the Second Part.

# PART FIRST

OF THE

# SECOND GEOLOGICAL REPORT

OF

# KENTUCKY.

# GENERAL REPORT.

---

## CHAPTER I.

## AGRICULTURAL GEOLOGY.

### GENERAL REMARKS ON SOILS.

The citizens of Kentucky are pre-eminently an agricultural people. In the Eastern and Northern States the wealth, influence, and intelligence of the population are, in a great measure, concentrated in cities, towns, and villages. Not so in Kentucky; the substantial patriarchal farmer forms by far the larger and most influential part of the Commonwealth. All, therefore, that relates to the cultivation of the soil is of very general interest. In prosecuting the geological survey of the State special attention has, therefore, been devoted to the Agricultural Department of the survey, especially during the two past seasons. Already upwards of one hundred and eighty soils have been collected, upon a systematic plan, with reference not only to the subjacent geological formation, but also to various local modifications of soils, derived from the same leading formation, where a difference or peculiarity in the lithological character of any underlying member indicated that there might be a marked character of the soils also. A large proportion of these soils have also been collected in sets of three or more from the same locality, with the following objects in view: Placing implicit reliance on the capabilities of chemical science to indicate, by the analysis of soils, the ingredients removed by the cultivation and harvesting of successive crops, it was hoped that by collecting samples of the virgin soil, and of the same soil from an adjacent old field, that not only the different substances asssimulated out of the soil could be ascertained, but also the exact proportion of these, so that the farmer might know precisely what must be restored to the land to bring it back to its original fertility. Hence, wherever a favorable opportunity offered, the virgin soil has been collected in connection with a soil from an adjoining field where the same timber formerly grew that now prevails on the site of the virgin soil selected. At the same time the sub-soil and under-clays were generally secured in order, if possible, to discover whether any or all of the materials required to enrich the land could be obtained from them, because, if so, that would present the most accessible source, and by far the cheapest means of restoring the exhausted elements to the soil.

A science which has extracted from the fixed alkalies metals lighter than water, that burst into flame the instant they come in contact with that fluid; which has reduced from clay a metal bright as silver, yet light as marble, that resists corrosion, that forms with copper an alloy having the color and brilliancy of gold; which distills from bones a body of the consistency of wax, so combustible that summer heat almost suffices to inflame it; that prepares from kelp a body whose vapors of the richest violet hues will render a silvered plate so sensitive to light that a few seconds suffice for impinging rays to paint their image on its surface; which compounds principles so subtile that a grain or two will impregnate the whole atmosphere of an apartment with the most deadly fumes, while the compound itself hardly loses any perceptible weight; which unites together the *same substances* so as to form, at one time, the most active poison, at an another, by varying only slightly their proportions, a substance altogether inert. A science, I say, which has accomplished wonders like these, is surely capable of disclosing the mysteries of the chemistry of agriculture. Indeed, it has already mastered and explained most satisfactorily many of the phenomena; and, although there remains much still to be accomplished, it has already laid down some of the most important principles for the guidance of the farmer, which, if rightly understood and carried out in practice, cannot fail to be of the most essential service.

Let us here briefly review some of the best established facts in agricultural chemistry, since, in the deductions which I am about to make from the investigations on Kentucky soils, I shall have frequent occasion to refer to them; and, as many of my readers may have had no opportunity of making themselves acquainted with the recent disclosures of this branch of science, it will be almost indispensable to precede this part of my report by the following remarks.

All plants obtain their nourishment partly from the atmosphere and partly from the soil.

The elements which form by far the greater part of the bulk and weight of plants are derived, either directly or indirectly, from the atmosphere, or from the carbonic acid condensed in atmospheric and spring water.

The mineral constituents which plants obtain from the soil, though always small in proportion to those assimulated from the atmosphere and water, being often only fractional parts of the whole substance of the plant, are, nevertheless, absolutely essential to their growth and maturity; and however important it may be that the farmer should aid nature in conveying to the roots of plants the atmospheric elements, it is yet more indispensable for him to see that his soil is not deficient in any one of the mineral constituents demanded, because plants can assimulate these only through their roots ramifying in the soil, and have no other source whence they can be obtained, whereas, if time sufficient be allowed, vegetation will itself, without assistance, appropriate the others, which always floating in the air are wafted by every breeze to the spot where they may be required, precipitated by each rain or fall of snow to the ground, or attracted to the earth by reason of certain affinities existing between them and a few substances usually existing in soils.

The most important substances derived from the atmosphere, which serve as nourishment for plants, are carbonic acid, water, and ammonia. Out of the carbonic acid, which is absorbed in part by the leaves and in part by the roots, previously condensed or dissolved in the water percolating the soil, the carbon is assimulated which forms more than one half the bulk and weight of the principal vegetable tissues and proximate principles, while at the same time the oxygen therewith combined is mostly restored to the atmosphere and there sustains the respiration of animals.

Atmospheric water not only furnishes oxygen and hydrogen required by plants, but is the principal solvent, especially when charged, as it always is, more or less with carbonic acid, of the various mineral constituents of plants, which must be carried into their organs in a state of solution.

Ammonia may be generated in soils from the decay of animal and vegetable substances, if they contain nitrogen; but this volatile alkali is chiefly condensed out of the atmosphere by rain and snow, since that derived from the decay of organic substances has been mostly received from the same source. However, the sulphate of ammonia and sal-ammoniac may be regarded as mineral products. Ammonia, both in its free state and combined with carbonic acid, is continually being evolved from all decaying animal and some vegetable substances, but being in both these states very volatile it rises and escapes into the air, unless it meets in its nascent state with certain substances which have the power of fixing it, either by chemical affinity or absorbing it into their pores, such as humus, vegetable mould, sulphuric acid, charcoal, or muck. These will arrest the ammonia generated, which is otherwise dissipated into the atmosphere, to be again precipitated to the earth, as we have already said, by every shower of rain.

This volatile alkali is one of the sources of the nitrogen of plants, which enters as an assential, though not most abundant element, into certain proximate principles of plants, viz: vegetable albumen, casein, fibrin, and glutin, and of all those vegetable products which are alone capable of producing blood and muscle. The proximate principles, woody fibre, sugar, starch, gum, oils, and resins constitute by far the most bulky parts of plants. For the production of these carbonic acid and water suffice, since they contain no nitrogen.

Although four-fifths of our atmosphere is composed of nitrogen, most vegetable physiologists contend that vegetation has little or no power to appropriate this nitrogen directly from the air. During thunder-storms the electric discharges may produce nitric acid by the union of one equivalent of the nitrogen of the air with five equivalents of its oxygen, but a renewal of these discharges may again decompose it. Some writers on agricultural chemistry believe that nitric acid produced in this way contributes a considerable amount of the nitrogen assimulated by plants; there are others, however, who, although admitting nitric acid, to a certain extent, as a source of nitrogen in agricultural processes, yet contend that that nitric acid is chiefly derived from the oxidation of the nitrogen in ammonia, generated by the decay of plants in the presence of fixed alkalies or alkaline earths. Still it appears more than probable that primeval vegetation may have derived its nitrogen directly by absorption from the air, more especially since M. Ville's experiments go to prove that plants do absorb notable quantities of nitrogen—far more than can be accounted for either by the floating ammonia or nitric acid; besides, the principal present source of atmospheric ammonia, from the decomposition of vegetable and animal matter, could not have supplied the volatile alkali before the existence of vegetables, whence all animal life derives its nourishment either directly or indirectly. This much, however, is certain, the great mass of vegetation receives its supply of nitrogen independent of manure. This is true even of domestic plants. The proof is found in the fact that various crops may be grown on soils, while the nitrogen of that soil may even be on the increase. Again: we know that though millions of hides of buffalo and cattle feeding on the prairies and pampas, secured for the use of man, remove many tons of nitrogen, the productiveness of these native pastures is not impaired, as far as the nitrogen and organic elements of vegetation are concerned. The cereals would undoubtedly disappear from Northern Europe if not fostered by cultivation; but it has been satisfactorily shown that this is not owing to any deficiency of nitrogen or the organic elements of plants, but from the exhaustion of the mineral constituents of the soil.

The mineral food of plants, derived mostly or altogether from the soil, are phosporic acid, sulphuric acid, potash, soda, lime, magnesia, oxide of iron, chlorine,* silica, and sometimes a little oxide of manganese. These do and must exist in all fertile soils, though in most cases, (with the exception of silica and oxide of iron,) in proportions less than one per cent.

There are certain chemical facts in relation to these two classes of vegetable food which deserve to be deeply impressed on the mind of the farmer, the misapprehension of which has greatly retarded the progress of agriculture: plants cannot assimulate the one without the other, for the weight of the crop will always be in proportion to the *atmospheric* and *mineral* food combined, and not to any one of them separately, since no plant can come to perfection unless it contains the due proportion of each, as indicated by the chemical analysis of both the *volatile* and *fixed* constituents.

If the ammonia, nitrates, or nitrogenized principle is supplied in sufficient quantity the plant will take up from the soil a full supply of its appropriate mineral food, provided, always, the soil contains these in a soluble condition; but if the soil do not contain the mineral ingredients, or if these be locked up in the soil in an insoluble condition,

*Chlorine and sulphur occur locally in the atmosphere.

the most profuse supply of ammonia will be of no avail in forcing the crop, unless, in the form in which it is applied, it first acts in rendering the mineral constituents soluble, as ammoniacal salts do to some extent. On the other hand, if there is an abundant supply of the mineral consituents in the available—i. e., soluble—condition, it is not indispensable to supply the ammonia or nitrogenous principle, because the plant can appropriate it from the atmosphere, especially in climates where the summer is long and hot, and the winter short. In warm climates the natural sources of nitrogen generally suffice, and in moderate climates an average crop can be obtained, if the soil have a full supply of mineral salts, without any addition of nitrogen but that which plants themselves can appropriate from the ammonia, nitric acid (or, perhaps, directly from the atmosphere, if M. Ville's experiments are to be relied,) which are continually conveyed to them through the medium of the surrounding air, especially if the soil be well aired by judicious tillage.

The artificial application of ammonia, in the form of farm-yard, guano, or other manures, is advantageous, particularly in cold climates, in *hastening* the assimulation of the mineral ingredients which proceed simultaneously with that of the nitrogen; and it is necessary, under the above conditions, in order to obtain the *maximum* production, since the more abundant the supply of nitrogen the more rapid will be the extraction of the soluble mineral ingredients from the soil; hence, in the usual acceptation of the term, there is nothing more exhausting to a fertile soil than the free use of nitrogenized manure; but if the farmer's object be to make money rapidly of course the heavier the crop, in the shortest time, the better; but he need not expect to do this without drawing largely on the resources of his land. Therefore, if he desires to keep his land in "*good heart*," he must return to the soil every year, or every few years, a portion at least of the mineral ingredients carried off by each successive crop Herein lies the art of successful and thrifty farming in all fertile districts.

The inference to be drawn from the preceding is, that ammoniacal salts *alone* will not keep up permanent fertility. The cheapest source of nitrogenized manures seems to be *good** guano, at least in all districts convenient to the seaboard.

*There is a great difference in quality and value of different varieties of guano, and a great deal of it is adulterated.

Plants can grow in a soil altogether destitute of organic and nitrogenized principles, as is evinced in volcanic districts, where the wild fig-tree springs up from the interstices of the cellular lava, while it is still hardly cooled, and from which, of course, every particle of organic matter must have been expelled, if it ever existed, by the molting heat to which it had been exposed. Yet the fig-tree must have organic food, to be assimulated and form the principal substance of the tree. Where else is it to be derived, in this case, but from the atmosphere?

Pines, too, grow on soils almost destitute of organic matter.

An acre of ground will yield 98,000 pounds of the fruit of the banana annually, year after year, which contain 17,000 pounds of carbon, and yet, at the end of half a century or more, the ground will be found richer in organic matter than at the commencement, from the decay only of the fallen leaves.

The cattle raised in France, according to good authority, consume in their food 76,789,000 pounds of organic matter, which is six times more than they restore to the soil; and on farms, generally, there is at least three times as much organic matter carried off as is ever restored. It is estimated that the annual destruction, or rather transmutation more properly speaking, of organic matter on the earth, is 140,000,000,000 of pounds, or upwards of 2,200,000,000 cubic feet. If vegetation, then, depends for its supply of organic matter on that which is in or on the soil, there would require to have been, 5,000 years back, in order to supply this consumption, an accumulation of organic matter on the ground of ten feet thickness; but three and a half grains of carbonic acid condensed in every pound of water would, according to Schubler, afford the amount of carbon required for the exigencies of plants, while one-thirteenth of a grain of ammonia in every pound of water would furnish all the nitrogen required, less than the amount found in any spring water. It is evident, from the above, that vegetable life must derive its principal supply of organic food from other sources than the soil.

It is well known that in poor sandy soils the crops raised are short and meagre, but it is not generally known what is the proper means of reclaiming such land. This was most satisfactorily shown by seven years experiment, undertaken by Liebig on ten acres of barren sandy soil, in the vicinity of Giessen, and conducted upon purely chemical principles, for the purpose of either confirming, modifying, or refuting his inferences derived from the chemical analysis of soils and the ashes of plants. These ten acres, at the time of the purchase, did not, as we are informed by the experimenter, support a sufficient growth of grass and herbs to sustain a single sheep. After the lapse of four years this poor sandy land was brought into such a condition as to supply the wants of a family who kept two cows, raised annually several oxen, besides the means necessary to erect buildings on what is now a small farm.

It is important here to remark, that this was all effected without the aid of ammonia, humus, vegetable mould, or nitrogenized manures, since there was little or none, either in the soil when first put into cultivation, or included in the mineral manures applied; which where composed of phosphates, fixed alkalies, soluble silica, and sulphates. With these alone he was enabled to raise trees and perrennial crops that acquire their nitrogen only by degrees, and therefore take up the mineral constituents proportionally slowly. He did not, however, succeed so well with the cereals, such as wheat, rye and oats, until he added, along with the mineral manure, saw-dust, which by decomposition furnished carbonic acid, which acting as a solvent over these mineral applications, brought them into solution faster, and consequently the assimulation of ammonia or nitrogen was proportionally hastened.

For the reason previously stated more abundant crops could have been obtained by the application of ammoniacal manures, but these were entirely dispensed with in this experiment, as its principal object was to test the correctness of the chemical reasoning which maintained the *indispensability* of the mineral constituents, and to prove that the barrenness of this poor soil was due to the absence of these in its composition.

Of course this experiment was attended with considerable expense, the outlay being $670, exceeding greatly the value of the crops harvested, but this only proved that the owner of barren land labors under great disadvantages, since the fertilizing of soil, almost destitute of the mineral food of plants, involves an expenditure which exceeds the price of the most fertile soil, consequently it is seldom, if ever, that land so completely barren as this is brought into cultivation; nevertheless it demonstrated most satisfactorily the important fact to the agriculturalist, that to be a successful farmer he must know and understand the mineral composition of his soil, and that all lands, however fertile, must, after a time, lose their fertility if the mineral constituents of plants, removed in each harvest, are not ultimately restored or developed by renewed disintegration of the soil. It is the duty of the farmer, therefore, if his land is deficient in any of the mineral constituents, to illiminate or add these, and to endeavor to keep up the supply of those that are in sufficient abundance in the original soil.

All of our Kentucky farmers are aware that land is improved by turning in green crops, but few understand the principle on which it depends; and being ignorant of this they are incapable of judging, unless it be by very long experience, which kind of green crops are the most applicable. It may be useful, therefore, in this place, to treat briefly on this subject.

Green crops, especially those which present an extensive surface of leaves, are able to appropriate, if there be a sufficient supply of its mineral constituents in the soil, a very large proportion of their weight and bulk out of the elements in the atmosphere. When turned under by the plow they are in the best condition to enter rapidly into a state of decomposition, and thus enrich the soil in the same manner as well preserved and well rotted manure; while, at the same time, the soil is not deprived of any of the *mineral constituents,* since these are restored. Besides these benefits those green crops which have deep-seated, searching roots, absorb and pump up from the substratum soluble mineral ingredients beyond the reach of the roots of any other plants, which are thus made available for the succeeding crop.

It follows, from the preceding, that those green crops will be the most servicable which have the most extensive leaf-surface and largest roots, provided, always, the soil is adapted to their growth—which means, contains the mineral food which they require. For instance: 1,000 parts of dried red clover will require 75 parts of mineral constituents, in the following proportions:

| | |
|---|---|
| Lime, - - - - - - - - - - | 37.1 |
| Potash, - - - - - - - - - | 26.7 |
| Soda, - - - - - - - - - - | 7.1 |
| Phosphoric acid, - - - - - - - | 8.8 |
| Sulphuric acid, - - - - - - - | 6.0 |
| Chlorine, - - - - - - - - - | 4.8 |

| | |
|---|---|
| Silica, - - - - - - - - - | 4.8 |
| Magnesia, - - - - - - - - - | 4.6 |
| Peroxide of iron, - - - - - - - - | .2 |

If the soil is deficient in any one of these chemical substances, it will not be suitable for clover, especially if lacking lime and potash, which are in large proportions. In such cases other green crops should be selected, better adapted to the composition of the soil.

Green croping is the least laborious way of manuring land, and the cheapest method when labor is high, but is, probably, not the most economical where labor is cheap, or near large cities and seaports, where guano and other manures can be obtained at a reasonable rate.

During last summer a soil was collected in Bullitt county, from an old field which has been fifty to sixty years in cultivation, and which will now no longer produce clover. I will venture to predict that when the analysis of this soil shall be completed it will be found to be deficient in some of these constituents, and the analysis will probably show what other green crop might succeed better for the renovation of such land.

If a soil contains a sufficient supply of the mineral constituents of plants, in a soluble condition, no benefit will be derived from the application of mineral manures to such land. This is one reason why many farmers have been disappointed in benefits they expected from their use.

Different crops require particular mineral ingredients, and in different proportions from others; therefore, when manures are to be added not only the chemical composition of the soil should be taken into account, but the kind of crop that is to be raised on the land. Potatoes appropriate from the soil potash and sulphuric acid in larger proportion than the other mineral constituents, hence soils rich in these mineral constituents are best for this crop, and ashes and gypsum are good applications, after repeated crops, of this tuber.

The same plant always contains the same mineral constituents. The only exception to this rule is, that in certain instances the alkalies replace lime, or the reverse. The first of these axioms in agriculture is strikingly exemplified in tobacco. Potash is so invariable and *essential* a constituent of this plant, that the government of France, who has a monopoly of the tobacco trade, caused an extensive series of analyses of this plant to be made, by which it was ascertained not only that potash is a constant ingredient in its ashes, but existed in *variable quantities*, corresponding to the quality of the tobacco. These analyses finally demonstrated that *the value of tobacco stands in a certain relation to the quantity of potash contained in its ashes*, and that this furnishes the best means, not only of estimating the value of the tobacco, but of distinguishing the different soils on which the tobacco has been cultivated. And it was further shown, that just in proportion as certain kinds of celebrated American tobacco deteriorated in *quality* the *quantity* of ashes diminished. This was, no doubt, due to the soil on which it grew becoming gradually exhausted of the potash and lime, which are the principal constituents of the ashes of tobacco; and these, fortunately, can be entirely replaced by mineral manure, if we include in the term sulphate of ammonia and sal ammoniac.

The waste of manure is, therefore, a serious loss to mankind. In the preservation of these the Chinese nation are much more particular than we are, the knowledge of the value of manures having, doubtless, been brought more immediately home to them, by reason of the density of the population, which necessarily gives to all agricultural produce a high price.

One load of *well preserved* manure is worth ten loads of manure that has long laid exposed to the weather. To prevent the liquid oozings from running to waste, and the ammoniacal vapors from volatilizing, the manure-pile should not only be placed under shelter, but disposed in a pen with a slightly inclined floor, so that the liquid part can be run into reservoirs placed to receive it, from which it can be either carried separately on to the land, or pumped up from time to time over the manure-pile which has been disposed, with proper absorbents to retain it, which may be such as will also fix the ammonia, such as alternate layers of muck, decayed wood, vegetable mould, clay, charcoal powder, or the proper quantity of diluted sulphuric acid, viz: just sufficient to prevent all odor from arising from the pile, without arresting the proper decomposition of the mass.

When ammonia or nitrogenous matter is exposed to the action of moist air or oxygen, in the presence of alkalies or alkaline earths, the nitrogen becomes oxidized with the production of nitric acid, which combines with the base to form a nitrate, and the process is known under the name of nitrification. In this way the nitrogen of fresh urine can be fixed, but if the lime be added to manure in a state of fermentation, where ammonia is being generated, it will have the effect to expel the ammonia with loss of its nitrogen, at least for this particular locality. A natural process of nitrificaction takes place in the open fields where the bases are present ready to combine with the nitric acid.

By a recent analysis of rain water, collected at the Paris Observatory, there was found rather more nitric acid than ammonia to be present. The nitric acid was most abundant after storms; and in the months when it was found most abundant there was least ammonia.

Dr. Gibert's investigations on the relative amount of nitric acid and ammonia in rain water give $\frac{1}{1,000,000}$ of nitrogen, in the form of ammonia, and $\frac{5}{1,000,000}$ in that of nitric acid. Ammonia he found most abundant in mists and dews. He found the quantity of nitric acid variable, at different seasons of the year, but most abundant after storms. He is disposed to think that nitric acid is probably equally efficacious with ammonia in supplying nitrogen to plants; and being more abundant, according to his observations, than ammonia, must furnish a larger supply to vegetation than ammonia, which had generally been supposed to be the principal source of that element. Gypsum, in a certain condition of humidity of the air, is an absorbant of carbonate of ammonia, by reason of the mutual chemical affinities exerted between the sulphuric acid and ammonia and the carbonic acid and lime. Burnt clay also attracts ammonia.

The insoluble mineral constituents gradually become soluble in time by the action of air and water on the soil. It is upon this principle that fallow, or what is improperly called rest, increases the fertilily of land.

The solubility of some of these mineral ingredients, as for instance, silicate of potash, is hastened by the action of lime on the soil. The presence of carbonic acid, and probably also ammoniacal salts, increases the solubility of the mineral constituents. Much of the benefit experienced by the decay of vegetable and animal matter is due to the formation of such compounds, and the subsequent solvent action they exert on the mineral constituents.

Frequent tillage, by exposing new particles of soil to the air, and rendering it finer, facilitates the development of the soluble mineral salts in the soil; but it has, at the same time, one deleterious effect, in causing the too rapid decay of the organic constituents of the soil, at a time when the crop cannot be benefited by it. It is upon this principle that Schleiden, in Germany, has recommended the less frequent use of the plow, and looks upon plowing as a "necessary evil—one to be employed only so far as necessity requires—because in the too frequent loosening of the soil the decomposition of humus is so rapid as to overbalance the benefit from exposure to the air."

It has been shown by S. Smith that in countries where labor is cheap the fertility of exhausted land can, to a great extent, be restored without any kind of manure, by a simultaneous system, on one and the same pieces of ground, of fallowing strips of three feet in width, and drilling in the grain on alternate strips of three feet, in rows one foot apart, sowing single grains every three inches in the drill, in connection with a system of spade husbandry, carried on while the crop is growing by which the alternate three feet of fallow ground is turned over, so as to expose, each season, from four to six inches of the subsoil. The planting of wheat took place the beginning of autumn, and the spade trenching as soon as the plants were fairly up. In spring the wheat rows were well hoed and hand-weeded, and the intervals stirred with a one-horse scarifier. In this way, on a field which had been severely cultivated for a century, he raised, from a little over one peck of seed, thirty-six to forty bushels of wheat to the acre, or rather half acre, (since only one half the field was in reality planted,) and the grain off each acre sold for a clear profit of $38 over and above the cost of the labor of production. He employed six men at twenty-four cents a day, who dug an acre in five days.

This system of renovating exhausted land is based altogether upon the system of bringing fresh particles of soil to the action of the air, by which the insoluble mineral constituents become soluble by degrees in sufficient quantities to meet the wants of vegetation, which proves that judicious tillage is equivalent, for a time, to manuring the ground.

M. Baudrimont has shown, that there are what he calls "insterstitial currents" in arable soil, which bring up, by a kind of capillary action, the soluble, saline constituents from the sub-soil to the surface. The advantages of fallow are in part due to this natural circulation in the soil.

Ground rich in organic matter or humus, in consequence of its absorbent power of moisture, stands drought much better than soil poor in these ingredients. It is also much warmer, for the darker the earth the better adapted it is to receive the suns rays. These are two of the principal benefits which agriculture derives for either natural or artificial organic manures, independent of the atmospheric fertilizers, which are absorbed into the earth simultaneously with the moisture.

The practice of a rotation of crops is founded upon the fact that different plants assimilate different proportions of the mineral constituents; for example, wheat requires a much larger proportion of silica, phosphoric acid, potash, and magnesia than oats; barley demands more silica and less potash and phosphoric than corn; and potatoes assimilate more potash and less lime than turnips; clover and grasses generally, as well as peas and vetches, require much lime.

Now since it should be an axiom in agriculture to endeavor to retain the due balance or proper proportion of the mineral constituents, the same crop should not be put successively on the same ground, nor should two crops follow one another, both of which require large proportions of the same ingredients; for instance, wheat should not be followed by rye, because they both assimilate large quantities of soluble silica, potash, magnesia, and phosphoric acid; but should be followed by clover, which requires much lime, and less of the other ingredients; then potatoes might follow, which require comparatively little lime and phosphoric acid; this may be followed by corn, which appropriates more phosphoric acid and silica and less potash. Thus not only is time given for those ingredients, removed in largest quantities, to be renewed by further disintegration of the soil, and the action of solvents, but the too rapid exhaustion of one or two of the constituents of the soil avoided.

Again: two plants do not succeed well together, or after each other, that have roots of equal or nearly the same length, because they draw their sustenance too much from the same stratum of earth.

If a cheap source is at hand, of such mineral manure as contains the ingredients removed by any given plant, then a rotation of crops becomes unnecessary, and the farmer may then raise, on the same land, in succession, whatever he finds most profitable. The establishment of the fact that rotation of mineral manures may be adopted as a substitute for a rotation of crops, is one of the great boons which agriculture has derived from chemistry.

In order to obtain a maximum produce there must be, as we have already remarked, a correspondence in the proportion of the available mineral and atmospheric constituents; an excess of either beyond the due proportion will not increase the harvest. It was for want of a proper understanding of this law of agricultural chemistry, together with that previously stated, that the addition of mineral manures alone to land, already containing enough for the necessities of the plant, would not increase the crop, that led some of the leading agriculturists of England to the most erroneous inferences, from experiments which they instituted on the application of mineral manures, as to their importance and value, and which caused them to reject and discountenance what they denominated *"Leibig's Mineral Theory."*

Leibig has, however, in an admirable little work published in 1855, under the title of "Principles of Agricultural Chemistry," conclusively demonstrated, not only the correctness of his views, but has shown most satisfactorily that the very experiments deduced to disprove his doctrine tend, on the contrary, to confirm and strengthen the chemical reasoning which led him to teach that no soil could be a fertile one that was deficient in the inorganic or mineral constituents found in the ashes of plants; that it was of no avail to add nitrogenized manures alone to land, where these were absent or very sparingly distributed, and that it was of more importance to restore to exhausted barren lands these mineral ingredients, found in the ashes of plants, than ammonical or other nitrogenized and purely organic manures; because, as has been already explained, these can, in due time, be appropriated from the atmosphere, spring, and rain water.

In promulgating this doctrine it was never intended to depreciate the importance of the nitrogenized organic manures, but to explain under what circumstances they might be applied with advantage; also, to correct an error which led many to estimate the value of a manure by the nitrogen, or at least only from the organic constituents which it might contain.

The chemical investigations of the soils of Kentucky bear ample testimony of the justness of Liebig's doctrines, as will appear in the sequel.

There is invariable enough of silica in the soil for the use of plants, but it is not always in the soluble form required for assimilation. When plants lodge, that is, when the stalks of plants are so weak and feeble as not to be able to support the weight of the head, it is an indication that there is not sufficient *soluble* silica for the use of the plant. In such case, it is not an application of silica that is indicated, but of potash or lime which will render the silica *soluble*. Fallow, however, will often effect the same end. When, on the contrary, the head does not fill out properly, it proves either that there is a deficiency of phosphoric acid in the soil, or that it is in an insoluble condition. When the leaves look fresh and green it is an indication that they have a full supply of ammonia; when they appear reddish it is from deficiency of this alkali.

Lime applied to soil not only hastens its disintegration, but it acts energetically on the organic matter of soil, rendering it soluble and available. In this way, it is true, it exhausts a soil, but only in a *necessary* way, if high farming is the object. Lime also corrects acidity of soils. For both reasons lime may be applied to much greater advantage, and in larger quantity to land rich in organic constituents, than to a soil nearly destitute of it.

The purest limestones are by no means the best as fertilizers of land, because accidental admixtures of clay, phosphoric and sulphuric acid, and alkalies are valuable additions to the soil.

If lime is deficient in a soil the application of gypsum sometimes produces a sour reaction by the liberation of sulphuric when there is no base ready to neutralize it. In such case caustic lime should be added along with the gypsum, which will effectually counteract this tendency. The more soluble the constituents of a manure the more valuable it is, since it acts promptly and its benefits are obtained without waste of time and consequently capital. In such a condition it is of course stronger and more concentrated, hence due caution is required in the quantity applied.

In urine the salts are in a more soluble condition and its nitrogenous principles enter more quickly into a state of putrescence than in solid dung, on this account the former is more forcing, but the latter contains more of the phosphates. For this reason urine and the soluble salts generally are leaf-producing, as these salts predominate in this part of the plant, while the dung and the more insoluble salts are grain-producing. Both must be united to produce a truly efficient manure.

Ashes are not always the best form to supply potash to land near large cities, as salts rich in potash can often be obtained. Ten to fifteen cents worth of which would be equal to thirty bushels of ashes. Nitrates of alkalies are good substitutes for ashes.

On stock-farms, where young animals are reared and sold, when they come to maturity; also on dairy-farms, where milk, butter, and cheese are prepared for market, there is necessarily a great drain upon the soil, principally in phosphate of lime or bone-earth. It has been estimated that there is removed one thousand pounds of this bone-earth in the shape of veal, butter, and cheese derived from twenty cows, pastured during one summer on a farm. Chiefly by this means, and by the successive crops of wheat exported from the Genesee and Mohawk vallies, the average crop of wheat has been reduced from thirty-five to forty bushels to twelve and a half bushels per acre, and this reduction has been proved to be, in most cases, due to the removal of phosphoric acid from the soil.

Sandy soils are greatly improved by the application of absorbents, such as muck, humus, vegetable mould, charcoal, coal-dust, and clay, because pure sand has comparatively a small absorbent power, and allows what ammonia and nitrates may have been brought down by rains to evaporate again into the atmosphere, or filter through it.

It is very essential that a soil be made by tillage porous, since all the rain that runs off, without soaking into it, carries away with it ammonia, nitrates, and other fertilizers, which would otherwise be absorbed and appropriated. There is another great advantage in rendering the soil porous: free access of air is admitted, and more carbonic acid admitted with the water, so that, independent of the service rendered by carbonic acid, as a source of carbon, it exerts a greater solvent power of the mineral constituents of the soil, and thus renders a larger supply available. There is, however, this drawback in frequent tillage, that should be borne in mind, as already hinted at, that it accelerates the decomposition of the organic principles.

The free access of air, and consequently of oxygen, also counteracts any deoxidizing effect which any of the constituents of the soil may possess, such as protoxide of iron, which, by absorbing oxygen as it passes into the state of peroxide, has an injurious influence on vegetation. In contact with air it gradually passes into the state of peroxide, which, on account of its absorbent effect on ammonia, is serviceable. The free access of air also corrects acidity by developing bases which neutralize the free acids. It promotes the formation of carbonic acid, and sometimes also of nitric acid, by the oxidation of the carbon and nitrogen of organic principles in the soil, both of which act beneficially, as previously explained. It hastens the decomposition of the soil, and therefore liberates fresh mineral fertilizers.

In thickly settled countries, where farming produce is high, and labor cheap great advantage has been obtained from under-draining. In the western country what little draining has been done has, for the most part, been effected by open ditches. Under-draining has many advantages over open ditches. It does not cut up the ground, nor foster weeds. All the water, in passing through the soil, gives up its condensed fertilizers, which are, in a great measure, lost in the flow of surface of water. In its passage downwards it removes or changes effete matters, secreted from the roots of plants, which are deleterious to vegetation. Under-draining also warms the surface-soil, and the cooler substratum condenses the moisture out of the atmosphere, which has free access to it; this, in a great measure, counteracts droughts, besides effecting all the good already indicated, as obtained by a free access of air to the soil. Under-draining is said, also, to prevent the dying out of grasses, by keeping their roots free from injurious influences; to distribute nutriment more effectually; besides removing excess of water, and thus deepening the surface-soil. It renders vegetation earlier in spring, and prevents the freezing out of the grain in winter; too rapid evaporation of the water, and the formation of a hard crust on the surface.

We are apt to regard draining as requisite only for wet and low swampy ground, but experience seems to prove that it is nearly equally beneficial in counteracting too great dryness. It is estimated by farmers, in districts where it has now become an almost universal practice, that it increases the profit on the land ten to twelve per cent.; and pays for expenses of underdraining in three years. It is said, also, and I have little doubt of the fact that it removes the malaria or whatever other cause produces intermittent fevers; so that these diseases have almost disappeared from districts affected with such disorders previous to its introduction. In addition, investments made in underdraining farms have proved to be amongst the safest now made in the farming districts of England, and the government of Great Britain encourages it by loaning money at five per cent. to aid in extending this system; and money can be obtained on easy terms from other sources for the same purpose; all of which is sufficient proof of its success in a pecuniary point of view in that country.

That it would be equally beneficial to our lands there is little doubt; whether it would pay is a question that still remains to be proved.

It may be well, in this connection, to mention that the stiff clay soils usually prevailing in Kentucky in the vicinity of the Black Devonian Shale are lands that can hardly be brought into successful cultivation without draining; and it appears that this shaly rock, when sufficiently hard and tabular in its structure, may answer as a tolerable substitute for the manufactured under-drain tile, where these cannot be obtained.

Our country is, probably, in a condition to be benefited, at present, more by sub-soil plowing; though to derive the full advantage from this system it should be combined with under-draining. According to the most approved method of subsoiling, at present practiced, the subsoil is only loosened—not turned up on the surface—and for this purpose Mapes' form of sub-soil plow is generally adopted, since it may be worked, in most cases, by a single yoke of oxen. It is considered advisable only to disturb a few inches of the sub-soil each season, as some sub-soils require considerable exposure to the air before their fertilizing effect is developed, and a mixture of too great a quantity of this at a time, with the surface-soil may, sometimes, injure the crop.

During the progress of the survey I have met with but one instance where the benefits of deep plowing and sub-soiling have been at all doubtful. That case was in Nelson county, on the waters of the Chaplin fork of Salt river, along the range of the out-crop of the silicious mudstone intercalated in the blue limestone, on Mr. Beauchamp's farm. If it really proved injurious it was probably because too large a portion of the subsoil was mixed, at one time, with the surface-soil. We shall, however, be better prepared to give an opinion on this subject in the second part of this volume, when the chemical analysis of this soil and sub-soil shall have been completed.

# CHAPTER II.

## AGRICULTURAL GEOLOGY.

### KENTUCKY SOILS.

On a former occasion I alluded to the distinctive characters which might be observed even in the external appearance of Kentucky soils, based on the principal geological formations. I shall now point out more precisely in what that difference consists, by presenting a comparative view of their chemical constitution, selecting, for this purpose, some of those specimens, at this time collected, and of which Dr. Peter has now completed the chemical analyses, placing them here for the sake of convenient reference and comparison, in apposition.

(*A*) is a virgin soil from the Leptaena, Pleurotomaria and Bellerophon beds of the Blue (or Lower Silurian) Limestone of Woodford county.

(*B*) is a virgin soil from the Magnesian Limestones, under the Coralline bed of the Falls of the Ohio, belonging to the Upper Silurian System, collected in the eastern part of Jefferson county.

(*C*) is a virgin soil, over the Lithostrotion beds of the Sub-carboniferous Limestone of Barren county.

(*D*) is a virgin soil from the Coal Measures of Ohio county.

(*E*) is a virgin soil from the Quarternary Formation of Henderson county.

In 100 parts of the soil, dried at 300°—

| | (*A*) | (*B*) | (*C*) | (*D*) | (*E*) |
|---|---|---|---|---|---|
| Organic and volatile matters, | 7.771 | 7.996 | 5.200 | 5.080 | 5.080 |
| Alumina, (the plastic earth of clay,) } Oxides of iron and manganese, } | 12.961 | 7.480 | 3.460<br>2.540 | 4.349 | 3.490 |
| Carbonate of lime, | 2.464 | .394 | .366 | .176 | 1.254 |
| Magnesia, | .173 | .240 | .205 | .166 | .447 |
| Phosphoric acid, | .319 | .205 | .159 | .101 | trace. |
| Sulphuric acid, | .170 | .082 | .197 | .413 | not de. |
| Chlorine, | not determined. | | | .060 | not de. |
| Potash, | .393 | .200 | .197 | .157 | .085 |
| Soda, | .130 | .043 | .090 | .015 | .034 |
| Sand and insoluble silicates, | 75.266 | 83.134 | 87.686 | 90.166 | 89.670 |
| | 99.648 | 99.774 | 100.100 | 100.683 | 100.060 |
| Moisture driven off at 300°, | 4.700 | 4.420 | 2.340 | 1.740 | 2.042 |

The Woodford soil (*A*) of this table supports a growth of sugar-tree, pig-nut, hickory, hackberry, ash, walnut, mulberry, buckeye, and box-elder, with an under-growth of pawpaw and elder. It was collected from the water-shed between Grier's and Clear creek.

The predominating growth, in that part of Jefferson county where soil (*B*) was collected, on gently rolling land near the sources of Floyd's fork of Salt river, Goose, and Harrod's creek, is beech mixed with hickory, ash, walnut, cherry, and poplar.

In the early settlement of that country, that portion of Barren county where the virgin soil (*C*) was collected, was emphatically a grass country—the "Barren grass" reaching as high as a man's head; now a large portion of the country of the Barren limestone region is grown up with small oaks.

The Ohio county soil (*D*) is only an average specimen of a Coal Measure soil, supporting a growth of white oak, hickory, ash, and poplar.

The Quarternary soil (*E*) of Henderson county, is characterized by a large growth of white oak, poplar, and walnut.

The Blue limestone soil of Woodford (*A*) is peculiarly well adapted for the growth of hemp. On some farms this crop has been grown, almost without interruption, for twenty successive years, and still the land is exceedingly productive.

The soil derived from the blue limestone and marls of the Lower Silurian Period, in Kentucky, are remarkably genial to the growth of grasses—to blue grass in particular—as is evinced in the remarkable development of the stock pastured on the Blue Limestone Region generally. For the most part they are almost a year in advance in bulk, weight, and form, to the stock raised on the soils derived from the Carboniferous group.

The acknowledged superiority of this soil is evidently due to the preponderance of the mineral constituents, lime, phosphoric acid, and the alkalies. It contains, as will be seen, fourteen times as much lime, three times as much phosphoric acid, and more than twice as much potash as soil (*D*,) from the Coal Measures. Its superior fertility bears no relation to the organic and nitrogenous constituents, for it has no more of these than many soils far inferior to it in productiveness. The large amount of alumina and oxide of iron, no doubt also greatly contributes to the permanent fertility of the blue limestone soils, since these not only materially aid in condensing ammonia out of the atmosphere, but the former retains and prevents the soluble salts from filtering away beyond the reach of the roots of plants, and by gradual disintegration supplies, by degrees, soluble salts of potash, which are being continually liberated from their original insoluble combination of silicate of alumina and potash, existing originally in the rocks and minerals from which clays have been derived.

Though the Woodford county blue limestone soil does not yield to carbonic acid water quite as much soluble saline matter as the subcarboniferous soil No. 234, of Wayne county, or as the quarternary soil No. 126, of the former report, yet it still stands third in the list, even in the quantity of soluble ingredients ready for the use of vegetation. It resembles, in its composition, some of the fertile soils near Tulln, in Lower Austria, except that in the Austrian soil the alkali seems to be almost entirely soda, while the Kentucky soil has both alkalies—the potash predominating over the soda. It has, also, much the same constitution as a very fertile soil of Hungary, near Esakang, except that in this soil, likewise the soda predominates over the potash, and it contains considerably more magnesia than the Woodford soil.

In this connection I would especially call the attention of the farmers of Woodford county, and those of the blue grass region generally, to the important fact that the sub-soil No. 552, of this report, taken from a foot beneath the surface, in an old field adjoining where the virgin soil (*A*) or No. 550, of Dr. Peter's Report, and therefore within easy reach of the sub-soil plow, is even richer in all the mineral fertilizers, except sulphuric acid and soda, than the virgin soil (*A*) itself, although the old field, from which this sub-soil was collected, has been in cultivation ever since 1808. Herefrom we have a sure guarantee of a cheap and easy means of restoring this variety of the blue grass land of central Kentucky, when its fertility diminishes, simply by gradually stirring up with the sub-soil plow, this invaluable store of mineral fertilizers, which, by judicious management, may be said to be almost inexhaustible.

The chemical analysis of a red under-clay from the east part of Fayette, No. 509, shows a corresponding richness in mineral fertilizers.

In my special report, made last February, the difference was exhibited between the Blue Limestone soil of Woodford, of the Lower Silurian Period, and a soil from the eastern part of Jefferson county, collected over the Magnesian Limestone of the Upper Silurian Period, that underlie the Coralline beds of the Falls of the Ohio, of which the analysis is given in column (*C*) of the foregoing table. These remarks are here inserted for the benefit of those who may not have had an opportunity of seeing that report.

"It will be perceived, from the foregoing comparative analyses, that the acknowledged superiority of the blue limestone soil does not depend, as is usually supposed, on its greater richness in organic matters, since, in fact, it contains 0.225 less of these principles than exists in the Jefferson county soil. This is also proved by comparison with the analysis of the Fayette county blue limestone soil, given on page 277 of the first Geological Report; for in that soil the organic and volatile matters are only .004 more than in soil (*B*) of the above table. On the contrary, these comparative analyses, as well as many others which will be hereafter furnished in the succeeding reports, give abundant evidence that it depends on the greater proportion of *inorganic* constituents, viz: the phosphates, sulphates, and alkalies; although these, as may be observed, do not amount, in either soil, to a large *per centage;* together with the much larger quantity of alumina and oxide of iron, which is 5.481 per cent. more in (*A*) than in (*B*). It is now well established that phosphates, sulphates, alkaline earths and alkalies are essential constituents of plants and must form a part of their food; in fact, the cereals cannot come to perfection and form a nutritive grain for man or animals, if the soil is destitute, or even very deficient, in phosphoric acid; and, though we are accustomed to view a soil proportionally rich to the larger or smaller quantity of vegetable mould it contains, yet the presence of the above inorganic constituents are as essential and more difficult to restore, when once exhausted, inasmuch as they cannot, like the organic, be appropriated out of the atmosphere, but must be renewed, either by disintegration of the rocks which contain them, and from which they are originally derived, which must necessarily be a slow and tedious process, or by the labor and expense attendant on the transportation of guano, bone-earth, and a few other varieties of manures in which these substances are contained. Indeed, some of the earthy fixed constituents of soils, especially the alumina and oxide of iron, i. e. ferruginous clays, are important vehicles, through the intervention of which ammonia is absorbed and fixed out of the atmosphere, and it is mainly from the presence of lime in the soil that nitric acid is produced; a remarkable fertilizer; either from the elements of ammonia, as evolved in the nascent state from decaying animal and vegetable matter and perhaps to some extent directly from the atmosphere during thunder-storms.

"The quantity of inorganic fertilizers which the Woodford county soil possesses over the Jefferson county soil, in one hundred parts, are therefore:

| | |
|---|---|
| Alumina and oxide of iron and manganese, | 5.481 |
| Carbonate of lime, | 2.070 |
| Phosphoric acid, | 0.114 |
| Sulphuric acid, | 0.068 |
| Potash, | 0.194 |
| Soda, | 0.087 |
| Total, | 8.014 |

"The total preponderance of these fertilizers in the Woodford soil is, therefore, a little over eight per cent.; and, of some of the individual ingredients, only a small fraction of one per cent.; this, at first sight, appears small and insignificant, but when calculated over a single acre of ground, only six inches in depth, the number of pounds becomes very considerable.

"Assuming, as found by trial, that the average approximate weight of these air-dried soils, in the condition used for analysis, to be about sixty pounds to the cubic foot, we obtain the following amounts in pounds, on each acre six inches deep, which the blue limestone soil of Woodford county contains over that in the soil of the eastern part of Jefferson county:

71,625 pounds of alumina, oxide of iron and manganese.
27,050 pounds of carbonate of lime.
1,489 pounds of phosphoric acid.
888 pounds of sulphuric acid.
2,535 pounds of potash.
1,136 pounds of soda.

104,723 pounds total.

"Hence to make the Jefferson county soil equally productive with the Woodford, for six inches in depth, there requires to be added to each acre of ground:

71,625 pounds of ferruginous clay.
27,050 pounds of limestone or 16,188 pounds of burnt lime.
3,175 pounds of bone-earth.
2,161 pounds of gypsum.
4,848 pounds of unleached ashes.
2,455 pounds of common salt.

111,314 pounds total.

"In place of the 3,175 pounds of bone-earth and 2,161 pounds of gypsum, there might be substituted, with advantage, 3,000 pounds of superphosphate of lime; that is bone earth which has been treated with sulphuric acid, by which results an acid phosphate of lime and gypsum; which has been found, in practice, an excellent form of application of the highly important inorganic constituents of soils. Soda and lime are also very advantageously employed in agriculture by dissolving common salt in as much water as is required to convert the caustic lime into a fine powder and slacking the lime with this brine.

"However, all these ingredients, except the ferruginous clay, can be obtained by the application of farmyard manure, of which there is always more or less on every farm; but not in the manure as it is usually found, after long exposure to weather, along side the stable, but in fresh stable and cow-house manure, or in manures which have been properly preserved *under cover* and in such a manner that neither the soluble portions have been drained away in liquid oozings from the manure pile; nor the ammonia volatized into the atmosphere; in manure, in short, which has been carefully heaped up under shelter in a slightly inclined plank pen, with tight floor constructed so that all the liquid part can drain into a tank or cistern, from which it can be pumped over the manure heap from time to time, or carried separately on to the land; while, at the same time, care has been taken to fix the ammonia which would otherwise escape into the atmosphere, either by the addition of alternate layers of muck, decayed wood, or other vegetable mould, charcoal powder, or, better than all, so much diluted sulphuric acid as shall prevent any odor being emitted. To such a manure pile the lime slacked with brine is also a good addition, as the sulphate of lime and chloride of calcium thereby formed are excellent absorbers of ammonia.

"In the same region of Jefferson county a soil was collected from a field which had been twenty-five to thirty years in cultivation; also, samples of the immediate sub-soil and red under-clay, which almost universally underlies these lands at the depth of a few feet.

"The following is the comparative analysis of the virgin soil of Jefferson, (*B*,) while (*A*) is of the soil from the adjacent field long in cultivation:

| *In* 100 *Parts.* | *A.* | *B.* |
|---|---|---|
| Organic and volatile matter, | 7.996 | 4.506 |
| Alumina, oxide of iron and manganese, | 7.480 | 6.204 |
| Carbonate of lime, | .394 | .316 |
| Magnesia, | .240 | .200 |
| Phosphoric acid, | .205 | .191 |
| Sulphuric acid, | .082 | .067 |
| Potash, | .200 | .158 |
| Soda, | .043 | .070 |
| Sand and insoluble silicates, | 83.134 | 88.318 |
| Loss, | 0.226 | 00.000 |
| | 100.000 | 100.066 |

"It appears from the above that there has been carried off from this field, by the succession of crops harvested, assimilation by stock, filtering and washing, a part of every fertilizing ingredient except soda, in the following proportions in 100 parts:

| | |
|---|---|
| Organic and volatile matters, | 3.490 |
| Alumina, oxide of iron and manganese, | 1.240 |
| Carbonate of lime, | .078 |
| Magnesia, | .040 |
| Phosphoric acid, | .014 |
| Sulphuric acid, | .015 |
| Potash, | .042 |
| Total, | 4.919 |

"This, calculated for one acre, six inches deep, gives:

45,607 pounds of organic and volatile matter.
16,204 pounds of alumina, oxide of iron and manganese.
1,019 pounds of carbonate of lime.
522 pounds of magnesia.
182 pounds of phosphoric acid.
196 pounds of sulphuric acid.
548 pounds of potash.

64,378 pounds total.

"Here then we have the amount in pounds which would be required to be restored to each acre of this field to bring it back to its original fertility, provided the influence of cultivation has only extended to the depth of six inches; but these amounts would require to be doubled, if the exhansting influence has extended to one foot.

"In Europe, where the different kinds of manure, both organic and unorganic, have a commercial value, the

| | | |
|---|---|---|
| 45,607 | pounds of organic and volatile matter would be worth, | $ 28 50 |
| 16,204 | pounds of alumina, oxide of iron and manganese would be worth only what it might cost to restore it, either by means of the sub-soil plow or the cost of hauling it on to the land, if beyond the reach of the sub-soil plow. | |
| 1,019 | pounds of carbonate of lime, worth, | 63 |
| 522 | pounds of magnesia, worth, | 25 |
| 182 | pounds of phosphoric acid, worth, | 1 82 |
| 196 | pounds of sulphuric acid, worth, | 24 |
| 548 | pounds of potash, worth, | 8 22 |
| 64,378 | | $ 39 66 |

"The importance of the information conveyed by these results is most manifest.

"By far the most expensive part of the above ingredients, if required to be purchased and hauled on the ground, would be the organic constituents; but fortunately there are other more economical alternatives of reclaiming the lost humus of a soil. The most abundant proximate principle of humus is vegetable fibre which, by decay, yields chiefly carbonic acid and the elements of water. It is by supplying these to plants that it is mainly efficacious in agriculture. Fortunately there is an inexhaustible store of these principles in our atmosphere, and the farmer has the power, if he knows how, to appropriate them to his use from that source, without seeking further. Strange as it may at first sound, land can be *manured* from the atmosphere; that is, it can receive from it the fertilizing elements of the organic constituents of manures. But this must be effected through the intervention of the mineral, inorganic or fixed constituents of the soil; that is, those earthy principles which cannot be burnt off by fire and are, therefore, found in the ashes of plants—such as the phosphoric and sulphuric acids, lime, clay and alkalies—for with an abundant supply of these and ammonia a luxuriant growth of leaves and roots overspread and penetrate the ground, having, during their growth, fixed a very large proportion of their weight and substance out of the atmosphere; it is upon this principle that the improvement of land by green cropping is based, which, when turned in, passes rapidly into a state of decay, furnish in this way an immediately available and abundant supply of carbonic acid, and oxygen and hydrogen in the proportions in which they exist in water. But these substances can moreover be condensed out of the atmosphere by good tillage, for the more porous and loose a soil is the more it is penetrated by air and rain water, in which more or less carbonic acid is always condensed. Thus, if the farmer takes care that his land is sufficiently supplied with these inorganic constituents above mentioned and a certain amount of the nitrogenous principles, he need not go to much expense in hauling humus, or its equivalent substances mainly consisting of woody fibre, as the atmosphere has always a liberal supply on hand. Indeed the nitrogenous principles can also be obtained to a considerable extent from the same source; since there are abundant emanations continually volatilizing ammonia and carbonate of ammonia into the air, which are returned to the earth by every shower of rain or fall of snow, besides what is absorbed by a porous, well tilled soil, particularly if that soil has a notable quantity of clay and peroxide of iron.

"Seeing then whence the organic and volatile matters of the soil may be derived, the next inquiry which presents itself in connection with the comparative analysis of the soil just given is, can any or all of the removed inorganic constituents be obtained from the sub-soil or underclay that underlie the soil? Because, if so, this is, undoubtedly, the most accessible and cheapest source, whence they can be restored to the soil.

"The following analyses of the immediate subsoil (*A*,) and the underclay (*B*,) give the answer to this question:

| | *A.* | *B.* |
|---|---|---|
| Organic and volatile matter, | 2.844 | 3.112 |
| Alumina, oxide of iron and manganese, | 6.235 | 17.020 |
| Carbonate of lime, | .356 | .194 |
| Magnesia, | .226 | |
| Phosphoric acid, | .099 | .477 |
| Sulphuric acid, | .082 | .088 |
| Potash, | .181 | .297 |
| Soda, | .028 | .111 |
| Sand and insoluble silicates, | 89.900 | 77.434 |
| Loss, | .049 | .881 |
| | 100.000 | 100.000 |

"The conclusion, from the preceding analysis is, that they can be supplied to a *limited* extent by the immediate sub-soil; but in much greater abundance by the red, ferruginous, under-clay which is found universally a few feet under the soil of this part of Jefferson county. This under-clay is not only rich in alumina and peroxide of iron, uncontaminated with *pro*toxide of iron, substances which have a remarkable power of absorbing ammonia from the atmosphere and yielding it by degrees to plants, besides retaining other manures and water, but in addition, this red under-clay, it will be observed, contains more than twice as much phosphoric acid, and nearly double the amount of alkalies, which are in the virgin soil. How important is this information to the farmer. He learns by these chemical analyses that he need not go to any other source, at present, for his supply of the inorganic food of plants; and that by the aid of powerful sub-soil plows, where this red clay is sufficiently near the surface to be reached by this operation, or where it lies too deep to be thus turned up, he can obtain it by only sinking with his pickaxe and shovel a few feet beneath the surface of his *own land.*

"Let those who have hitherto had little faith in the powers of chemistry to reveal to agriculture invaluable truths contemplate these results.

"I have heard farmers, and even those professing to be chemists, express their doubts that that science could ever disclose the mysteries of vegetable assimilation, or the way in which plants received their nourishment and the transposition of the elements thereto contributing; but he who has closely watched the rapid strides of discovery in chemistry in the last quarter of a century cannot fail to have most implicit confidence in this noble science."

All of the varieties of the blue limestone soils are not equally rich, in the mineral fertilizers, with the Woodford and Fayette soils; but, so far as the chemical analyses have yet been carried, they exhibit properties far above the average in fertility. Considerable variation will, no doubt, be observable when all the different varieties already collected, and others which it is intended hereafter to select, shall be analysed; for both the lithological character of the various strata of which this formation is made up, and the difference in the growth of timber and shrubs, indicate this. For instance: there is interstratified, in the blue limestone formation, a peculiar silicious mud-stone along a belt of country which will be more particularly described hereafter, ranging through Jessamine, east part of Fayette, Scott, Grant, and a part of Owen, Harrison, Boone, and Carroll counties, which not only stamps a marked character to the soil, but a peculiar disease follows its range, which I have found no where else on this formation.

I am able at this time to call attention to the chemical analysis of one soil, No. 504, of Dr. Peter's Report, collected from a narrow strip of beech timbered land, ranging through the eastern part of Fayette county, where silicious mud-stones and shales, No. 505 and 506, are superimposed on the Isotelus and Leptaena beds, in the blue limestone. It is, perhaps, not as characteristic a sample as I shall be able to supply hereafter from Grant county, where these strata, known as the "rotten sandstone," are more developed, and extends over a wider belt of country. In Fayette it only occupies a strip of so called "sobby beech flats," of about a quarter to half a mile wide, on the head waters of Elkhorn creek. This soil is not nearly so productive as the true blue limestone soil. The analysis shows that it contains less than half as much of most of the mineral fertilizers as the best blue limestone soil of Fayette county.

It is particularly worthy of note that the only milk-sick region which I have, as yet, become acquainted with, on any part of the range of the blue limestone formation, follows the out-crop to the surface of this so called "rotten sandstone," and its accompanying characteristic soil.

Again: an intimate connection is evidently traceable, in a portion of Nelson county, between the soil and underlying substratum of shell limestone in which the fossils are all converted into silex. This bed of rock, of no great thickness, by disintegration gives rise to a silico-calcareous reddish-brown earth, in which the silicified fragments of the characteristic *orthis* are distinctly traceable. This earth is coincident with the so called "blue ash lands" of Nelson county, whereas the upper member of this formation, in Spencer county, give rise almost exclusively to beech.

The chemical examinations of the soils and sub-soils from Woodford county—Nos. 550, 551, 552, and 553—are attended with peculiar interest, inasmuch as this, and a few of the adjacent counties of Kentucky, are regarded by our citizens as the "Garden Spot" of the West, and not without justice; for although there may be other blue limestone soils in the state, as well as in Ohio and Indiana, that will compare favorably with it, still it is doubtful whether a tract of equally fertile *upland soil* can be found, having the surface so level and unbroken, and therefore so favorably situated for cultivation, in consequence of their being located toward the sources of small streams. The county of Woodford borders, it is true, on the valley of the Kentucky river, but since its bluffs rise very abruptly, and its eastern tributaries rising in the county are very short in their course, because the summit level between the Kentucky and Licking rivers lies, in this part of the state, comparatively close to the former river, the configuration of the land is, in consequence of this peculiarity of its physical geography, very gently undulating, and comparatively level, even within a mile or two of the deep gorge of the principal water-course of the state.

The peculiar adaptation of the soil to the growth of hemp, as already remarked, has caused the early settlers to put large tracts of their land, for many years in succession, in this staple. As the inorganic base assimilated in largest quantity by this plant is lime, being eight to ten times more than of the other mineral fertilizers which the Woodford soil and sub-soil has in the greatest abundance; and since so much of the plant is usually restored to the soil which nourished it, it will be apparent why it is by no means an exhauster of the soil, like wheat, corn, and tobacco; hence, comparatively speaking, the resources of the lands of this part of the state have not been so severely taxed, as in the tobacco and more exclusively grain growing counties. These advantages, together with the practice of retaining large portions of the farms in pasture for stock, has retained, in a great measure, the primitive fertility of her soil.

A comparison of the composition of soils Nos. 550 and 551, shows the amount of ingredients which have been removed by nearly fifty years of cultivation, chiefly in hemp, with occasionally corn and wheat. In 100 parts there has been a loss of 2,258 organic and volatile matters; 0.013 of phosphoric acid; 0.113 sulphuric acid, 0.189 of potash; while, notwithstanding, the demands of hemp for lime, there has been a gain both of this alkaline earth and magnesia, viz: 0.270 carbonate of lime and 0.160 of carbonate of magnesia, which proves how rich the substrata must be in calcareous matter. Assuming a cubic foot of this soil, in its dry state, to weigh seventy pounds and that cultivation has extended to the depth of one foot, it would require to be added to each acre to restore this land to its native fertility only three hundred and ninety-one pounds of phosphoric acid or nearly double that of bone-earth; 3.441 of sulphuric acid or a little less than twice that quantity of gypsum, and 5.762 pounds of potash or four tons of unleached ashes.

In this connection, permit me in this place, again to call particular attention to the chemical analysis of the sub-soil of this old field, No. 552, and the red under-clay, No. 553, prevalent throughout Woodford county. These will be found amply capable of supplying all the lost ingredients; for even the sulphuric acid and soda, which are not in quite so large quantities as in the soil, have been found in great abundance in the red under-clay. In fact, taking the sub-soil and under-clay together they are richer in all the essential fertilizers but soda, and even that alkali falls but little short in the red under-clay. As the removed organic and volatile matters may be restored by green crops, the land proprietors of Woodford county have, within easy access of the sub-soil plow or shovel, an inexhaustible store of agricultural wealth. Who will pretend to set a monied estimate on the value of this information to the farmers of Woodford? By judicious management this soil may be considered good for centuries to come. It is much to be desired that some of the farmers in Woodford or Fayette counties would test the result of the action of these red under-clays on the poorest land they can find, and communicate the result, since chemistry points so decidedly to this as a cheap and convenient natural manure. They need not, however, expect to observe as much effect in the first as the succeeding years, when the mineral fertilizers become more soluble and, consequently, more available. Three or four inches of these under-clays will be enough to stir up or mix with nine inches or one foot of the surface soil at one time. They should also try it in connection with green crops turned in or by addition of ammonical salts, nitrates or some organic manure; because, as we have elsewhere stated, the assimilation of the mineral constituents is in proportion to the supply of the organic elements received into the substance of plants, and, if there be already as large a quantity of the mineral constituents in the soil as will balance the amount of organic food derivable, in their climate and locality, from the atmosphere, water, and what may be also in the soil itself, little or no benefit may be experienced from the application of these earths; this is very apt to be the case in rich soils like the blue limestone soils and this is the reason why I recommend the trial on the poorest and oldest land that can be found.

Upon the same principle, in the best lands of the blue limestone region, where the mineral fertilizers are in large and available quantities, if a maximum produce is desired to be obtained on the *"high farming system,"* it will be necessary that ammonical salts, guano, a nitrate, or some nitrogenous manure be added to the soil to be so cultivated, so as to force the crop to take up more of the soluble mineral constituents than it would be otherwise capable of doing with only such quantity of the organic elements as it could appropriate from natural sources.

No more conclusive proof can be deduced, of the intimate connection between the composition and quality of the soil, and the subjacent rock formation, than the inspection of the chemical analysis of the Leptaena limestone, No. 547, and the Bellerophon limestone, No. 549, which underlie the country south of Versailles, in comparison with the Woodford soils and sub-soils, Nos. 550, 551, 552, and 553, all collected from the same locality. The source of the large amount of the great mineral fertilizers—phosphoric acid, sulphuric acid, and the fixed alkalies—in these soils and sub-soils, is most apparent by taking note of the per centage of these acids and bases in the above underlying limestones, No. 547 and No. 549.

Contrasting the composition of these limestones with those of other limestones of which the analyses are given, both in the first, and this report, it will be observed that they contain a larger amount of phosphoric acid and alkalies *combined*, than any of the other limestones except No. 484, from Anderson county, and 507 and 508, from Fayette, all of which rock specimens are from the blue limestone formation. Nos. 507 and 508 underlie the region where the fertile blue limestone soils and sub-soils, No. 27, of the first report, and Nos. 509 and 510, of this report, were obtained. The only other limestones analysed, up to this time, which approach to these, in the large amount of the above acids and bases, are some of the black bituminous ferruginous limestones of the Coal Measures; which, however, on account of the circumscribed area which they occupy, can have but a very partial and limited influence on the soil.

This leads to a very important deduction, the principle of which was already hinted at in the first chapter; and serves, moreover, to correct a very erroneous idea which prevails: that the purer a limestone is, which is to be applied as a mineral manure, the better; when, in reality, it is those limestones which are most argillaceous, and which are replete with organic remains, that are by far the most valuable for that purpose, since they furnish not only the lime which is required, but also the phosphoric acid and the alkalies.

I will venture to say, if any of the farmers living in Hopkins county will try the application of the black bituminous limestones, No. 132, 134, or 154, either burnt or in the raw condition ground, on any of their soils that may be deficient in phosphoric acid and alkalies, they will be astonished to witness the fertilizing effect it will have in a few years after its application—say thirty to fifty bushels to the acre.

Though it will not answer, in a pecuniary point of view, at present to transport Lower Silurian Blue Limestones and Marls to any great distance, the time will come, as our means of transportion shall be more extended and cheaper, when many of the inferior sandy soils and stiff cold tenaceous clay-lands of Kentucky will be manured with these fossiliferous limestones, even if they have to be carried hundreds of miles; because these limestones will be found far more valuable, as mineral manures, than the generality of limestones from the other formations of the state.

Again: in the centre of Washington county there is a marked soil of the blue limestone formation, producing good hemp, but on which tobacco grows too rank and coarse to afford a fine quality of that plant. Where this soil prevails remarkably large, yellow poplar flourish; and it appears to be of different properties, not only from the preceding, but, also, from that which furnished a soil found around Springfield, in the same county, where white oak is the prevailing timber. Samples of soils have been collected from these and other localities in the Blue Limestone Formation, that presented marked characteristics. When the chemical analyses of these shall have been completed, we hope to be able to deduce therefrom some interesting results from the chemical peculiarities they may present.

The Barren limestone soil (*C*) contains, as will be perceived, very nearly the same amount of carbonate of lime, magnesia, phosphoric acid, and alkalies, as the magnesian limestone soil (*B*,) of Jefferson county; rather less alumina and oxide of iron, but more than twice as much sulphuric acid, which rather exceeds that in soil (*A*). This is undoubtedly the reason why this soil is so well adapted, both for grasses, oats, and corn crops, which require a large supply of this chemical substance.

The Coal Measure soil (*D*), contains still smaller quantities of carbonate of lime and phosphoric acid, but more sulphuric acid; indeed, more than has been found in any soil yet examined, except another Coal Measure soil, No. 236, from Union county. This is no doubt due to the frequent occurrence of minerals, containing sulphur, in rocks of the carboniferous era. These soils ought, therefore, with the occasional addition of unleached ashes, nitrate of potash, or some combination of that alkali, to be better adapted than any of the other soils for potatoes, turnips, oats, and with the addition of common salt and bone-earth, or its equivalent, for buck wheat.

In the Coal Measure soils, Nos. 10, 126, 138, and 155, the silica and insoluble silicates range from eighty-six to ninety per cent.; but a large proportion of their salts are, in that soluble condition, available for the immediate use of plants. Soil No. 126, of the first report, gave up no less than seven parts of solid extract in 1,000, and No. 155 six parts in 1,000. The former, therefore, contains more, and the latter nearly as much, soluble ingredients as the rich Woodford virgin soil, No. 550.

The Barren limestone soil has usually, a large proportion of peroxide of iron in its composition—three to five per cent.—which gives the lands of this formation generally a deep reddish brown color. This, together with the considerable quantity of clay derived from the partings between the limestones and the interstratified argillaceous marly beds, contributes, doubtless, greatly to their fertility, for the reasons already given, when adverting to the action of the same admixtures in some of the blue limestone soils. This is one reason why these soils have proved much more productive than was at first supposed, in the early settlement of the country, and which rapidly raised the price of these lands, after they began to be appreciated, from five dollars to forty dollars per acre in some portions of the state.

There is another variety of soil derived from the silicious strata of the lower division of the sub-carboniferous group, which is greatly inferior to that which prevails in the upper calcarious and marly division of this group of rocks, which prevails on the knobs; fortunately, however, it is of very limited extent.

The analysis of three soils from this formation have been completed; No. 232, from Cumberland county; No. 228, from Monroe county, and No. 226, from Russell county. In these the sand and insoluble silicates range from 87.110 to 90.786. The quantity of argillaceous earth is very small in all of them, and the lime and phosphoric acid in No. 226 and 228, is far below the average. They would, therefore, be greatly improved by the application of some of the fat marls and stiff red clays of the adjacent upper division, of the same formation, which is often sufficiently accessibly; or even of the clays produced from the black Devonian shale, that often forms the base of the knobs.

No. 232 afforded a large amount of soluble extract—five parts in 1,000—and contains the largest amount of sulphuric acid of any soil yet analysed. It should be remarked, however, in regard to this soil, that it was collected from the bottom land of Sulphur creek, and has therefore received the washing from adjacent knobs, and reposes on the clay derived from the underlying black slate, and is therefore of superior quality to the soils from this formation, on the summits and slopes of the knobs. The frequent occurrence of iron pyrites, in the strata towards the base of the knobs, accounts for the large proportion both of sulphuric acid and iron.

Where the encrinital beds of limestone crop out, which are interstratified locally with strata of the knobs, a much more productive calcarious soil results.

As yet no analyses have been completed of soils derived from the black Devonian slate.

Besides the various shades of difference which will no doubt be exhibited hereafter in the different quaternary soils, I will mention, at this time, two marked varieties. The earthy ingredients of both of these are in a very fine state division; but one is of a mouse color, and is derived from a fine silicious, and more or less calcarious loam—No. 3, of the section given on page 22, of the first report; the other remarkable for its whiteness, as well as its lightness, due to exceedingly fine rounded particles of hyaline and milky quartz. This variety originates from the disintegration of the white silicious and silico-magnesian earth, No 6, of the same section. The analysis given under column, (*E*,) is of the first variety, which predominates greatly over the second, of which an analysis is given on page 334, No. 128, of Dr.

Peter's first report. Both varieties are usually tolerably rich in lime, and the first seems to have, in the specimen examined, a large proportion of its mineral salts, in a soluble condition. Specimen No. 126, or (*E*,) of the column, gave up no less than seven parts of solid extract, to water impregnated with carbonic acid. This is the largest amount of ingredients, dissolved by the same means, in the sixty-three soils already analysed, except one—No. 234—a sub-carboniferous soil, from Wayne county. It contains, therefore, a very large amount of mineral salts, in a state fit to enter into the circulation of the plant, and already prepared for its immediate nourishment.

It is probably due to this condition of these quaternary soils, that they are so well suited for the growth of fine silky tobacco.

Soil No. 230, from Daviess county, also remarkable for its large growth of tobacco, and which receives its material chiefly from the Quaternary Formation, contains more than the average quantity of soluble saline ingredients. But neither this soil, nor No. 126, contain, on the whole, as large a proportion of potash as some of the other Kentucky soils; however, No. 218, of this report, which is also a quaternary soil, will be seen to possess a larger amount of alkalies, viz: 0.120 potash, and 0.020 soda. Indeed, there is no doubt that a considerable amount of potash has been removed from soils, No. 126, and 230, since they are soils that have been under cultivation in two of the principal tobacco growing counties in the state, and there was a mature tobacco crop on the ground when No. 126 was collected.

Some of the specimens of quaternary soils, which have been analysed, contain more phosphoric acid than (*E*); for instance, No. 218, of the present report, from Ballard county, has 0.410 per cent., which is nearly as much as the rich virgin Woodford soil. The sand and insoluble silicates together amount, in the specimens subjected to chemical examination, to from 89.670 to 91.71; of this, from 48 to 80 parts are in an exceedingly fine state of division. This condition of the silicious earths indicates that they will absorb moisture, and retain it, together with the soluble organic and inorganic manures, much more effectually than soils containing an equal amount of ordinary palpable sand; while, at the same time, they are soils that require much less power to break them up than more tenacious clay land.

These kind of soils prevail, to some extent, in those parts of Henderson and Daviess county bordering along the principal water-course, and even in the adjacent hills, to the hight of one hundred or one hundred and fifty feet above low water of the Ohio river; but more especially in Ballard, Hickman, Fulton, and most of the counties of the Jackson Purchase.

It is probable that there will be found as much, if not more, variation in the chemical composition of the Coal Measure soils, than in those derived from the other formations, since the lithological character of the strata composing this group varies, perhaps, more than that of any of the others. Those, however, derived from any of the principal geological systems of Kentucky, will, undoubtedly, also show many shades of difference, corresponding to modifications in the lithological character of the various members, of which each formation is made up.

The largest amount of phosphoric acid or phosphates, obtained from the sixty-three soils and sub-soils analysed up to this time, are from the blue limestone formation of Woodford, Fayette, and Clark counties, viz: in Nos. 27, 28, 500, 501, 550, 551, 552. The phosphate of lime in No. 27, is 0.56, as stated on page 277, of the first report, but incorrectly transferred to the table on page 370, of same volume where it stands only 0.26.*

The largest amount of sulphuric acid, as already remarked, have been found in soil No. 232, from the bottoms of Leek creek, in Cumberland county, at the base of the Knob Formation; next to it, the Coal Measure soils of Union and Ohio counties—Nos. 236 and 223.

The largest quantity of alkalies yet obtained have been from the blue limestone soils and sub-soils of Clark, Jefferson, and Woodford—Nos. 500, 501, 509, 510, 552 and 553—and the sub-carboniferous Wayne county soil, No. 234, of which I shall speak in the sequel. It is worthy of note in this connection, that it is the same soils which contain, as a general rule, the largest of alumina, from which we may infer that these are chiefly liberated by the disintegration of this earth, derived, no doubt, originally, from rocks allied to the potash and soda felspars.

The prevailing member of the blue limestone formation, where soil No. 517, of Dr. Peter's report, was collected, in Franklin county, is an

*The phosphate of lime of No. 155, of the same table, should be 0.08 as on page 352, of that report, in place of 0.68, in the table.

encrinital layer of this formation, and not far from a fine locality of *Cytherina*, allied to, but probably distant from, *C. Baltica*. The predominating growth of timber on the land is sugar-tree, mixed with honey-locust, mulberry, and poplar. New ground will produce from fifty to sixty bushels of corn to the acre. The comparative analysis of the virgin soil, from this part of Franklin county, and that of a field where the same kind of soil existed, and which had been twelve years in cultivation, shows a loss of 0.151 of phosphoric acid, 0.38 potash, but a gain of .026 soda, and 1.443 of carbonate of lime. The latter has been derived from the turning up of a sub-strata of earth, richer in lime than the immediate surface soil; the former, perhaps, from the salt consumed by the stock on the farm, or been also brought from a more impervious layer beneath, into which it had filtered from the surface by the action of rain. Assuming that a cubic foot of this dry soil weighs seventy pounds, and that the influence of cultivation has extended to the depth of one foot, then there would require to be added to each acre of this field, to restore it to its original fertility, 4,604 pounds of phosphoric acid—or if added in the form of bone-earth, nearly five tons; 182 pounds of sulphuric acid—or if added in the shape of gypsum, nearly 400 pounds; and 1,158 pounds of potash, or 16,000 pounds of unleached ashes, besides 81 tons of organic matter; but, as the greater part of this is vegetable fibrin, which resolves itself, for the use of plants, into carbonic acid and water, good tillage, and a few green crops turned in, would be a good substitute, and the least expensive plan of effecting the restoration of the lost organic and volatile matters. Since stable manure contains only three tenths to a half per cent. of phosphoric acid, it would require from four hundred to seven hundred tons of such manure to each acre to furnish the lost phosphoric acid from this field. This shows how inadequate the usual supply of stable manure would be to restore the exhausted constituents of soils. Allowing a ton of stable manure to be worth $2, it would require the expenditure of $800 to bring this soil back, by this means, to its original condition; hence the necessity for the farmer to be acquainted, himself, with other sources of the mineral fertilizers. If the geological survey is able to point out a cheap and easily accessible source of this, and other mineral fertilizers, it will be conferring an invaluable boon upon the commonwealth, compared with which the cost of the survey is a mere trifle. It has already done so in the case of Woodford and Fayette counties, and I believe it will be able to do so for Franklin county also; for the appearance of red under-clays, observed in portions of Franklin county, are so similar in their aspect to Nos. 552 and 509, that it is highly probable they will show a corresponding richness in phosphates and other mineral fertilizers, when they come to be analysed, and be capable of affording a supply amply sufficient to compensate for the loss, from cultivation, which could only be *fully* restored, at so ruinous a cost, by ordinary manure.

I would also in this place call the attention of the farmers of Wayne county to the analysis of the sub-carboniferous soil No. 234, of Dr. Peter's report.

It will be perceived that this contains the largest proportion of organic and volatile matters of any soil yet analysed; yet it has been found by the ordinary system of cultivation, unproductive; proving, most conclusively, that blackness and richness in organic constituents is not the only requisite to fertility. It will be seen, moreover, that this soil is also rich in the essential mineral constituents. There can be no doubt that the cause of its unproductiveness is the want of draining and access of air, so that the standing water has, in all probability, produced a sour condition of the soil. For this reason, notwithstanding the considerable proportion of carbonate of lime present, and, also, on account of the large quantity of organic matter in this soil, burnt lime will, undoubtedly, be found beneficial, in order chiefly to counteract that acidity, and then bring the organic matter into a proper condition to resolve itself into carbonic acid, water, and ammonia. With this treatment, and a proper and efficient system of drainage, I will venture to predict that it will become one of the most productive soils in the state. I believe there is a good deal of such land in the valley of Meadow creek, in Wayne county; therefore, it is of considerable moment that it should be reclaimed.

Iron exists in soil in two conditions: in the state of protoxide, with eight parts of oxygen united to twenty-seven to twenty-eight of iron, and as per or sesqui-oxide, containing twenty-four parts of oxygen, and fifty-four to fifty-six of iron. The former of these is continually abstracting oxygen from air, water, and other substances containing oxygen, with which it may come in contact, and therefore exerts a deoxidizing influence injurious to vegetation, which requires all the oxygen it can obtain from the air and water in the soil. Where it exists to any great extent in soils, it appears to exert a baneful influence. The remedy for this is free access of air to the soil, by frequent working, so as to cause this protoxide to pass as rapidly as possible into the state of peroxide. Protoxide of manganese acts in the same way.

There are some facts which seem to lead to the conclusion, that a notable quantity of protoxides of iron and manganese may be the cause of rust. Two soils in the vicinity of Brunswick, containing a considerable portion of these oxides, produced grain which was attacked by rust. One of these was a heath soil, on which the rust was not counteracted even by the application of lime, marl, potash, wood-ashes, bone-dust, ashes of the heath plant, common salt, and ammonia. The other was a fine-grained loam, which did not suffer from want of drainage, as it was well exposed to the sun, in an elevated situation. "In order to ascertain whether the rust was due to the constituents of the soil, or to certain fortuitous circumstances, unconnected with their operation, a portion of the land was removed to another locality, and made into an artificial soil of fifteen inches in depth. Upon this barley and wheat were sown; but it was found, as in the former case, that the plants were attacked by rust; whilst barley growing on the land surrounding this soil was not at all affected by the disease. From this experiment it follows, that certain constituents of the soil favor the developement of rust."

It was thought it might be from the presence of phosphate of iron, but I think it more probable that it was due to the deoxidizing effect of the protoxide of iron; for we know that in our country it is apt to occur, when there is a superabundance of rain, accompanied by hot weather; now the effect of this will, undoubtedly, be to produce acids in the soil, which is almost always accompanied by the absorption of oxygen, an effect quite analogous to that of protoxides of iron and manganese. The per or red oxide of iron, on the contrary, seems to be a valuable constituent of soils; at least it is almost universally the case that red soils are fertile soils, and the redness is always due to the presence of peroxide of iron. It is probable its beneficial effect depends on the property it possesses of condensing ammonia, either from rain water or the atmosphere, and perhaps also to peculiar relations which it bears to certain organic matters, which both it and alumina seem to possess of holding them, so that water alone has little power to remove them.

The red lands of Cheshire and Somersetshire, derived from the disintegration of rocks of the trias system, as well as the red lands of Devonshire and Herefordshire, on the Devonian or Old Red system, are amongst the most productive lands in Great Britain. It is true, that where the members of these formations are, over limited tracts, *exclusively* sandstones—as in parts of Pembrokeshire—the derivative soil cannot be considered fertile; but this is because it is on that account altogether too sandy, and hence so porous that it allows every thing soluble to filter through it; or, in the language of the farmer, "*eats all the manure and drinks all the water.*" But the most loamy of the red lands of Hereford afford the finest crops of wheat and hops, and bear the most prolific apple and pear-trees, and the sturdiest oaks, of all England; and the *red* clay and *red* loamy soils of Cheshire, Somerset, Staffordshire, and Leicester, are capable of producing the most luxuriant growth of almost any crop that may be put on them.

Again: the fertility of the red lands of Texas is well known.

All my experience so far, in regard to the red soils of Kentucky, goes to show that they are very productive soils.

The red sub-soils of Woodford and Fayette, on the blue limestone formation; and of Jefferson county, on the magnesian limestone formation; and of Simpson and other counties, on the sub-carboniferous limestone formation, all show large amounts of phosphoric acid and alkalies. For some further remarks on this subject see Dr. Peter's report, under the head of Simpson county, No. 480.

The blackness of a soil, as we have said, is by no means a sure indication of fertility, at least not of permanent fertility; and, indeed, even not always a criterion of the amount of organic and volatile matters, as is shown by the analysis of a remarkably black soil, No. 237, which only contains 4.38 of organic and volatile matters, while we have many instances of soils, comparativly light in color, that contain more than twice as much of these substances.

# CHAPTER III.

## CHEMICAL ECONOMIC GEOLOGY.

### COALS.

Several of the coals, of which proximate analyses were made, and recorded in the former report, have been subjected since to ultimate chemical analysis, showing the total amount of carbon in coke, and volatile matter united, together with the total amount of hydrogen, oxygen, sulphur, nitrogen and ashes. As such investigtion as contribute greatly in affording an insight into the character and quality of a coal, they are of great importance; and are especially valuable as giving further insight into their gas and oil producing qualities--since the nearer they approach in composition to the atomic proportion of carbon and hydrogen in olefiant, or heavy carburetted hydrogen, or pure oil gas, viz: eighty-six of carbon and twelve of hydrogen, the more valuable they become for these purposes.

All other things being equal, the less water (hygrometric,) the less sulphur, the less ashes, the less oxygen and nitrogen, and the more carbon and hydrogen, the better adapted they are for the production of illuminating gas. These ultimate analyses of coals, (formerly reported only proximately,) are: Mulford's main five-foot coal, No. 185; Casey's coal, No. 166; Union coal, No. 240; Ice House coal, No. 188; Haddock's cannel, No. 166; Eades' coal, No. 157; Airdrie coal, No. 156; Robert's coal, No. 191; Wright's mountain coal, No. 135; Ashland coal, No. 101; Wolf Hill coal, No. 189; Sneed's coal, No. 25; Breckinridge, No. 243.

The results will be seen by consulting Dr. Peter's report, under the heads of the counties in which they occur; also, in the comparative table of coal analysis at the conclusion of that report.

The last of these coals, the Breckinridge, No. 243, has been subjected, by Dr. Peter, to investigations by various methods for its oily products, by which he obtains, on an average, very nearly the same results obtained by myself, viz: thirty-two per cent. of the weight of the coal. For a full account of these results see Dr. Peter's report, under the head of Hancock county, No. 243. Under the same head will also be found an interesting comparative statement of the results in crude oil, gas, coke, and ammoniacal water, obtained from the

Breckinridge, Haddock's, Union, Mulford's five-foot, Robert's, Ice House, and Youghiogheny coals.

The first workable coal under the Anvil Rock, where it has, in some parts of Union county, a peculiar birdseye appearance, and assumes a compact structure, and presents a peculiar glimmering lustre on the oblique faces of fracture, has been analysed in this laboratory for its oily products, with the following results, in 1,000 parts:

264 of crude oil.
645 coke.
48 good illuminating gas.
43 ammoniacal water.
———
1,000

On the large scale this coal will yield, from a ton of 2,000 pounds, 500 to 528 pounds of crude oil; equal to 60 to 66 gallons, or 30 to 33 gallons of purified oil; 1,290 pounds of coke; 2,238 cubic feet of good illuminating gas.

From the best information, at present, in my possession in regard to the cost of manufacturing, though the quantity of oil which it will yield is seven per cent. less than that from the Breckinridge, still this birdseye coal ought to afford a handsome profit to the manufacturer of coal oils, &c.

Since the publication of the first report, both the proximate and ultimate analysis of seventeen other coals have beem completed, as follows:

Woolich coal, Christian county, No. 462;
Garrard coal, Clay county, No. 460;
Triplett's coal, Daviess county, No. 502;
Hawesville Little Bed coal, Hancock county, No. 468;
Judge Mayhall's coal, Hancock county, No. 519;
Pate's coal, Hancock county, No. 520;
Hall's coal, Hopkins county, No. 463;
Samuel's coal, Hopkins county, No. 465;
McHenry's coal, Lawrence county, No. 466;
Keener's coal, Lawrence county, No. 469;
Walker's coal, Muhlenburg county, No. 464;
Pitchener's coal, Ohio county, No. 459;
Barrett's coal, Ohio county, No. 470;
Jackson's coal, Ohio county, No. 461;

Haddock's cannel coal, Owsley county, No. 161;
Lear's coal, Pulaski county, No. 467;
Lower Cumberland, Pulaski county, No. 471.

Of these the Haddock cannel coal, from Owsley county, proves to be a coal producing 246.50 parts in a 1,000 by weight of crude oil, which is only 17.50 in a 1,000 parts less than the Birds-eye coal of Union county, and 77.50 less than the Breckinridge. It will yield, therefore, from fifty-five to sixty gallons of crude oil, or about twenty-seven to thirty gallons of purified oil to the ton. These three coals are the best oil producing coals that have yet been examined from Kentucky.

For further remarks on these coals consult Dr. Peter's report, under the respective numbers above stated, who has has taken unusual pains in the developement of their chemical properties. He has also presented, in connection with the Breckinridge coal, under the head of Hancock county, the comparative chemical composition of the oil producing Boghead coal of Scotland.

For comparison with Kentucky coals, as to their gas producing qualities, I here subjoin some proximate and ultimate analyses made in this laboratory of four varieties of coal used, in 1856, by the Manhattan gas company, in New York, for supplying that city with gas.

*Albert coal, Hillsboro', Albert county, New Brunswick.*

Proximate analysis—

| | | |
|---|---|---|
| Specific gravity: | | |
| Total volatile, | | 61.4 |
| Coke, | | 38.6 |
| | | 100.0 |
| Moisture, | 1.5 | |
| Carbon, | 84.972 | |
| Hydrogen, | 9.144 | |
| Sulphur, | 0.000 | |
| Oxygen and nitrogen, | 4.284 | |
| Ashes, | 0.100 | |

Calculated for 100 parts of dry combustile matter we have:

| | |
|---|---|
| Carbon, | 86.250 |
| Hydrogen, | 9.292 |
| Sulphur, | 0.000 |
| Oxygen and nitrogen, | 4.458 |
| | 100.000 |

This an excellent oil-producing coal, as well as a fine gas coal.

*New Castle, England.*

| Proximate analysis: | |
|---|---|
| Total volatile matter, | 30.1 |
| Coke, | 69.9 |
| | 100.0 |

| | |
|---|---|
| Moisture, | 1.2 |
| Volatile combustible matter, | 28.9 |
| Fixed carbon in coke, | 69.3 |
| Ashes, | 0.6 |
| | 100.0 |

| | |
|---|---|
| Moisture, | 1.2 |
| Carbon, | 80.063 |
| Hydrogen, | 4.944 |
| Sulphur, | 1.000 |
| Oxygen and nitrogen, | 12.193 |
| Ashes, | 0.600 |
| | 100.000 |

Calculated for 100 parts of dry combustible matter this gives:

| | |
|---|---|
| Carbon, | 81.530 |
| Hydrogen, | 5.034 |
| Sulphur, | 1.018 |
| Oxygen and nitrogen, | 12 418 |
| | 100.000 |

*English Cannel Coal.*

| Proximate analysis: | |
|---|---|
| Total volatile matter, | 43.6 |
| Coke, | 56.4 |
| | 100.0 |

| | |
|---|---|
| Moisture, | 1.2 |
| Volatile combustible matter, | 42.4 |
| Fixed carbon in coke, | 54.0 |
| Ashes, | 2.4 |
| | 100.0 |

Ultimate analysis of the same with ashes and moisture included in the per centage:

| | |
|---|---|
| Moisture, | 1.200 |
| Carbon, | 78.982 |
| Hydrogen, | 5.711 |
| Oxygen and nitrogen, | 11.707 |
| Ashes, | 2.400 |
| | 100.000 |

Calculated for 100 parts of dry combustible matter this gives:

| | |
|---|---|
| Carbon, | 81.931 |
| Hydrogen, | 5.924 |
| Oxygen, | 12.145 |
| | 100.000 |

*Scotch Cannel.*

Proximate analysis:

| | |
|---|---|
| Total volatile matter, | 45.9 |
| Coke, | 54.1 |
| | 100.0 |

| | |
|---|---|
| Moisture, | 3.2 |
| Volatile combustible matter, | 42.7 |
| Fixed carbon in coke, | 42.1 |
| Ashes, | 12.0 |
| | 100.0 |

Ultimate analysis of the same with ashes and moisture:

| | |
|---|---|
| Moisture, | 3.200 |
| Carbon, | 70,582 |
| Hydrogen, | 9.788 |
| Sulphur, | 1.400 |
| Oxygen and nitrogen, | 3.030 |
| Ashes, | 12.000 |
| | 100.000 |

Calculated from 100 parts of dry combustible matter

| | |
|---|---|
| Carbon, | 83.233 |
| Hydrogen, | 11.542 |
| Sulphur, | 1.650 |
| Oxygen and nitrogen, | 3.575 |
| | 100.000 |

Four of the best gas coals used in England and Scotland are: Lesmahago Cannel coal; Ramsay's New Castle coal; Weymss Cannel coal; Wigan Cannel coal. The specific gravity, and proximate analyses, of these coals I here append, for the sake of comparison:

Specific gravity, - - - - - - - - 1.222 to 1.228

*Two proximate analyses of Lesmahago Cannel coal:*

| | *A.* | *B.* |
|---|---|---|
| Volatile matters, - - - - - - - - | 49.6 | 49.34 |
| Carbon, - - - - - - - - - - | 41.3 | 40.97 |
| Ashes, - - - - - - - - - - | 9.1 | 6.34 |
| | 100.0 | 96.65 |

A ton of this coal yields of

| | | | |
|---|---|---|---|
| Coke, - - - - - - - - - - - | 1091. | to | 1064. |
| Pounds of gas, - - - - - - - - - | 463. | to | 483.5 |
| Tar, - - - - - - - - - - - | 594. | to | 603. |
| Ammoniacal water, - - - - - - - | 4.5 | to | 4.5 |
| Loss, - - - - - - - - - - - | 87.5 | to | 85. |
| | 2240. | | 2240. |

This is equal to 11.681 to 9.875 cubic feet of gas, of specific gravity 0.361, 0.54, to 0.65; having an illuminating power of 2.33; gas of specific gravity 0.36, being considered one or unity.

*Ramsay's New Castle Coal.*

| | | |
|---|---|---|
| Specific gravity, - - - - - - - - | 1.29 | |
| Volatile matter, - - - - - - - - - - | | 36.8 |
| Fixed carbon in coke, - - - - - - - - - | | 56.6 |
| Ashes, - - - - - - - - - - - - | | 6.6 |
| | | 100.0 |

A ton of this coal yields of

| | |
|---|---|
| Coke, - - - - - - - - - - - - - | 1435. |
| Gas, - - - - - - - - - - - - - | 410. |
| Tar, - - - - - - - - - - - - - | 295. |
| Ammoniacal water, - - - - - - - - - | 6.72 |
| Loss, - - - - - - - - - - - - - | 93.28 |
| | 2240. |

This is equal to 9,016 cubic feet of gas, of specific gravity 0.604, and having an illuminating power of two, by the same standard as the preceding.

*Weymss Cannel Coal.*

| | | |
|---|---|---|
| Specific gravity, - - - - - - - - - | 1.1831 | |
| Volatile matters, - - - - - - - - - - | | 58.52 |
| Fixed carbon in coke, - - - - - - - - - | | 25.28 |
| Ashes, - - - - - - - - - - - - - | | 14.25 |
| | | 98.45 |

A ton of this coal yields of

| | | |
|---|---|---|
| Coke, - - - - - - - - - - | 1124. | to 1188 |
| Gas, - - - - - - - - - - | 551.4 | to 528 |
| Tar, - - - - - - - - - - | 224. | to 197 |
| Ammoniacal water and loss, - - - - - - | 341. | to 327 |
| | 2240. | 2240 |

This is equal to 10,976 to 10.192 cubic feet of gas, of specific gravity 0.670 to 0.691, having an illuminating power of 2.47, according to the same standard.

*The Wigan Cannel Coal.*

| | |
|---|---|
| Specific gravity, | |
| Volatile matter, - - - - - - - - - - | 37 |
| Fixed carbon in coke, - - - - - - - - - | 60 |
| Ashes, - - - - - - - - - - - - | 3 |
| | 100 |

A ton of this coal yields of

| | |
|---|---|
| Coke, - - - - - - - - - - - - | 1326 |
| Gas, - - - - - - - - - - - - | 338 |
| Tar, - - - - - - - - - - - - | 250 |
| Ammoniacal water and loss, - - - - - - - - | 326 |
| | 2240 |

This is equal to 9,408 cubic feet of gas, having a specific gravity of 0.478, and an illuminating power of 1.5, according to preceding standard.

At present I am only able to furnish the *ultimate* analysis of one of these coals—the *Wigan Cannel:*

| | |
|---|---|
| Carbon, - - - - - - - - - - - | 79.23 |
| Hydrogen, - - - - - - - - - - - | 6.08 |
| Sulphur, - - - - - - - - - - - | 1·43 |
| Nitrogen, - - - - - - - - - - - | 1.18 |
| Oxygen, - - - - - - - - - - - | 7.24 |
| | 95.16 |

The investigations of the past two season have fully proved the fact that, in the western coal field at least, the same bed of coal varies considerably, both in its thickness and chemical composition; for instance, the first workable coal under the Anvil Rock ranges, in thickness, from twenty-one inches to four feet, and that in the distance of ten miles. At the most southwestern of the above localities it has a beautiful "birds eye" structure, possesses the properties of some varieties of cannel coal, and contains nearly ten per cent. more volatile matter, over fifteen per cent. less fixed carbon, and eight per cent. more ashes, than it does ten miles to the north-west, where it posseses no cannel structure. Hence coal locations, even on the same bed, a few miles apart, require special examinations, both as to thickness and chemical composition; and we are not justified in deciding, because a coal is good or bad at one mine, that the same coal may not be better or worse at any other, although the *probabilities* may be in favor of analogy in the same bed, throughout its area.

The subject of the equivalency of our Coal Measures is of peculiar interest, as well in a scientific as in a practical point of view, both as regards their horizontal extension through the different counties in the western coal field, as, also, by drawing comparisons between the western and eastern coal field, now separated by an axis of older rocks, occupying nearly one hundred and fifty miles of the surface, from the present limits of the western to the eastern coal field. At this time I am only able to offer a few suggestions on this subject.

Previous to the survey of Kentucky, the only state which had made the detailed stratification of the coal formation a special investigation is Pennsylvania. Unfortunately the public has not yet had the final report of the state geologist laid before them. Until its appearance we shall not be put in possession of all the facts, especially the palaeontological data, collected during the survey of that state. A very valuable work has, however, recently been issued by J. P. Lesley, the topographical assistant of that survey, containing some excellent generalizations on this subject, which affords us an opportunity of drawing some comparisons from lithological grounds and order of superposition, and, perhaps, something from topography; though these features, in western Kentucky, are, by no means, as conspicuous as in Pennsylvania.

It appears, from Lesley's "Manual of Coal and its Topography," that the Pennsylvania Coal Measures are about 1,300 feet in thickness, 900 feet of which lie from the Pittsburg coal down to the conglomerate, and four hundred feet above that bed. The four hundred feet immediately underlying the Pittsburg coal are destitute, except locally, of any workable beds of coal, and have, on this account, received the name of the "*Barren Coal Measures*," these being limited downwards by a prominent sandstone, known as the "*Mahoning Sandstone*." This sandstone is sometimes pebbly, though it lies five hundred feet above the true conglomerate, at the base of the Coal Measures, and, for this reason, as suggested by Lesley, would very properly form the base of the *Upper* Coal Measures, as the true conglomerate is the base of the *Lower* Coal Measures.

Six principal beds of coal are enumerated between the true conglomerate and the Mahoning sandstone, forming the Lower Coal Measures. The two lower of these are sometimes united into a thick bed, known as the "Mammoth Bed." The third in the ascending series, (or properly the second since A and B, of Lesley's section, may regarded as one,) is the great depository of the cannel coal of Pennsylvania. The fifth and sixth are known as the Lower and Upper Freeport coals.

From the Mahoning sandstone to the Pittsburg coal, a distance of about four hundred feet, there are but two beds of coal specified, and these, usually, only a foot or thereabout in thickness, with some four or five beds of limestone in the upper one hundred feet, immediately underlying the Pittsburg coal.

Less is known, as yet, of the details of the four hundred feet above the Pittsburg coal than of the measures below it. It includes, however, the "Great Limestone" of the Pennsylvania Coal Measures, separated into two members by a sandstone, more or less developed, which lies about one hundred feet above the Pittsburg bed, in which space there is one bed of coal between two limestones, measuring, with its shale roof, about three and a half feet. Above the "Great Limestone" there is but one good workable, reliable coal, lying about sixty feet above the "Great Limestone," characterized by a shale roof full of ferns: *Neuropteris flexuosa, Cyclopteris obliqua, and Neuropteris loschii.*

Taking, for the present, merely lithological and chemical characters as the ground-work of comparison between the Kentucky and Pennsylvania Coal Measures, we are disposed to regard the Mahoning sandstone as the equivalent of our Curlew sandstone, of Union county, since it forms the second principal sandstone, in the ascending order, above the conglomerate, and has a coal, with limestone under it, in the Curlew hill, occupying, apparently, a similar position to the Upper Freeport or coal E of Lesley's section, with limestone (E) under it; while his description of the contorted Sigillaria, Lepidodendron, and Calamites sandstone, underlying his coal D, (the Lower Freeport,) with its lenticular deposites of coal, agree very precisely with the mass of the Finnie Bluff, of Union county, which is the second principal sandstone above the conglomerate.

If the equivalencies here suggested should prove, by subsequent investigations, correct, then coal A and B, of the Pennsylvania Measures, would occupy the place of the Bell and Cook, resting on the Caseyville sandstone conglomerate, and being the equivalent of the so called Mammoth vein, near Johnstown and Tangastootac, where coals A and B are united into one bed, practically speaking. If this be the case, then in the place of the lower hundred feet of the "Barren Coal Measures," of Pennsylvania, we appear to have two good workable coals in Union county, and only a space of one hundred and eighty feet between these and the main five foot Mulford bed, to represent the three hundred upper feet of the Barren Coal Measures of Pennsylvania. The main five-foot Mulford would, then, be the equivalent of the Pittsburg bed. A circumstance which lends probability to this suggestion is found in the fact, which I would beg, in this place, to call particular attention to, that, of all the Kentucky coals examined chemically, up to this time, the main five-foot Mulford comes nearest, in its composition, to the Pittsburg coal.

In the one hundred feet above the main five-foot Mulford, and between it and the Anvil Rock, in Union county, we can trace, as yet, but little analogy to the one hundred above the Pittsburg, with its "Great Limestone Bed," which seems to be absent, in Union county at least. In the three hundred feet of the upper division of the Upper Coal Measures we can trace some analogy with the borings reported below the level of Gruudy's ridge; but until these strata can be viewed in natural sections, above the drainage of the country, one can hardly come to any definite conclusion in regard to the equivalency.

If any part of the western Kentucky coal field deserve the name of *barren*, it seems to be the two hundred and fifty feet of space from the Ice House to the four-foot bed, the first above the Curlew sandstone; but we doubt the propriety of any such distinction, since the Curlew coal, below the Curlew sandstone, may prove to be workable in some of its area, as at Freeport, in Pennsylvania, where it is six feet thick.

Partly from differences in the chemical composition of several of the upper coals of the Henderson section, and those under the Anvil Rock, in Union county, and partly from variations in the lithological characters of the associate strata at each locality, together with some relations of dip and position in the coal-field, the two sections were supposed, in the first year of the survey, to occupy lower and upper divisions of the Coal Measures. During the progress of the survey facts have come to light which render it possible that the same measures which occur in Union county, under the Anvil Rock, are repeated in Henderson county, in the Holloway borings, under the level of the Ohio, on the east side of the Bald hill and Highland creek axis. In consequence, however, of our knowledge of the lower two-thirds of the Henderson Measures being derived from the reports of borings, it has been difficult to draw a satisfactory parallelism. If, however, the Great Salt-bearing sandstone occupies a constant geological horizon, at the base of the Coal Measures, as appears probable from the experience in Pennsylvania, and if the western Coal Measures are of a corresponding thickness to the eastern, then the four foot bed of coal, one hundred feet below the bed of the Ohio river, in the Henderson shaft, would be the equivalent of the Pittsburg coal; and the "Twin Vein," near low water mark, would, in that case, represent coals near the top of the Upper Measures, near the "Great Limestone," while the thick coal reported on the top of the one hundred and fifty-six feet of Salt-bearing sandstone, at the base of the Holloway borings, would occupy the place of the "Mammoth coal," reposing on the millstone grit, at the base of the Pennsylvania coal-field. We are led, therefore, to suspect that the Holloway borings of 1,037 feet, may include members of the lower, as well as the upper, division of the Coal Measures, as recognized by Lesley's sections, in Pennsylvania; the more so, since we have some evidence of considerable masses of limestone in the water-shed between Green river and the Ohio, not far from the Holloway borings; nevertheless we are, at present, unable to recognize, in the space of three hundred feet reported in the Holloway borings, above the Salt bearing sandstone, either the lower cannel coal, (C,) of Lesley's section, or the Ice House coal, (D,) of the Union county section, which should come in respectively—the cannel at sixty to eighty feet, the Ice House at one hundred and fifty feet above the lowest six-foot coal, placed on the top of the Great Salt-bearing sandstone.

However, to established satisfactorily these identities or equivalencies between the eastern and western coal-fields, as well as between different portions of the same connected basin, it is necessary to bring all the palaeontological evidence to bear on the subject, especially that relating to the vertical range of certain species of fossil plants, in connection with the shells and fishes, and the relation they bear to the shale over each bed of coal. In the absence of any botanico-geological report of the Pennsylvania Coal Measures, we have to rely upon other sources for this desired comparison.

Before the whole of the second part of this report goes to press I hope to be able to lay before you some reliable and valuable information, touching this branch of the survey, since I have been fortunate in securing the aid of the highest authority now living, on the identification of the specific characters of the fossil flora of the United States—M. Leo Lesquereux, the well known Bryologist, who was, for two years, charged with the botanical branch of the geological survey of Pennsylvania; whose early life first led him to study minutely the flora of peat bogs, and investigate the production of that kind of vegetable fuel, of comparatively recent origin, in all its various phases; and who, from this study, was very naturally induced to turn his attention to the examination of the extinct tribes of plants that flourished during the carboniferous era. One who has made these two subjects a speciality, during a life of laborious research, must be peculiarly well suited to supply the desired information, particularly as a considerable share of his experience was gained in Pennsylvania, the stratigraphical relations of the Coal Measures of which state we desire, at this time, to bring in comparison with those of Kentucky.

I am unable to say at present, whether, before the second part of this report must go to press, sufficient materials shall have been brought together to enable M. Lesquereux to make a satisfactory preliminary report on this subject, as the large amount of the appropriation required to be expended on the topography of the state will only permit me to retain M. Lesquereux's services, in the geological corps of Kentucky, during a limited part of the present season; but the importance of this branch of the geological survey of Kentucky demands that it should hereafter be carried out to its fullest extent, not only for the interests of Kentucky, but for that of other states embraced in coal districts, which may follow in detailed surveys of their resources in fossil fuel.

## ORES.

Three iron ores, from the Knob Formation of Bullitt county, have been analysed—Nos. 488, 489, and 493. Two of these—Nos. 488 and 493—belong to the carbonates; one—No. 489—is a limonite. By consulting the tables appended to Dr. Peter's report, the composition of these ores will be seen; the two carbonates yield—No. 488, 32.62 per cent. of iron; No. 493, 31.32 iron. The limonite—No. 489—gave 43.46 of iron. This latter is in an irregular band and segregations, locally found in the equivalent of the Knob building stone, and is but little used at the Bullitt county furnace. Where these ferruginous segregations occur in the Knob building stone, they greatly deteriorate its quality for constructions.

For further remarks in regard to the Bullitt county ores, and limestones used as a flux, slag, &c., consult the body of Dr. Peter's report, under Nos. 488, 489, 490, 491, 492, and 493.

The limonite ore of Carter county, which is associated with the subcarboniferous limestone, proves to be a rich ore; the analysis shows 54.93 per cent. of iron. It contains, as will be seen by consulting Dr. Peter's report, under No. 473, more than the usual quantity of oxide of manganese, (3.17,) which will most likely have the effect of producing a hard white iron suitable for the manufacture of steel; not perhaps from the quantity of manganese which may enter into the composition of the steel produced from such cast iron, but from the peculiar condition and union which the manganese induces in the iron and carbon.

The chemical analyses of twenty-eight ores, of the limonite variety, have been completed up to this time, since the first report was made; these are from the Kenton, Caroline, Pennsylvania, Buffalo, Bellefonte, Raccoon, Greenup, Laurel, and Mt. Savage ore banks.

Some very interesting results will be found in the chemical report, under the head of Greenup county, in regard to some of these ores.

There have, besides, been completed ten analyses of ores of the carbonate variety, from Greenup county, from ore banks in the vicinity of the above furnaces.

The ore No. 406, submitted for analysis, from Laurel county, by General Jackson, is an earthy carbonate of iron, containing 19.10 per cent. of iron, and 3.60 of manganese. No other metal, useful in the arts, exists in this ore in quantities over one and a half per cent. The carbonate of iron—No. 410—from Craig's creek, proves to be richer in iron than No. 406, since it gave 35.45 per cent. of iron. No. 411, from the same county, is also carbonate, yielding 33.05 per cent. of iron, and no other valuable metal, in sufficient quantity to make it valuable for any other purpose but as an ore of iron.

Ore furnished from Thos. Helm's land, Lincoln county, No. 407, has also been examined. It is a carbonate of iron, containing 30.77 per cent. of iron.

No. 418, the Monroe county ore, from Malone's land, Cole's fork of Mill creek, is a limonite, containing 53.85 per cent. of iron. No. 405, from Ohio county, 7 miles north of Hartford, near the top of the hill at Mr. French's, is also a limonite, yielding 27.64 per cent of iron.

The ore near high water-mark of Green river, at the Livermore landing, in Ohio county, No. 413, contains 42.14 of iron, which is a large per centage for that variety of ore.

Two ores of iron have been analysed from from Pulaski county: one—No. 452—from the head waters of Indian and Rock House creek, and Grassy gap, is a carbonate, yielding 35.60 per cent. of iron; and No. 546, from Rockcastle river, which is an impure limonite, containing only 27 per cent. of oxide of iron.

Two ores have been analysed from Trigg county: one—No. 420—from the waters of Little river, which is a limonite, possessing a fine honey-comb structure, which yields 39.28 per cent. of iron; the other, from the same vicinity—No. 421—is a hæmatitic variety of limonite, having a compact texture, which yields 55.60 per cent. of iron. There appears to be a considerable body of this ore in Trigg county, at the above locality.

The ore from Warren county, collected in a ridge watered by Clay lick creek, above the conglomerate, is also a hydrated oxide, containing 47.02 per cent. of iron.

No. 450 is a bog ore, from Meadow creek, in Wayne county, which only contains 16.59 per cent. of iron; and No. 453 is the ore formerly used at the old Iron Works of Wayne county, which is a limonite, containing 40.82 per cent. of iron.

Ores from the Log Mountain, in Whitley county, and from the falls of the Cumberland river, in the same county, have been subjected to further examinations by Dr. Peter, since the first report, with similar results. See his report under the head of Whitley county.*

An ore from the head waters of Mud creek, in the same county—No. 448—is a rich limonite, yielding 56.37 per cent of iron.

The Poplar creek ore—No. 449—of Whitley county, is a carbonate of iron, containing 37.60 per cent. of iron; that from south part of the Pine Mountain is a limonite, affording 44.53 per cent. of iron.

Four ores have been analysed from Edmonson county; three of these from near the base of the Coal Measures, in the ridges adjacent to Nolin creek—Nos. 414, 415, 416, and 419. Two of these—Nos. 414 and 415—are limonites; No. 414 contains forty-two per cent. of iron; No. 415, from the shale above the coal, yields 52.31 per cent. of iron; the third—No. 416—is a carbonate, from the shale† above the sandstone, gives 37.04 per cent. of iron; No. 419 is a dull yellowish-brown limonite ore, obtained in shales above the first coal over the conglomerate; it yielded 43.50 per cent. of iron.

Two varieties of pig-iron have been analysed, produced at the Laurel Furnace, in Greenup county—No. 435 and 434.

Five different furnace slags have been analysed: two produced at the Bellemont Furnace, in Bullitt county—Nos. 491 and 492; one produced at the Buena Vista, of Greenup—No. 330; two at the Caroline Furnace, in Greenup, No. 423 (*A*,) a granular variety, and No. 423 (*B*,) a glassy variety. For details and inferences in regard to these the reader will please consult Dr. Peter's report.

Six limestones have been examined, chemically, from Greenup county, since the publication of the first report, viz: No. 481, used as a flux at the Buffalo Furnace; No. 477, under the limestone ore, used

*For the results obtained previously, see first report, page 225.

†See the Edmonson county section, in the first report.

as a flux at Pennsylvania Furnace; No. 433, used as a flux at the Laurel Furnace; No. 432, ferruginous limestone, running upwards into limestone ore, Oldtown creek; No. 426, ferruginous limestone, on which the limestone ore rests, Caroline Furnace; No. 427, four-foot ferruginous limestone, under the limestone ore, Caroline Furnace.

Three limestones from Bullitt county have also been analysed, one of which—No. 490—is used as a flux at Bellemont Furnace, intercalated in the black shale at the base of the Knobs.*

These analyses of Kentucky ores, supply ample evidences of the abundance of rich ores which exists in the state, both interstratified in the Coal Measures, conformable and associated with its coals, disseminated and associated with the sub-carboniferous limestone, as well as in the shales of the Knob Formation, at the base of the sub-carboniferous rocks.

The best ores collected, up the present time, in the Coal Measures proper, have been found in the *Lower* Coal Measures; there is, however, some excellent black-band ore high up in the *Upper* Measures of Muhlenburg county, but, so far as we have seen, only six inches thick; and there appears to be a considerable quantity of iron stones in the Upper Coal Measures, in the shaly beds, lying some distance under the Bonharbor coal.

A specimen of ore was sent to Dr. Peter for analysis, by O. C. Winburn, said to be obtained from near Irvine, in Estill county, which proved to be an ore containing 21.13 per cent. of copper. The locality of this ore has not yet been examined. If ore of this per centage exists in a regular vein it is rich enough to work, being above the average of copper ores in its yield of that metal. I propose to visit this locality in June or July of the present season, examine, if possible, its situation, and ascertain its geological position.

The zinc ore of Monroe county—No. 454—which we found running in slender veins through limestone belonging to the Devonian Period, in the bed of Sulphur Lick creek, has been analysed. It is essentially a sulphuret of zinc, containing 51.77 per cent. of zinc. The sulphuret is combined in this ore with 17.48 per cent. of silica, besides 5.19 per cent. of carbonates of lime and magnesia, and a little disseminated sulphuret of lead.

*For further particulars in regard to these limestones consult Dr. Peter's report, under the heads of Greenup and Bullitt county.

Of all the limestones examined up to this time, the pure magnesian limestones are, probably, the most durable building material, at least amongst the limestones. There are two or more beds of this kind of rock interstratified, in the Birds-eye limestone, forming the cliffs of Boone creek, in Fayette county. Grimes has opened a quarry into these magnesian limestones, which has supplied some of the best building stone of central Kentucky. The same beds are seen occupying a similar geological position near Roger's mill, a mile and a half from Clay's ferry, on the Kentucky river. Dr. Peter has analysed three varieties of this rock: one—No. 512—collected at Grimes' quarry, on Boone creek, which proves it to be a very pure magnesian limestone—the carbonate of lime and carbonate of magnesia being very nearly in the proportion of their equivalents, while the insoluble earthy silicates only amounts to 2.79 per cent.; the alumina and fixed alkalies exists in it only to the amount of fractions of one per cent. The other variety—No. 511—which lies in layers from five inches to one foot above the last, is not as pure a magnesian limestone; it contains a little more than ten per cent. less carbonate of magnesia, and near ten per cent. more earthy silicates. This is not as good a building stone as the other. The third specimen—No. 513—was taken from a corner stone placed as a monument of the boundary of the city of Lexington; it is intermediate in quality, containing three and a half per cent. less carbonate of magnesia, and 0.59 per cent. more insoluble earthy silicates, than No. 512, with a much larger per centage of potash (2.35,) than either of the other two, the origin of most of which may possibly be extraneous, as suggested by Dr. Peter, in his remarks following the analysis of the specimen No. 513; to which article of his report I would call the attention of those interested in the building materials of the State.

In this place I would beg particularly to direct to the notice of the citizens of Jefferson county a magnesian limestone, belonging to the Upper Silurian Period, lying near the surface in many places in the eastern part of that county, and quite convenient to the Louisville and Frankfort railroad. The cellular variety—No. 528—probably contains a little too much magnesia, and is rather too porous to be as good a building material as the best of Grimes' quarry, but the compact variety—No. 530—differs but very little in its chemical composition from the best magnesian limestone of Fayette county; and will undoubtedly be far superior in durability to many of the building stones in use in Jefferson county.

Two of the limestones before mentioned as having been analysed from Bullitt, are also magnesian limestones, belonging to the Upper Silurian system. In them the magnesia falls from eight to ten per cent. short of the best proportion for a building stone; and No. 495 has 10.32 per cent. of earthy insoluble silicates. The variety No. 494 has, however, but 2.18 per cent. of earthy silicates, and will probably make a good building stone.

Five limestones have been analysed, which were supposed to possess hydraulic properties from their appearance, fracture, and mode of weathering. Of these, No. 521 is the hydraulic limestone of Jefferson county, of which such large quantities were excavated out of the Louisville canal, and have since been extensively manufactured into water cement, and sold throughout the western country, for all purposes for which such cement is applicable. It is an earthy limestone, of a slightly bluish-green ashen tint, with an earthy flat conchoidal facture. It contains, as its principal and characteristic constituents, 28.29 parts of lime, 8.89 of magnesia, and 25.78 per cent. of earthy insoluble silicates, of which 22.58 is pure silica. It is particularly worthy of note that the lime and silica in this celebrated and well known cement rock, are exactly in the proportion of their equivalents, proving, most conclusively, that its hydraulic properties are due to this definite chemical relation of these substances, which, after the rock is properly burnt and ground, unite, in connection with the water, to form a hydrated silicate of lime, in which there is one equivalent of silicic acid united to one equivalent of lime, which acts as a powerful cement, to agglutinate the grains of sand added in the mixed mortar, which is usually three times the bulk of the hydraulic lime employed. In some hydraulic limestones magnesia seems to form an important constituent, and it is possible that in this cement it may enter into chemical union, and act a *subordinate* part. It has been remarked, by the investigators of hydraulic limestones, that they contain more than the ordinary proportion of fixed alkalies. This Kentucky cement rock contains 0.32 of potash, and 0.13 of soda, which is rather above the average amount in the limestones analysed, though not quite as much as the *Leptaena* limestone of Woodford, No. 547. These alkalies probably only act in connection with the lime in the kiln, after the expulsion of the carbonic acid, in bringing the silica into that condition favorable for its subsequent prompt combination with lime, when mixed with water, and in the act of cementation. But it is not likely that the alkalies enter, as an essential constituent, into the cement, because it seems to be expelled with the surplus water which oozes away as a slippery ley; and, because they would, probably, impair the durability of the cement, as they no doubt contribute to the disintegration of the original rock in place, so remarkable for the facility with which it cracks, splits, and turns to calcarious mud, where exposed to the vicissitudes of the weather.

In No. 456, the Grayson county "gravestone," the lime is 26.28, and the insoluble silicates 20.78. The pure silica was in this specimen not isolated and determined; but it is more than probable that it will be found but little under the definite chemical proportion in the Jefferson county hydraulic limestone. In the Grayson stone the alkalies are in still larger proportion: 0.50 potash, and 0.37 soda. There is little doubt, therefore, that it will be found to have hydraulic properties, and is altogether unsuitable for building purposes.

The four-foot bed, near the base of the Upper Silurian system, of Trimble county, Kentucky, and Jefferson county, Indiana, and from seventy-five to eighty feet above the Murchisonia marble, of the Lower Silurian system, has very nearly the same constitution as the Louisville water limestone, according to an approximate analysis which I made of that rock in 1853. One variety contains 28 per cent. of lime, and 25 per cent. of insoluble silicates, and 10.06 per cent. of magnesia, which is within a fraction of the same amounts of these substances found in the Jefferson county hydraulic limestone. There is hardly a doubt that this rock has also hydraulic properties, which of course is sufficient to condemn it at once as a building stone, though the same rock has been largely used in the construction of one of the principal public buildings in Louisville.

Of the two limestones from Trigg county, supposed from their appearance, to possess hydraulic properties, one, No. 458, contains 28.61 lime, and 13.68 of earthy insoluble silicates, and 14.77 per cent. of magnesia. In this the silica is considerably under the equivalent number, but the lime is nearly in the same per centage as in No. 521. It is somewhat remarkable that this rock has 0.92 per cent. of phosphoric acid—even a larger proportion of this ingredient than has been found in any other limestone except No. 484, a limestone from the Lower Silurian system of Anderson county, and some of the dark ferruginous, bituminous, limestones occupying the place of the black band horizon in some parts of Hopkins county; as for instance, No. 132, in the first report; it would, therefore, be a valuable limestone for agricultural purposes also. There is considerably more magnesia (14.77,) than in No. 521; this may be no disadvantage, because it appears that magnesian limestones, with proper additions, make excellent hydraulic mortars—better than limestones without magnesia; because silicic acid seems to manifest a preference to combine with two bases rather than one; for instance, the Tarnowitz limestone, which contains 10 per cent. of magnesia, and only 3.3 per cent. of silica hardens very well under water, even without any addition of silica; indeed, Fuchs "has shown that the two bases—lime and magnesia—can combine and harden under water, without the addition or presence of any silica; and Dumas, also, has shown that nearly equal equivalents of burnt lime and magnesia—i. e. 44 lime to 36.8 magnesia—when slaked and made into a paste with water, become tolerably hard after being left under water for nine days." In fact the investigations of modern science all go to show that the reaction of silicic acid on lime, and that exerted between lime and magnesia, are the two principal causes of the solidification of hydraulic limestones; but the degree of solidification depends much upon the molicular state of the silica, and the amount of base already in combination with it. It appears, moreover, that when the proportion of lime rises to 48, and the silica to 52—i. e. 3 equivalents of lime to 2 of silica—there is no disposition in such a compound to harden under water, either in the burnt or unburnt condition. In the case of No. 457, from Trigg county, in which the lime is 43.91, the magnesia 7, and the silica 8.36, the proportion is one equivalent of lime to combine with one equivalent of silica, to form silicate of lime, and another equivalent of lime to combine with one equivalent of magnesia; hence it is highly probable that this rock will be found also to possess valuable hydraulic properties.

Of the five sandstones, of which the analysis is recorded in the first part of this report, three are from the Knob Formation of Bullitt. They are not as durable rocks as the fine pure gritstones of the Coal Measures, but, by a proper selection, rejecting those beds that contain segregations of oxide of iron, and too much argillaceous matter, they afford a building material that works free under the chisel, is susceptible of fine carving, and is of tolerable durability. Those of Bullitt county, that have been examined, contain from 93.68 to 94.78 of fine sand, and 2.85 to 3.95 of alumina, with small portions of oxide of iron, only traces of lime, and 0.7 to 2.29 of carbonate of magnesia.

Two varieties of the mudstone have been analyzed, which is interstratified in the blue limestone underlying the beech flats, spoken of in the second chapter, in connection with the soil of the beech flats of Fayette. This is a more argillaceous rock than the preceding sandstone, containing from 8.65 to 10.25 alumina, with a little oxide of iron, only a trace of lime, 1.40 to 2.30 of carbonate of magnesia, and the largest proportion of sulphuric acid in any sandstone yet examined—the very ingredients, therefore, calculated to produce, by its decomposition, astringent salts. It is worthy of especial note, in this connection, that the only regions in the range of the blue limestone formation, where I have found milk-sickness prevail, follows most remarkably the out-crop of this mudstone, as I have elsewhere stated.

### MINERAL, SPRING, AND WELL WATERS.

Since the first report was made, thirty-eight mineral and other waters have been examined and analyzed. Of these, twenty-five have been tested, qualitatively, at the fountain head; while quantitative analyses have been made of thirteen in the laboratory, which will be found tabulated and reported by Dr. Peter, under the head of Lincoln county, in the first part of his report. The twenty-five which were tested at the fountain head I now proceed to report on.

Yelvington Spring, Daviess county, showed the presence of
Free carbonic acid;
Bi-carbonate of lime;
Bi-carbonate of magnesia;
Chloride of sodium;
A trace of oxide of iron.

There is perhaps a trace of carbonate of alkali, which may give the flat taste to this water, after it has stood for some time; but the reaction of the reagents which detect these salts is so indistinct, in the unconcentrated water, that it is difficult to decide. Part of the magnesia present may be united with chlorine as chloride of magnesium,

If sulphates are present they are in too small quantities to be detected with any degree of certainty, without boiling the water down.

The Oliver Spring, in the same county, on the waters of Blackford, gave nearly the same results, except that there is more oxide of iron in that water.

During July of 1856 the Paroquet Springs were tested at the fountain head, *qualitatively;* since that time the principal spring has been analyzed, *quantitatively*, by Prof. J. Lawrence Smith, of Louisville, with similar results as to the principal bases and acids. I subjoin Prof. Smith's analysis as it gives the *quantities* of the saline ingredients in a gallon of the water.

Sulphuretted hydrogen, 30 cubic inches;
Carbonic acid, 6 cubic inches;
Chloride of sodium, 309.6000 grains;
Chloride of calcium, 67.7100 grains;
Chloride of magnesium, 48.0300 grains;
Chloride of potassium, 0.4860 grains;
Sulphate of soda, 2.4120 grains;
Sulphate of lime, 2.2800 grains;
Sulphate of alumina, 0.4920 grains;
Carbonate of soda, 0.3780 grains;
Carbonate of lime, 2.4000 grains;
Carbonate of magnesia, 1.5060 grains;
Carbonate of iron, 0.1800 grains;
Iodide of sodium, 0.1560 grains;
Iodide of magnesium, 0.2460 grains;
Bromide of sodium, 0.1800 grains;
Bromide of magnesium, 0.3120 grains;
Silica, 3.9000 grains;
Organic matter, 2.1360 grains.

When examining the Paroquet Spring it was a question with me whether a portion of the magnesia found was not combined with sulphuric acid as sulphate of magnesia. I perceive Prof. J. Lawrence Smith has given the most of it to the chlorine, considering it to exist mostly as chloride of magnesium. A difficulty has arisen in my mind, as to the compatibility of that salt with carbonate of soda. It is possible, that in *small* proportions these salts may exist *together* in solution; though in larger proportion the carbonic acid would certainly go over to the magnesia, while the soda would combine with the chlorine. As there does not appear to have been sufficient sulphuric acid found to saturate more than the soda and lime, it is inferred that the most of the magnesia is united with chlorine, the only element which existed in sufficient quantity to saturate the surplus of that alkaline earth, over and above that which is combined with carbonic acid, iodine, and bromine.

The presence of chloride of magnesium, in *large quantities*, in either well, spring, or mineral waters I regard as objectionable. The proportion in this water may not, materially, detract from its other virtues, derivable from the presence of the sulphur, chloride of sodium, and iodine, which act beneficially on diseases of the skin, digestive organs, and glandular system.

The qualitative examination of the "Alum Spring," at the base of Burdett's Knob, on James Richardson's place, gave as its principal constituents,

Sulphates of alumina and protoxide of iron;
Bi-carbonate of lime;
Bi-carbonate of magnesia.

It has strong astringent properties, and is, undoubtedly, deleterious to health.

This water issues from near the out-crop of the hydraulic limestone and black slate, at the base of an outlier of the Knob Formation, a a mile and a half from the forks of the turnpike leading from Danville to Lancaster.

A well water was also tested at the forks of the above road, on Mrs. Hoskins' place, which was found to contain chiefly,

Bi-carbonate of lime;
Bi-carbonate of magnesia;
Sulphate of soda;
Sulphate of magnesia.

The nearest underlying rock on this farm is a subcrystaline, blue limestone, with cherty segregations belonging to the lower silurian system, and containing *Atrypa modesta*, and a coarse ribbed *Orthis* like *O. Calligramma*.

Yates' mineral water, in Boyle county, near the base of "Knob Lick," contains principally,

Bi-carbonate of lime;

Bi-carbonate of magnesia;

Chloride of sodium;

Sulphate of soda;

Sulphate of magnesia;

and, perhaps, a small quantity of carbonate of soda, as it has an alkaline reaction on litmus and georgina paper, and gives some indications of the presence of carbonate of alkali with tincture of champeachy wood, and addition of chloride of calcium and sulphate of copper.

The well from which this water was taken is sunk in the ash-colored argillaceous shale of the Knob Formation, and goes four feet into the black shales of the Devonian Epoch.

In Lincoln county, at the sources of Salt river, a sulphuretted water, issuing from the above black shale, gave to chemical reagents the indication of the same saline ingredients as in the Yates well, with the addition of free sulphuretted hydrogen. This water is known as the "Nevien's Sulphur Spring."

The Rochester Mineral Spring is in the blue limestone formation of the western part of Boyle county. This water contains a large quantity of magnesia, mostly in the state of sulphate.

Its principal ingredients are,

Sulphate of magnesia;

Sulphate of lime;

Sulphate of soda?

Probably a *small quantity* of alumina?

and a trace of sulphate of protoxide of iron.

Janes' mineral water, of Washington county, on Road run, four miles from Springfield, is a sulphuretted saline water, containing

Free sulphuretted hydrogen, strongly impregnated;

Chloride of sodium;

Sulphate of soda;

Sulphate of magnesia;

Bi-carbonate of lime;

Bi-carbonate of magnesia.

This spring issues through the *Lynx* beds of the blue limestone, which prevail along the bed of Road's run.

The examination of the water of the public well, at Bloomfield, in Nelson county, indicated, as its principal constituents,

Chloride of sodium;

Bi-carbonate of lime;
Bi-carbonate of magnesia;
Sulphate of soda;
Sulphate of magnesia.

The water has a feeble alkaline reaction, and may contain a trace of carbonate of soda, and a trace of alumina. The rock of this part of Nelson belongs to the Lower Silurian Period, and is of a blue color and subcrystalline texture. There are several mineral waters in this formation, in the vicinity of Bloomfield.

The examination of R. B. Grigsby's "White Sulphur Mineral Water" indicated,

Free sulphuretted hydrogen, (tolerably strongly impregnated);
Chloride of sodium;
Sulphate of soda;
Sulphate of magnesia;
Bi-carbonate of lime;
Bi-carbonate of magnesia;

It has a slightly alkaline reaction, and the log-wood test, with chloride of calcium, indicates a small quantity of carbonate of soda. There is also a trace of bi-carbonate of iron. The water is a tolerably strong sulphuretted saline water, with slight tonic properties. A peculiar shell-earth, derived from the disintegration of a bed of the blue limestone, of which I had occasion to speak in the chapter on "Agricultural Geology," is well developed on this property, under the soil.

The "Mammoth Well," on the west branch of Simpson creek, in this county, was found to contain

Chloride of sodium;
Bi-carbonate of lime;
Bi-carbonate of magnesia;
Sulphate of soda;
Sulphate of magnesia;

and a trace of iron.

It is a feeble tonic, and mild aperient and alterative.

The Creel "White Sulphur Mineral Water," in Marion county, gave

Free sulphuretted hydrogen;
Chloride of sodium;
Sulphate of soda;

Sulphate of magnesia;
Bi-carbonate of lime;
Bi-carbonate of magnesia;
A little carbonate of soda.

This water is distinctly alkaline.

The reaction of the Campbellsville "Sulphur Water," with reagents, indicated, as its constituents,

Free sulphuretted hydrogen;
Sulphate of soda;
Sulphate of magnesia;
Chloride of sodium;
Chloride of magnesium;
Bi-carbonate of lime;
Bi-carbonate of magnesia;
Trace of carbonate of soda.

This water is from the table land of the sub-carboniferous limestone of Taylor county. There used to be a noted lick at the place where the gum enclosing this water is sunk.

The water of the well, at the hotel in Campbellsville, was tested, which showed it to be a hard limestone water.

There are fine springs of water in Taylor county, between Campbellsville and Saloma.

Examination of Washington Bell's Sulphur Spring of Marion:

Free sulphuretted hydrogen, and trace of sulphuret of alkali;
Chloride of sodium;
Sulphate of soda;
Sulphate of magnesia;
Bi-carbonate of lime;
Bi-carbonate of magnesia;
Carbonate of soda.

This water has a distinct alkaline reaction on reddened litmus paper. This water issues from the black shale of the Devonian Period, on Sulphur Lick creek, and is now owned by Ex-Governor Wickliffe.

At the Plantation Lick, three to four miles from Sulphur Lick, where this water issues, some salts have been made, probably a mixture of sulphate of magnesia and chloride of sodium.

In Nelson county, in the precincts of New Haven, James T. Weathers' well water was examined, which gave

Trace of sulphuretted hydrogen;
Sulphate of magnesia;
Sulphate of soda?
Small quantity of chloride of sodium;
Bi-carbonate of lime;
Bi-carbonate of magnesia;
Trace of carbonate of soda and iron.

This water has a feeble alkaline reaction. It is a mild aperient, and feeble tonic.

The ash-colored and black shale of the Devonian Period are the strata exposed in the vicinity of Mr. Weathers'.

The "Howell Mineral Spring" is an alkaline saline chalybeate, which issues in a copious flow from the sub-carboniferous limestone of the "Barrens" of Hardin county. This is a remarkable mineral water, having more iron, and a stronger alkaline reaction, than most of the saline waters hitherto tested. Its constituents are

Sulphate of magnesia;
Sulphate of soda?
Chloride of sodium;
Bi-carbonate of lime;
Bi-carbonate of magnesia;
Bi-carbonate of protoxide of iron;*
Carbonate of soda.

The Bedford Spring, owned by Noah Parker, in Trimble county, is collected in a basin of the blue limestone, which has a hydraulic layer of four to five feet, interstratified, from which a part of the water seems to come. Its constituents are

Chloride of sodium;
Sulphate of soda;
Sulphate of magnesia;
Bi-carbonate of lime;
Bi-carbonate of magnesia;
Carbonate of soda.

It has an alkaline reaction to test paper.

The Epsom Spring, owned by Thomas Rolland, in this vicinity, contains the same ingredients, only the sulphate of magnesia seems to be in larger proportion.

*Some of the iron may be in the state of sulphate.

In Gallatin county, at "Big Lick," on Lick creek, a branch of Eagle, a fine sulphuretted saline water was tested, boiling up in a constant flow, and blackening the ground around. It contains

Free sulphuretted hydrogen;
Sulphate of magnesia;
Sulphate of soda;
Chloride of sodium;
Bi-carbonate of lime;
Bi-carbonate of magnesia;
Carbonate of soda.

It has a slight alkaline reaction, and indicates a trace of sulphuret of alkali, with the nitroprusside of sodium.

Examination of a water, two miles from Dowlingsville, near Clark's creek, in Grant county, where the silicious mudstone rock, associated with the Blue Limestone Formation, prevails, and where, in dry seasons, milk-sickness is prevalent:

Bi-carbonate of magnesia;
Bi-carbonate of lime;
Chloride of magnesium;
Sulphate of soda, (small;)
Sulphate of magnesia?
Suspended, and perhaps dissolved, alumina.

The most remarkable feature in this, and some other waters to be noticed, occurring along the same range of the geological formation, is the proportion of magnesia, which is considerably more than is usually found in water used for domestic purposes, and which, probably, exists partly in the state of chloride, which is an objectionable form of magnesian salt for waters in habitual use, in proportions above the normal standard. To determine whether the magnesia is mostly in this condition would require a careful *quantitative* analysis in the laboratory which would be a very important investigation, and which I propose hereafter to have instituted; because the therapeutic effects of chloride of magnesium, taken in considerable quantities, border closely on those prominent in milk-sickness, as will be perceived by the following description:

*Action of chloride of magnesium on the animal economy.* Acid eructations, with regurgitation of the injesta. Nausea, with accumulation of water in the mouth. *Great weakness in the lower limbs, with symptoms*

*of weariness. Constipation*, with urgent desire for stool, with shuddering after each effort. Fæces, when expelled, are scanty, either soft or in knots, like *sheeps dung*, with an appearance as if burnt; sometimes covered with mucus. This follows by a diarrhea.

Some of these last effects are perhaps not observed in milk-sickness, but the first are, so that it is highly probable that an extra proportion of this salt, together with other astringent salts, such as sulphate of alumina and iron, and perhaps suspended clay in water, may combine to produce that mysterious disease.

At all events, it is worthy of particular notice, that this complaint follows precisely the geological range of the silicious mudstone or "*rotten sandstone*," interstatified in the blue limestone formation of Scott, Owen, Grant, Boone, and other counties of central Kentucky; and, so far as my observations have yet extended, no where else within the range of that particular geological formation; in connection with the fact that the waters, which I have so far had occasion to test, qualitatively, in the worst milk-sick regions of these middle counties of the state, contain more than the normal proportion of magnesia. This is the case with the so-called "Poison Spring," which I tested, in the north eastern part of Grant, and elsewhere along the belt of country ranging with the out-crop of this peculiar member of the rocks of central Kentucky.

The "Hardinsville Sulphur Spring," which issues from encrinital layers of the blue limestone of Franklin county, contains

Free sulphuretted hydrogen;
Chloride of sodium;
Bi-carbonate of lime;
Bi-carbonate of magnesia;
Trace of sulphate of soda;
Trace of sulphate of magnesia;
Carbonate of soda?

This water seems to have a feeble alkaline reaction.

Reuben Jesse's mineral water, of Woodford county, near Versailles, struck by sinking a deep well in the blue limestone formation, contains

Free sulphuretted hydrogen;
Chloride of sodium;
Bi-carbonate of lime;
Bi-carbonate of magnesia;

Trace of sulphate of soda;

Chloride of magnesium?

The Harrodsburg "Saloon Spring," in Mercer county, is a saline water, in which magnesia is in large quantity, combined, in all probability, with sulphuric acid. Its principal constituents are

Sulphate of magnesia;

Sulphate of lime;

Bi-carbonate of lime;

Bi-carbonate of magnesia;

Trace of chlorides, and a small quantity of carbonate of protoxide of iron.

The "Greenville Spring," at Harrodsburg, has only a trace of sulphuretted hydrogen and iron; otherwise, it is very similar, in its composition, to the Saloon Spring. There is perhaps, a small quantity of alumina in both, but it is difficult to appreciate it in the unconcentrated water.

The beds of the blue limestone, which are exposed near these mineral springs, are characterized chiefly by *Chætetes*. There are, however, blocks of birdseye textured limestone, lying in blocks around and in the masonry of the wall built around the Saloon Spring, and which were quarried near by.

The mineral waters, at the celebrated locality of Big Bone Lick, in Boone county, now owned by Mr. McManning, are truly fine sulphuretted saline waters, of which there is the most abundant supply from the several springs which burst forth, in various directions, from the boggy flats forming the sources of Big Bone Lick creek, enclosed by an amphitheatre of hills composed of *Leptæna*, *Orthis*, and *Chætetes* beds of the blue limestone, amongst which the *Leptæna Sericea* is conspicuous, towards the base of the surrounding slopes.

The qualitative examination of these waters, at their fountain head, indicated, as the principal constituents,

Sulphuretted hydrogen, which escapes in intermittent volumes, proving the water to be saturated with this gas;

Chloride of sodium;

Sulphate of magnesia;

Sulphate of soda;

Sulphate of alumina?

Bi-carbonate of lime;

Bi-carbonate of magnesia;

Carbonate of soda.

This water has an alkaline reaction.

It was only in a few instances where any attempt was made to test for iodides, bromides, and other ingredients that may exist in these waters in minute quantities, because it requires large quantities of the water to be boiled down in order to arrive at any satisfactory conclusion as to their presence or absence; but many of these waters deserve a careful quantitative analysis; but for this purpose a special journey must be made to collect the water in sufficient quantity, and with the necessary precautions, to fix the evanescent constituents on the spot, and transport the water to the laboratory. The proportion of the *fixed* constituents can, however, be determined, if the water is put up well, by the proprietors of the springs, in clean bottles, and forwarded to Dr. Robert Peter, at Lexington, the Chemical Assistant of the Survey, if the legislature provide for the continuance of the survey.

In connection with this subject, it would be a matter, not only of scientific interest, but of practical utility, to have the waters of all the principal rivers of Kentucky analysed, quantitatively, in order to judge of the effects they may have upon the system, and when used for various industrial applications.

# CHAPTER IV.

## STRATIGRAPHICAL GEOLOGY.

### DAVIESS COUNTY.

Some further examinations have been made in this county, chiefly in the vicinity of Big and Little Blackford creeks. The predominating rock is a soft and rather ferruginous sandstone, passing downwards into a grey shaly clay, with imperfect seams of coal; below which is darker shale, with streaks of coal; under this is a seam of coal the thickness of which has never yet been ascertained.

North of Yelvington, on J. M. Robinson's place, just above his spring, there is a brownish freestone, which shows itself half a mile to the south. On Mr. Beauchamp's place, adjoining, limestone occurs towards the top of the ridge; and north of his house three to four feet of ferruginous limestone occurs, about half way up the ridge, near Robert's line. It seems to sink towards the west, as a limestone, apparently the same, has been burnt for lime at a lower level. Below this is a shaly fire-clay, where a spring issues. Here Wm. Beauchamp made a partial examination for coal, but never proved the ground.

One and a half miles east of Yelvington, at Mason Lee's, there is three feet of coal, without any regular covering, as far as it has been entered.

Near the head of Big Blackford, on a high oak ridge, a hard quartzose sandstone is exposed.

The sections hitherto obtained, in this part of Daviess county, are too obscure, and the exposures of out-crop too few and far apart, to enable me to make a satisfactory section of the succession until the detailed survey is carried over this and the adjoining counties.

### SUB-CARBONIFEROUS ROCKS IN BRECKINRIDGE, MEADE, AND HARDIN COUNTIES.

Near the head of the west fork of Clover creek I obtained the following section of the limestone, shaly limestones, and marly shales, with some alternations of sandstone, underlying the Coal Measures and upper conglomerate at the top of the hill, or its sandstone representative, since, on the eastern edge of the western coal field, it has only locally pebbles disseminated.

*Feet.*

230. Hard, thin-bedded sandstone, with red ferruginous subsoil near the top of the ridge, with perhaps a ten inch coal in this space.
170. Ferruginous sandstone.
160. Ferruginous sandstone.
150. Ferruginous sandstone.
142. Grey limestone, with chætetes abundantly disseminated.
120. Sandstone—red externally, white internally.
112. Light-colored limestone.
100. Sandstone and light grey limestone alternating.
85. Sandstone and light grey limestone alternating.
83. Marl, with some sandstone.
71. Schistose impure limestone and sandstone.
67. Earthy, shelly, limestone.
65. Marl.
63. Brownish sandstone.
61. Marly earth.
60. Top of bench of light-colored limestone, in benches with marly interstices.
54. Light grey limestone; thin streak of coal?
53. Bench of buff calcarious rock.
    Marl.
47. Bench of light reddish-grey earthy schistose limestone, two feet nine inches.
    Marl.
41. Bench of rugged calcarious rock nine feet.
    Marl.
32.5 Bench of rugged calcarious rock.
32. Ash-colored marly shale.
19. Ash-colored earth, and ferruginous calcarious rock.
15. Ash-colored earth, less ferruginous.
10. Ferruginous marly shale.
6. Light buff marl, with earthy light-colored schistose limestone.
5. Green and ferruginous shale.
0. Dark marly shales extending to the foot of the glade, where the measurement was made.

In the last cut of the railroad about half a mile from Cloverport, the base of the sandstone for two and a half to three feet, is strongly impregnated with petroleum; so much so, that it oozes out from its pores, and trickles on the ground. The rock has a dip 50° west of north, at an angle of about 5°.

In Hickory Hollow, up a ravine to the west of the railroad, the Archimedes Limestone is exposed, with other fossils to be hereafter described. It lies here from fifteen to twenty feet below a sandstone, which is undoubtedly the equivalent of that which forms the vertical cliff above the Tar Springs. These underlying limestones are the exact equivalent of the beds which were observed in 1845, at the falls of Rough creek, in the northwest corner of Grayson county.

The Tar Spring sandstone, near where the railroad intersects the line of separation between Breckinridge and Hancock counties, is very nearly one hundred feet above low water of the Ohio river, at Cloverport. At this point the sandstone is about fifty-five feet in thickness, and the top of it one hundred and fifty feet approximately above low water on the Ohio river. This sandstone is remarkable in Breckinridge county on account of the fine springs of petroleum which issue at its base, close to its junction with the Archimedes limestone. It oozes in the form of a thick oily fluid, nearly of the color and consistence of molasses; collecting more copiously in wet than dry weather; and, apparently, where the sandstone rock is comparatively thin. At Thomas Jackson's about ten barrels are collected in a year, but much more might be obtained if proper precautions were taken to save it. From the fact of its oozing more abundantly in wet weather, it is probable that the rain water filtering through the porous sandstone carries the oil with it to the base of the rock.

In some places in the same cliff of sandstone, above that part impregnated with petroleum, is of snowy whiteness, fit for the use of the glass manufacturer, and forms a most remarkable contrast to the dark umber of the same rock, saturated with the oil. Where it runs off on the ground it becomes inspissated into masses having a hardness and appearance intermediate between pitch and asphaltum, incrusted and incorporated more or less with earthy materials and small gravel. Cattle seem to have a predilection for it, and it is undoubtedly a valuable remedy in milk-sickness, for burns, and obstinate cases of rheumatism, for which it is used both externally, as a rubefacient, and internally. This "tar sandstone" occupies the same geological horizon as the Rock House sandstone, on the head of Shotpouch creek It marks the limit in connection with the underlying Archimedes limestone, of the productive Coal Measures, since no coal, of more than six to ten inches, has ever yet been found under it, and probably never will; because the circumstances favorable to the production of thick beds of coal do not appear to have existed, in the region of the western country, at least, until after the deposition of this bed of sandstone.

Near the head of Sinking creek a lower bed of sandstone than the Tar Spring sandstone, of thirty-eight feet in thickness, was observed about twenty-five feet above the bed of the creek; at sixty-five feet buff Archimedes limestone, associated with white oolitic beds, suitable for making fine white lime, and, if fairly opened in a quarry, might, perhaps, afford an oolite marble, such as is now worked by some of the marble cutters of the west. At eighty feet, in the same hill, is another bed of buff, encrinital and Archimedes Limestone. Below the sandstone cliff is a dark, marly shale, underlaid by reddish-grey productal limestone.

The succession of the sub-carboniferous rocks, in Breckinridge county, west of Sinking creek, from the Tar Sandstone downwards, is as follows:

| Thickness in feet. | |
|---|---|
| 50 to 70 | Tar Spring and Shot Pouch sandstone. |
| [illegible] | Upper Pentrimite and Archimedes limestone. |
| 30 to 40 | Grey shale and place of six to ten inches of coal.<br>Locally marly shales, Archimedes and oolitic limestones. |
| 20 to 40 | Sandstone 20 to 40 feet, softer and more schistose than the upper. |
| 10 to 20 | Marly shale 10 to 20 feet.<br>Productal and Archimedes Limestone.<br>Cherty limestone and chert. |
| 15 to 25 | Sandstone 15 to 25. |
| | Thick bedded limestone with cherty segregations extending down to Licking creek. |

It is to be observed, however, that there are considerable modifications in the relative thickness of the limestones, shales, and sandstones, at different localities sometimes not far apart.

On the east side of Sinking creek, in the vicinity of the Big Spring road, there is little or no sandstone visible, and the soil becomes of a redder cast as the middle cavernous beds of the sub-carboniferous limestone rise from beneath the drainage of the country. In consequence of the cavernous nature of the limestones in the eastern part of Breckinridge, a great deal of the surface-water sinks away into subterranean passages, and comes out in bold springs at the Big Spring and "Good Spring," near the eastern confines of the county. In some places water may be reached in wells, even at the depth of thirty feet; but many wells have been sunk fifty to sixty feet without getting water.

The characteristic fossil coral of the cherty beds of the sub-carboniferous limestone, in their southern and western range, does not seem to be abundant in this part of Breckinridge county; indeed, organic remains generally are few compared with what are usually found in the equivalent beds elsewhere.

Near the confines of Breckinridge and Meade counties, some of the beds of the sub-carboniferous limestone have the fine grain and smooth texture of lithographic limestones, and it is very probable that by opening quarries valuable beds might be found, which, if free from flaws and dry cracks, would fetch in the market twelve to fifteen cents per pound, and command ready sale.

Rock, possessing the structure of lithographic limestones, were also observed about half way down the descent to Otter creek, in Meade county, on the road from Good Spring to Garnettsville.

The following sections, taken in Meade county, along the bluffs on the Ohio river, will convey an idea of the succession of the beds above low water of the Ohio river.

*Section near Concordia.*

*Feet.*

240. Red limestone, with Archimedes.*

225. Oolitic limestone.

220. Base of bench of these limestones.

190. Top of heavy ledge of sandstone, ten feet in thickness.

180. Finely laminated shaly sandstone, underlaid by dark shale and shaly sandstone.

155. Top of limestone cliff of thirteen feet.

142. Base of reddish-grey solid limestone.

140. Light grey shale passing into shaly sandstone.

135. Ash-colored shale, three feet, and base of five feet ledge of sandstone.

90. Top of bench of limestone, slope concealed.

*In the spaces between the rocks noted the rocks are concealed from view by debris.

*Feet.*

74. Bottom of a limestone bench.
63. White and yellow sandstone.
35. Top of white and brown sandstone speckled with red, irregular in structure,
33. Bottom of sandstone and top of ledges overhanging the river bank.
26. White limestone, with dendritic streaks.
20. Brown sandstone and Euomphalus brown limestone.

L. W. Compact white limestone.

*Section one mile above Concordia.*

*Feet.*

270. Top of ledges of sandstone.
250. Base of ledges of sandstone.
 Slope of strata hidden.
230. Top of bench of upper Archimedes limestone.
225. Red limestone on weathered surface.
220. Oolitic limestone.
 Slope, strata hidden.
180. Top of sandstone.
140. Shales interlaminated with red speckled sandstone, and top of reddish-grey hard limestone, forming together a cliff of fifty feet space where strata concealed from view.
70. Pentremital limestone.
55. Red limestone, underlaid by brown, rugged, sandstone.
25. Top of white limestone.
5. Base of white limestone.

*Section at Glen's, Adit between, above Boonsport.*

*Feet.*

275. Upper Archimedes limestone.
250. Calcarious sandstone.
 Limestone.
230. Top of argillaceous marly shale with alternating thin bands of limestone in which adit run in search of coal.
210. Bottom of Glen's Adit.
 Vertical cliff of heavy beds of limestone fifty to sixty feet.
150. Base of overhanging ledges of limestone.
 Soft decomposing marly limestone.
130. Oolitic limestone of which white lime is made.
 Slope of one hundred and twenty feet where strata are concealed.

L. W. Light grey compact limestone at low water of the Ohio river.

*Section a few miles higher up.*

*Feet.*

300. Top of ridge and top of sandstone ledges.
 Slope where strata concealed.
255. Upper Archimedes limestone.

*Feet.*

245. Lower ledges of a bench of limestone.

100. Base of heavy ledges of limestone.

One hundred feet nearly concealed by talus.

L. W. Light grey compact limestone at low water.

*Section on Kentucky shore, opposite Leavenworth.*

*Feet.*

300. Sandstone and shale at the base of the coal measures, near the top of the ridge.

Red and grey sandstone.

280. Top of Pentremital and Archimedes limestone associated with a rich Productal and Terebratula limestone.

240. Buff Archimedes and Pentremital limestone.

Long slope where strata are concealed.

150. Limestone in thick beds forming a bench.

L. W. Aulopora and Cyathophyllum limestone.

*Section of bluff below Indian creek.*

*Feet.*

340. Brown sandstone near base of the coal measures.

300. Top of limestone.

285. Bed of white interstratified sandstone.

200. Base of overhanging ledges of limestone.

Decomposing ledges below.

70. Base of rocks visible.

L. W. Ohio river.

*Section two miles above North Hampton, near lime kiln.*

*Feet*

300. Highest limestone observed, with oolitic limestone beneath.

255. Oolitic limestone.

220. White oolitic limestone.

190. Rugged weathering limestone, of the Barren limestone series.

80. Base of cliff.

52. River bottom land a little above high water mark.

L. W. Low water of the Ohio river.

*Section five and a half miles above Brandenburg.*

*Feet.*

265. Top of hill and second cliff composed of close-textured, grey sonorous limestone, passing downwards into a granular limestone, with buff magnesian layers near the

160. Base of the cliff.

150. Slope with schistose and more earthy-textured limestones.

125. Top of first cliff or bench.

121. Top of quarry, composed of sub-crystalline stratified limestone.

Dark grey and buff magnesian limestone.

100. Base of quarry.

L. W. Low water Ohio river.

Several other sections have been made, on the Ohio border of Meade and Hardin counties, but those given will be sufficient, at this time, to show the nature of the rocks belonging to the sub-carboniferous limestone group, underlying these two counties. It will be observed that the strata at the very base of the coal measures only cap the tops of the highest hills, which would have to be at least fifty to one hundred feet higher to bring in any workable coal.

I would also call attention to the fact of the numerous repetitions of sandstones interstratified in the sub-carboniferous limestones of Meade and Breckinridge.

The most valuable materials under the soil of Meade and Hardin counties, so far as I have yet seen, are the materials for constructions, oolitic marble, lime, and lithographic stones, and the marly beds, some of which would be excellent for improving the more sandy soils of this and the adjacent counties.

A large area of Hardin county is based on the cavernous members of the sub-carboniferous group, proved by the disappearance of streams from the surface for many miles and the gushing out, locally, of subterranean springs. There is, also, evidence of the existence of very numerous caves, in this county, especially on the waters of Linder's creek, perhaps as extensive as in the eastern part of Edmonson county.

About two miles west of Elizabethtown a bed of earthy-textured limestone crops out near the road leading to Howell Springs, which has all the appearance of possessing hydraulic properties. It is associated with other varieties of close grey limestones with argillaceous partings, which approaches, in its character, to lithographic limestone.

Six miles west of Elizabethtown freestone crops out, which is one of the few localities where this kind of rock reaches the surface, since sandstones have only been observed in this western portion of Hardin, near the Grayson line, on the waters of Nebo, Brushy, and Meeting House creeks, and the argillaceous freestone of the Knob Formation, in the extreme northeast corner of the county.

The most common rocks seen on the surface are different varieties of chert, which have originated from segregations and interstratification, washed and weathered out of and from between the limestone which, resisting decomposing agencies longer than the matrix, lie in blocks on the surface.

## LOWER SUB-CARBONIFEROUS ROCKS AND BLACK DEVONIAN SLATES OF THE KNOB FORMATION, OF BULLITT COUNTY.

The descent from the table lands of the sub-carboniferous limestones of Hardin county, over the slopes of the sub-carboniferous freestones of the Salt river or Muldrow hills, is for the most part abrupt and precipitous.

About seventy feet above the base of these hills, near Key's Ferry, the fossiliferous, burhstone textured chert, and earthy encrinital limestones, are in place on the west side of Salt river. The total height of the Salt river hills, on the Bullitt side of this stream, is from three hundred and fifty to four hundred feet. On this side, the encrinital limestone is one hundred and fifty-five feet above the ferry house, and two hundred and ten feet above low water.

The summit level, where the road to the Bullitt Lick passes over, is nearly four hundred feet above the bed of Salt river, giving near two hundred feet of strata, soil and sub-soil, covering the geodiferous entrochital limestones which lie, therefore, nearly midway of these hills. The upper two hundred feet includes part of the sub-carboniferous limestone which overlies the Knob sandstone in this locality, but their relative thickness is not well seen here.

The Black Devonian slate, doubtless, exists here near the base of the hills, but it was not exposed on the immediate line of observation.

The iron ores which have been made mention of in the third chapter on chemical geology, are most abundant in the grey or ash-colored shales overlying the Black Devonian slate, in the south-east part of the range of the Knobs of Bullitt, extending along the waters of Cane Run south-eastwardly, into Nelson county.

A section will aid in illustration of its position:

| | Feet. | Inches. | |
|---|---|---|---|
| | | | Knob Freestone, enclosing locally ore No. 489.* |
| | 60 | | Grey or ash-colored shales, 50 to 70 feet, enclosing Kidney and Sheet ores, Nos. 488 and 493.† |
| | 15 | | Black shelly shale, locally enclosing limestone. |
| | 60 | | Black shelly shale, locally enclosing lime stone, greyer and more leafy in its structure from exposure. |

It is in the grey and ash-colored shales and clay, lying between the black shales and Knob freestones, that the ore of Bullitt county is found in greatest abundance, and mostly as carbonate of the protoxide of iron, except where it has been oxidized by partial exposure to air and permeating water. It varies in thickness from three to eight inches.

The two varieties, as recognized by the ore diggers, are the "Kidney ore" and the "Blue Sheet ore," because the former generally lies above the latter in more detached hemispherical masses possessing a concentric structure; whereas, the latter is more continuous or in the form of a pavement, and less oxidized by exposure. But neither of these ores are in as high a state of oxidation as the more irregular nests and bands in the overlying freestone, because the argillaceous matrix of the Kidney and Blue Sheet ores protects them more effectually from atmospheric influences.

The greater per centage of iron in a given weight of the ore,† enclosed in the freestone, is to be attributed chiefly to the loss of carbonic acid during the progress of oxidation.

In the south-east part of Bullitt county, these ores of the Knob Formation prevail with considerable regularity; the principal difficul-

*See Chemical Report.

†See Ore No. 489.

ty experienced by the ore digger in obtaining it being the abruptness of the declivities of the Knobs, which only occasionally form benches by which he is enabled to reach the ore with a reasonable amount of stripping. Nevertheless, it can be mined and delivered at the furnaces from $1 50 to $1 75 or $2, depending on the distance from the furnace, and the amount of stripping, which at present is from five to twenty or twenty-five feet. Upon the whole, considering the character and uniformity of the ore; the rate at which it can be mined and delivered; the quality of the iron it produces, which is both soft and tough, and in great request by the nail makers; the abundance of well timbered land, and the proximity to the line of the Louisville and Nashville railroad, this part of Bullitt county must be considered a favorable position for making charcoal iron.

The limestone which is at present mostly used as a flux, is included locally in beds interstratified in the Black Shale underlying the ore bearing shale, and contains, according to analysis No. 490, a little over thirteen per cent. of magnesia, but only about one and half per cent. of insoluble earthy silicates, which is much less than its external appearance conveys to the eye. It lies quite convenient to the Bellemont Furnace, the only one that was in operation in Bullitt county in 1856. More or less ore, of the carbonate variety, has been found disseminated in grey shales occupying the same geological position, within the range of the Knobs of this and adjoining counties, but not always in sufficient quantities to support a furnace.

Both at Bullitt Lick and Button Mould Knob considerable ore has weathered out of the shale, but still neither of these localities would alone justify the erection of a furnace; so that, before capital is expended in machinery and constructions, it is always proper to prove thoroughly the ore ground.

At "Bullitt Mould Knob," a celebrated locality for encrinites, in Bullitt county, there are three or more encrinital beds, interstratified with the ash-colored shale, which form a remarkable steep glade on the south-west side of the Knob; the glade commencing one hundred and twenty-five feet below the summit of the Knob:

*Feet.*

250. Summit of Knob.

235. Top of second benchof sandstone, in quarry.

225. Top of ledge of first bench sandstone.

*Feet.*
200. Slope with sandstone?
162. Lowest exposure of sandstone.
110. Top of bare glade.
105. Orthis Michellina bed.
Ash colored shale.
100. Orthis Michellina bed most abundant.
Ash-colored shale.
97. Weathered out carbonate of iron.
95. Weathered out carbonate of iron.
Ash-colored shale.
90. Weathered out carbonate of iron.
Ash-colored shales.
80. Branching corallines.
75. Weathered carbonate of iron.
65. Encrinital limestone.
60. Weathered carbonate of iron.
Ash-colored shale.
49. Encrinital limestone.
Ash colored shale.
35. Encrinital limestone.
Ash-colored shale at base of bare glade.
25. Black sheety Devonian shale extending to bed of creek.

Here we have nearly one hundred feet of ash-colored shales exposed, in a bare glade, with repeated alternations of thin bands of carbonate of iron, encrinital, argillaceous, and shell limestones, forming a remarkable feature of the landscape in the northern part of Bullitt county, adjoining Jefferson county. The position of this Knob is incorrectly laid down in Milne and Bruder's map, since it lies nearly one mile to the east of the railroad leading from Louisville to Shepherdsville.

The total thickness of the black shales, under the ash-colored shale, was not seen in any of the sections yet observed in Bullitt county. The best section was seen in the cut of the Louisville and Nashville road, about a mile from Shepherdsville, where forty feet is seen in one vertical cut.

In the bed of Salt river, below the bridge, several ledges of rugged, cherty, limestone can be seen, in a low stage of the river, these must doubtless be referred, in part at least, to the Upper Silurian Period, as the chain coral (*Catenipora escharoides*) is found here, associated with *Favosites spongites, Cyathophyllum helianthoides*, and *Polymorpha?* and other fossils of that Period, very imperfectly preserved however.

The coralline and shell-beds of the falls of the Ohio seem to be very thin and obscurely marked, or at least difficult to detect. The only place where I was able to recognize them satisfactorily, as yet, was along the bed of a creek near Mr. Benjamin Sanders' place, not far from Button Mould Knob, at a little lower level than the zero of that section. Beneath these a cellular magnesian limestone, which weathers in a very rugged manner, is the prevailing rock in the ravines, on both sides of the old road from Shepherdsville to Louisville.

### JEFFERSON COUNTY.

The Knob Formation, very similar in its component members to that described at Button Mould Knob, extends into the southern part of Jefferson county, forming the range of Knobs on the waters of Pond and Mill creek, their summits being capped with soft freestone, while the ash-colored shales, with intercalations of encrinital limestones, form their principal mass, resting on black Devonian shale.

Jefferson county affords the best exposures of the calcarious rocks, under the black slate belonging to the Devonian Period, yet seen. The projecting ledges on the bank of the Ohio river, that appear in connected succession between the head and foot of the falls, afford, probably, the best sections of these rocks in the western states.

We observe there the following succession and superposition:

1. Black bituminous slate or shale.
2. Upper crinoidal, shell, and coralline limestones above.
3. Hydraulic limestone.
4. Lower crinoidal, shell, and coralline limestones.
5. Olivanites bed.
6. Spirifer Gregoria and shell and coralline beds.
7. Main beds of coral limestones.

These beds rest upon a limestone containing chain coral, which is seen just above the lowest stage of water, at the principal axis of the falls, where the waters are most turbulent. Only a portion of the lower part of the black slate is seen immediately adjacent to the falls. Its junction with the upper crinoidal bed, No. 2, of the above section, can be well seen below the mouth of Silver creek, on the Indiana side, where there is a thin hard pyritiferous band between the black slate and limestone, containing a few entrochites.

Three sub-divisions may be observed in the upper coralline bed, No. 2, of this falls section:

(*A*). White or yellowish white, earthy fractured layers, containing, besides *Crinoidea* a *Favosite*, a large *Leptæna* and *Atrypa prisca*, with a fringe.

(*B*). Middle layers containing also a few Cystiphyllæ.

(*C*). Lower layers, containing most Cystiphyllidæ, and on Corn Island, remains of fishes. This is what has been designated as the Upper Fish Bed.

These crinoidal beds contain a vast multitude of the remains of different species of encrinites, mostly silicious, or more so than the imbedding rock, so that they often project and appear like black concretions. Remains of the *Actinocrinus abnormis*, of S. S. Lyon's report, is the most abundant. There is also a Syringapora; and short truncated Cyathophyllium. The Cystiphyllum is long, slender and vermiculiform, sometimes extending to the length of fifteen inches or more; also, a coralline, referrable either to the genus *Porites* or *Astrea.**

The hydraulic bed, as has been shown in the third chapter, is an earthy magnesian limestone, in which the lime and silica are in the proportions of their chemical equivalents. It is variable both in its composition, thickness, and dip. In the upper part of the bed, where it contains many *Spirifer euratines* and *Atrypa prisca*, it is more silicious than that quarried for cement. At the head of the falls it is eight feet above low water. At the foot of the falls it is only four feet above low water; and at the quarry, on the Indiana shore, eleven to thirteen feet. Here there is twelve feet of it exposed; but only a foot to eighteen inches of it quarried for cement. At the Big Eddy it is twelve to thirteen feet above low water; and at the middle of the falls as much as thirty-five feet above low water.

From the head to the foot of the falls, the Ohio river falls nineteen to twenty-one feet, depending on the stage of the water, and the distance on the general line of dip—west by south—one and a half miles. Hence there is an anticlinal axis about the middle of the falls, not uniform but undulating, amounting on the whole, to upwards of thirty feet in three quarters of a mile west by south. In the distance of four hundred and fifty yards from the quarry, on the Indiana shore,

*When these fossil corals are more thoroughly studied, we shall be able to give more definite references.

down stream, the strata decline fifteen to sixteen feet. It is at the anticlenal, above mentioned, where the steamboats so frequently scrape the rocks in gliding over the most turbulent portion of the falls. It is thickest at the foot of the falls, where it is twenty-one feet, it thins rapidly out in a north-east direction. At a distance of two and a half miles nearly east, where it is seen in the north-west end of the Guthrie's quarries, it is eighteen inches; and in a distance of three hundred yards to the south-east from this, it divides into two beds, and thins away to a few inches. Where it is divided an earthy limestone is interposed, not considered to possess hydraulic properties. It would seem, therefore, that the principal source of the hydraulic material was to the north-west of the main axis. At Hardin's in Clarke county, Indiana, cherty masses are interposed, which contain a variety of shells, amongst which the *Lucina proavia;* and very large *Atrypa prisca* are the most conspicuous.

The limestone which lies below the hydraulic limestone, composed, in a great measure, of comminuted remains of crinoidea, affords also *Spirifer cultriguqatus;* a very large undescribed species of *Leptæna,* which has been referred by some of our geologist, to the *Euglypha;* also, *Atrypa prisca,* and remains of fishes. This limestone is obscure on the middle of the falls; to the east it is better defined. On Fourteen Mile creek it is eleven feet thick; near the mill, on the east side of the Ohio, it is only three feet to three feet eleven inches. At Big Eddy the place of this limestone is six feet above the top of the Lower Fish Bed, but it is very obscurely marked at this point. To the east, in Jefferson county, Indiana, it passes into a well developed cherty mass of four or five feet in thickness, and is almost blended with the aforementioned cherty interpolations of the overlying beds.

Under the *cultraguzalus* beds succeeds the *Olivanites* bed, which is only six inches thick, near the mill on the south side of the Ohio, but attains a thickness of six or seven feet on Fourteen Mile creek, and runs down to a few inches at some places on the falls.

The next layer which is recognizable, is a cherty band charged with *Spirifer gregaria* of Dr. Clapp, and many small hemispherical masses of *Favosites spongites,* as at the foot of Little Island: one foot thick. Then comes a layer containing *concardium sub-trigonate,* of D'Orb., large hemispherical masses of *Stromatopora* and a *Ceriopore?* three to five feet.

Next come the Lower Fish Beds, nineteen feet in thickness, consisting of limestone, containing a large and beautiful species of undescribed *Turbo;* a large *Murchisonia;* a *Conocardium, Spirifer gregarea;* some small *Cyathophyllidæ,* and a *Leptæna.* The *Conocardium* layer is light grey, and more granular than the upper part, and esteemed the best bed for lime on the falls. The *Leptænæ* lie mostly about two feet above the *Conocardium.*

Next come chert layers, underlaid by coral layers, containing *Favosites maxima* of Troost, and *Favosites basaltica,* Goldfuss, which repose on a very hard layer.

The most of the remains of the fishes are found about three feet above the Turbo bed, but are more or less disseminated through the different layers, which have been designated as the Lower Fish Beds, and may, therefore, be subdivided thus:

1. Shellbeds.
   *A.* Conocardium bed, seven inches.
   *B.* Leptaena bed, (also with some conocardium,) six feet.
2. Parting chert layers, three feet.
3. Coral layers, seven feet.
4. Very hard rock, two feet.

The principal mass of corals on the falls of the Ohio, which must probably be grouped in the Devonian system, underlie these shell and fish beds, just mentioned, and repose upon a bed which can just be seen above the water level, at the principal axis, at extreme low water, which contains the chain coral, and which appears to be the highest position of this fossil.

Amongst the main coralline bed of the Devonian Period of the Falls may be recognized—

1. Dark grey bed, containing large hemispherical masses of *Favosites maxima* of Troost; *Zaphrentis gigantea;* and immense masses of *Favosites basaltica,* sometimes as white as milk; *Favosites* allied to *polymorpha,* but probably a distinct species, generally silicified and standing out prominently from the rock.

2. Black coralline layers, being almost a complete mass of fossilized corals; amongst which, a *Cystiphyllum, Favosites cronigera* of D'Orb., and *Zaphrentis gigantæ,* are the most abundant. These black layers contain also large masses of *Syringapora;* a large *Turbo* different from the species in the shell beds; also the large *Cyathophyll form Favosites*

allied to *polymorpha*, with *star-shaped* cells opening laterally on the surface of the cylinder, in pores visible to the naked eye; some *Cystiphyllum* carved into a semi-circle; large *Astrea pentagonus?* of Goldfuss, silicified, prominent, rugged, and black; this is the so called "Buffalo Dung."

The termination of these coralline beds of the Devonian system probably mark the place of the conocardium calcareous grit of the falls of Fall creek, Madison county, Indiana, and which is, undoubtedly, the equivalent of the Schoharie shell grit, near Cherry Valley, in New York, which underlies the Onondaga limestone of the New York system. No vestige of this calcareous grit has yet been found on the Falls, but there is reason to believe that it may be found in Jefferson county, about six miles above the Falls to the north east, on the farm of the late Dr. John Croghan, on the head of the Muddy Fork of Beargrass; and if so, though the Devonian and Silurian are apparently, at first view, so blended together on the Falls of the Ohio, this horizon between the black coralline beds above, and the chain coralline bed below, marks most satisfactorily the line of division between these two systems of rocks in Kentucky.

Time has not yet permitted a thorough investigation into the specific character of the numerous beautiful fossil-shells, corals and fish remains which occur at this highly interesting locality. Hereafter it is proposed, if occasion offers, to give more full and specific details of these rocks, and their imbedded organic remains.

As yet we have no good detailed sections of the Upper Silurian beds of Jefferson county, lying between the upper chain coral bed and the magnesian buildingstone. In the eastern part of Jefferson county, on Harrod's creek, a good section was obtained, showing the junction of the upper and lower beds with some of superior and inferior stratification.

The following is the section presented in the cut of Harrod's creek:

*Feet.*

240. Sneider House.

235. Magnesian limestone below house.

220. Red chert, with *Spirifer gregaria*, Porites, and other fossils.

180. Top of third bench of magnesian limestone.
Slope, with rocks concealed.

163. Base of third bench or offset of magnesian limestone.

*Feet.*

16?. Top of second bench of magnesian limestone.

154. Base of second bench of magnesian limestone.
Slope, with rocks concealed.

115. Base of overhanging ledges of cellular magnesian limestone.

110. Thin grey and reddish layers weathering and undermining the overhanging magnesian limestone, perhaps hydraulic in its properties.

107. Base of upper bench under the fall.
Earthy rock with some magnesia, perhaps with hydraulic properties.

100. Earthy rock with less magnesia?

95. Earthy reddish and green layers weathering with rounded surfaces like hydraulic limestones.

91. Hard grey silicious limestone, projecting from the bank.

90. Soft argillaceous layer decomposing under overhanding ledge above, partly hydraulic, upper two feet most earthy.

85. Hard layer on top of a little fall in bed of creek.

84. Ash-colored easily decomposing layers; lowest layer with nearly vertical fracture at right angles to the bedding.

86. Top of ash colored earthy hydraulic layers.

80. Top of lowest layer, with vertical cross fracture.
Junction of Upper and Lower Silurian Formations.

79. Limestone, with *Orthis Lynx.*

78. Brown layer of limestone, with branching Chætetes.

76. Layer with Cyathophylum?

67. More marly.

65. Hard thin layers of Leptæna limestone, with branching Chætetes.

59. Hard thin layers of limestone, containing *Leptæna alternatæ,* and *Atrypa capax*, of Con.

58. Hard layer, with irregular surface, four inches thick.

52. Hard layer, six inches thick.

50. Concretionary marly layer, containing *Leptæna planumbona.*

41. Irregular light colored layers, with remains of *Isotelus?*, Orthis, &c., five inches thick.
Dark marly regular layer, containing branching *Chætetes*, nine inches thick.

40. Ash-colored irregular layers, containing small branching *Chætetes.*

25. Fossiliferous slabs, with *O. Lynx*, and *O. formosa.*

22. Concretionary and marly ash-colored layers, with O. Lynx.

0. Slabs, with *Atrypa capacx*, and *Modesta*, at the junction of Harrod's creek with its Sneider branch.

The gregaria chert-bed lies on the Falls of the Ohio, about thirty feet above the base of the rocks of Devonian date. In this Harrod's creek section they were observed at 2?0 feet, where the junction of the Upper Silurian and Lower Silurian occurs at eighty feet; hence, if the rocks of Devonian date have the same thickness in the eastern part of

Jefferson county, as on its northern confines, the Upper Silurian rocks have a thickness, on Harrod's creek, of one hundred and ten feet. It is probable, therefore, that the upper chain coral bed, which marks the top of the Upper Silurian strata, is concealed ten feet up the slope, above the upper bench of protruding magnesian limestone in the above section.

Near the boundary between Jefferson and Oldham counties the cellular beds of the magnesian limestones of the Upper Silurian Period form the surface stratum, which is reached in sinking wells and found, on account of its spongy character, very difficult to blast.

### OLDHAM COUNTY.

The cuts in the hollows near Floydsburg very nearly expose the junction of the Upper Silurian with the Lower Silurian rocks.

The succession is as follows:

*Feet.*

95. General level of the country.
85. Bed of magnesian limestone.
75. Bed of buff magnesian limestone.
70. Ferruginous earth with entrochites.
68. Ash-colored earthy material.
65. Thin-bedded building stone.
60. Base of bench of buff and reddish rugged magnesian limestone, resting on a bed of buff, soft, crumbling, magnesian rock.
50. Bed of compact magnesian limestone used for gravestones.
45. Base of bench of magnesian limestone.
Argillaceous soft beds, associated with schistose magnesian layers, forming a slope of twenty-five to thirty feet.
15. Hard bluish-grey and buff limestone, which forms the top of the cliff at the Old Mill, on Garry's fork.
12. Resting on hydraulic looking earth limestone, of which twelve feet is seen.
0. Garry's fork of Salt river.

This branch of Salt river, where it meanders near the confines of Oldham and Jefferson counties, marks the boundary between the rocks of the Upper and Lower Silurian Period. In its bed lie strewed numerous specimens of the *Favistella stellata*, derived from an earthy hydraulic looking bed, that forms one of the uppermost layers of rocks referrable to the Lower Silurian Period, in this part of Kentucky. Almost immediately under it is a bed containing *O. Lynx*; then a bluish-grey bed of ten to twelve inches, which is the best rock for masonry on this creek. It reposes on a four to six-inch layer, containing Chætetes, under which, in the bed of the creek, is a bluish-grey limestone of eight to ten inches thick, of a sub-crystalline structure.

There is a layer of the blue limestone formation of this part of Oldham, near Hawley's and Ballard's, that admits of a tolerable polish, is variegated, dark, and reddish grey, and makes a tolerable marble.

About one mile from Lagrange, on the Sligo road, one of the earthy beds of the Upper Silurian system gives origin to a very peculiar white, ashey looking soil.

Lagrange is situated just at the junction of the upper and lower divisions of the rocks of Silurian date.

Between Lagrange and Harrod's creek is an oak country founded on the earthy beds at the base of the Upper Silurian system. Descending the hill at Lagrange you pass over buff and bluish-grey magnesian limestones, resting on hydraulic limestones and earthy beds, charged with *Favistella stellata*.

Between Ballardsville and Lagrange, in the deepest hollows, there is about eight feet exposed of grey marly limestones, containing a small nearly globular variety of *Chætetes lycoperdon*. This is overlaid by very rugged weathering limestone. The *Ambonychia carinata* is also characteristic of the rocks at this locality.

East of the head waters of Harrod's creek, the growth is chiefly beach, and the surface of the country less broken than west of this stream, in the oak land towards Lagrange. The ravines have seldom cut down into the Favestella beds.

Near Burrow's tavern the blue limestone shows itself under the magnesian and earthy strata; also, at the Sligo meeting house, near the line between Oldham and Trimble counties.

## TRIMBLE COUNTY.

In the southern part of this county chert beds, associated with a reddish earth, is frequently exposed in the hollows all the way to Bedford, overlaid by magnesian limestones, which occupy the higher grounds; and the country between Bedford and Milton is very much of the same geological character, the blue limestone appearing only in the lower cuts of the stream. On the points and slopes where this rock appears the land is much more productive and durable than on the higher grounds, and yields fine crops of tobacco and small grain.

The saline water, described in the third chapter, was first struck over the earthy hydraulic layers on high ground; the second pool, where the water is now collected, is excavated in the upper beds of the blue limestone formation. Lower down in the ravine is a fine building stone capable of receiving a polish.

The Epsom Spring, adjoining, issues from a bed (of calcarious?) clay, derived no doubt from the disintegration of the earthy beds above the blue limestone.

In the neighborhood of Bedford the rocks contain a large hemispherical fibrous-structured coral, resembling the Chætetes, but in concentric layers like the *Stromatopora*.

Descending Scott's hill the varigated banded earthy magnesian limestone, occupying the same geological position as the four-foot banded Madison building stone, of which I have previously spoken under the head of Jefferson county, is conspicuous, in step-like projections, overlying the blue limestone, the upper layer of which contains a large coral like the species just alluded to. This is forty to fifty feet down the descent. Under these beds of the blue limestone protrude layers containing *Atrypa capax, Leptæna planumbona, Streptelasma crassa, Orthis sub-quadrata;* and about forty to fifty feet still lower the Murchisonia marble rock is in place in considerable force.

The country descends very abruptly, over the beds of the blue limestone, to the little Kentucky river—almost too abruptly to admit of extensive farming, but the steep hill sides would, undoubtedly, make good vineyards.

The top of Scott's hill is approximately four hundred and sixty-six feet above the Ohio river; of this rather more than four hundred feet is all blue limestone.

## CARROLL COUNTY.

In this county my route led me along the extensive river bottoms, of which there are two terraces: one near high water mark, which is of a sandy character; the other, twenty to twenty-five feet higher, is a soil of fine quality, laid out extensively in orchards, which supply a large proportion of the fruit consumed in the towns on the lower Ohio river.

The hills in Carroll county have been only very partially explored as yet; enough is known, however, to place upwards of four hundred feet of the blue limestone, extending from the Ohio river to the general surface of the upland. Many of the beds of this formation are full of the *Orthis Lynx* and *occidentalis,* which are particularly abundant at an elevation of about three hundred and fifty feet above high water of the Ohio river.

Two varieties of soil were collected in Carroll county, for chemical examination, which are now in process of analysis, viz: one from the second bottom, and one from the upland, three hundred feet or thereby above high water of the Ohio river. The principal body of level upland seen in this county lies between Lick and Eagle creeks.

Slabs of the blue limestone, near the tollgate, and not far from New Liberty, contain *Orthis testudinaria,* and a *Leptæna* allied to the *deltoidea.*

### OWEN, GALLATIN, GRANT, AND BOONE COUNTIES.

The prevailing rocks which give character to a considerable area in these four counties, are different varieties of the peculiar earthy silicious mudstone, which I have had occasion to notice elsewhere in this report. The different varieties of this rock, in this part of Kentucky, will probably average about one hundred feet in thickness, and its elevation above the Ohio from two hundred to three hundred feet. It is usually of a buff color. Fossils can seldom be recognized in these beds of the Lower Silurian system of central Kentucky; those that have been observed are casts of *Leptæna.* Its composition will be seen by referring to the chemical report in the first and second parts of this volume, where from it will be seen to differ most decidedly from the composition of the calcarious beds of the same formation; this being, in fact, a rock not containing palpable grains of sand, but a large per centage of silex, and small proportion of lime, and a large per centage of sulphuric acid. It gives rise to a soil entirely distinct in its character from that resulting from the decomposition of the underlying and overlying blue and grey fossiliferous limestones, and stamps a marked feature to the whole country, where it exists in any great force. Some varieties give origin to sobby beech land; the better quality of soil, based on this formation, supports a growth of hickory, oak, poplar, and sugar-tree, and some walnut and hackberry on the hill sides, which may be considered the prevalent growth.

The waters of this district—even those habitually in use—appear to contain, as far as I am able to judge, from the *qualitative* examinations that I have made of them in the field, more than the normal proportion of magnesia found in natural spring water, not regarded as mineral, and that, probably, mostly in the state of chloride; but this is a matter that requires a careful *quantitative* analysis in the laboratory, in order to be able to pronounce, with confidence, as to proportional quantity and state of combination; and is a subject which demands farther and closer investigation. The beds of limestone which underlie this mudstone, and crop-out occasionally in the lower cuts of the hills in these counties, contain *Leptœna sericea, Orthis testudinarea, Atrypa capax and modesta.*

The hills are usually about four hundred feet above the level of the Ohio river; the mudstone occupying the higher grounds under the arable lands, which is often considerably broken; but the hillsides are more rounded off and less abrupt than where the limestones prevail, with a considerable thickness of marly shales.

Four different varieties of soil may be noticed in Boone county:

1. The beech, sugar-tree, white and blue ash lands.
2. The buckeye land.
3. The oak land, which is a superior soil, good for tobacco.
4. The wet or "sobby" beech land.

Between the towns of Union and Florence, in Boone county, at an elevation of two hundred and fifty to three hundred feet above the Ohio river, the shell beds of the blue limestone formation are in place, containing abundance of *Leptœna alternata, Orthis Lynx,* and branching form of *Chœtetes lycoperdon.* Some of the silicious mudstone is speckled with dark spots. This character of the rock was observed particularly near the head of Arnold's fork of Eagle creek, in Grant county, and on the waters of Cedar creek, in Owen county, which appears to be due to the presence of manganese.

Some of the springs near Dickey's branch of Cedar, in Owen county, are feebly impregnated with common salt, forming weak brines from which, in the early settlement of the country, salt has been made.

## KENTON, CAMPBELL AND PENDLETON COUNTIES.

The remarks which have been made in the preceding section, in regard to the geological formation of the counties adjacent to these on the west, apply to a considerable area of Kenton, Campbell and Pendleton counties, especially in the south and west portions of these counties.

In the north part of Kenton county, where the soils were collected for chemical analysis, the growth is chiefly beech mixed with sugar-tree, walnut, buckeye, large and small varieties of wild cherry, and black locust. About twelve feet below the general level of the arable up'and the beds of blue limestone are charged with *Leptæna alternata, Orthis Lynx, Chætetes lycoperdon.* The soil is best adapted to corn and rye. Wheat often freezes out; this could be avoided by underdraining and liming, and improving the mechanical texture of the soil by a proper system of cultivation.

The surface is very much broken in Campbell and the northern part of Pendleton; the predominating rock, towards the tops of hills, being the aforementioned silicious mudstone. About the centre of Pendleton county the hills are about three hundred and twenty-five feet above the Licking. Their base is composed of yellow marly shales and rough weathering limestones, containing *Orthis testudinaria, Leptæna sericea,* and *plano-convexa,* alternating with marly argillaceous shales. Above these come in shell-beds, full of broken fragments of *Leptæna alternata.* Owing to the large amount of marly argillaceous shales the ravines are very much gullied out, and for the same reason the soil is quite marly and argillaceous in its character. Vestiges of the silicious mudstone are still seen on some of the hill tops. The soil of this part of Pendleton is derived chiefly from the *Leptæna* and *Chætetes* limestone and marly argillaceous shales—the sub-soil being a yellow clay. The growth is black walnut, large white and red oak, black ash, black locust, and some wild cherry and shell-bark hickory. Between six and seven miles north of Falmouth the soil is derived more from the buff silicious mudstone; here the growth is mostly white oak and small scrub oak.

In the northern part of Campbell county the soil of the table land is chiefly derived from the silicious mudstone, resting on a deep yellow subsoil, supporting a growth of white oak, beech, sugar-tree, and black-walnut. Under the mudstone are yellow marly shale, with thin beds of hard (limestone?) insterstratified, containing branching forms of *Chætetes lycoperdon* and *Leptæna sericea.* The base of the hills are composed chiefly of blue marly argillaceous shales, with bands of limestones containing *Orthis testudinaria*, remains of *Encrinites*, and *Calymene senaria*. These two last mentioned members occupy together about one hundred feet of the base of the hills.

The southern part of Campbell county is much broken, and the geological formation very similar to that above described, the growth being white and red oak, shell-bark hickory, some poplar, with locally scrub and black-jack oaks; on the north slopes sugar-tree, black walnut, hickory, and beech prevail.

In the south part of Campbell county, about three miles from the county line, there is a considerable bed of bog iron ore. Masses of from two to four superficial feet are strewed on the surface, in a field on Mr. Yelton's farm, and large masses of the same description of ore also occur in the northern part of Pendleton county. If the specimens prove to be rich enough in the per centage of iron, on analysis, it is possible that enough might be obtained in this part of Kentucky to support a furnace. Near the top of the ridges, north of Falmouth, beds of white rounded quartz pebbles, as large as partridge eggs, were observed. These seem to be derived from some conglomeritic rock, which very likely may occupy the place of the Oneida conglomerate of the New York system; but, as yet, the rock itself has not been seen satisfactorily in place. If it prove, from future investigations, that these pebbles are derived from such a member of the rocks of Lower Silurian date in Kentucky, it will, I believe, be the first time that it has been observed in the western states. These pebbles can be detected occasionally, for a distance of ten miles, near the Cynthiana and Falmouth road.

### HARRISON AND SCOTT COUNTIES.

The blue limestone formation of Harrison county seems to be traversed by veins containing some sulphuret of lead, accompanied with sulphate of barytes, as near the Kentucky river, on the southern confines of Woodford county The soil of the southwestern part of Harrison county is a dark crumbling soil, based on a mulatto sub soil derived from rough weathering sub-crystalline. close-grained, light-grey limestones, containing *Leptæna cincinnatiensis*, under which are limestones containing fragments of *Asaphus* (*Isotelus*) *gigas* in large quantities and *Pleurotomaria* (Turbo?) *bilex*, of Conr. This description of soil commences four miles north of Cynthiana, and extends to the southern and western limits of Harrison county, at an elevation of about one hundred feet above the south fork of the Licking river.

Very little of the silicious mudstone was observed in Harrison county. There are some weak brines that rise in springs along the head waters of Mill creek, at Lee's lick, in the south western part of the county.

The soil in the neighborhood of Georgetown, in Scott county, is derived from a grey sub-crystalline member of the blue limestone formation, weathering reddish-grey, and containing *Atrypa capax* and *modesta*. The surface of the county is level, and the farms in a high state of cultivation; this character of soil and country extends for about four miles north of Georgetown, when the country becomes more broken, the hills assuming a peculiar rounded contour.

The growth, three quarters of a mile north of Georgetown, is sugar-tree, thick and shell bark hickory, black locust, wild cherry, and pin oak. Little or no poplar. This is on the second bottom of Elkhorn creek. Four miles north, where the surface becomes more broken and hilly, the principal timber is beech, white oak, small and large hickory; the rocks are more shaly, and the intervening layers more argillaceous.

Fourteen miles north of Georgetown the silicious mudstone appears, and gives more or less character to the country north, even as far as the Ohio river, as heretofore described in my remarks on Owen, Grant, Gallatin, and Boone counties.

### FRANKLIN COUNTY.

At Benson creek the beech timber of the eastern part of Shelby county terminates, and a growth of small sugar-tree, white and black hickory, black ash, and walnut sets in, with only very few beech trees. The undergrowth is dogwood, small black hickory, and oak, with only very few pawpaws except on the rich bottoms. On the summits and backbones of the ridges white oak prevails, while the hillsides have the deversified growth just mentioned, with red-bud and honey-locust in addition.

For special remarks on the soil of the western part of Franklin county the reader is referred to the second chapter of the first part of this report, containing remarks on the soils of Kentucky.

Near Bridgeport, on the banks of Benson, there are alternations of thick and thin bedded sub-crystalline buff, grey, and bluish-grey limestone. The thin layers are of irregular fracture, and rather concretionary. One stratum, six to seven feet above the bed of the creek, and twelve to eighteen inches thick, would probably make tolerably good dimension stones, requiring, however, some fifteen feet of stripping, mostly of rock, in order to reach the bed.

Lower down Benson, in the neighborhood of the Riffle, near Bright's mill, is an interesting geological locality from the occurrence, in the rocks, of very perfect specimens of a singular fossil belonging to the tribe of sponges, and referable, probably, to the genus *Scyphia*, of Schweigg, at least one of the forms, which is most abundant. This fossil consists of a cup-shaped central body, having a ring-shaped base, by which the animal, probably, fixed itself to objects. A variable number, (eight to eleven observed,) of large cylindrical tubes, radiate from the circumference of the body; the first inch of each of these passes off at right angles or horizontally, from the periphery; turning then at right angles they rise nearly vertically upwards and out-wards. The texture seems to be analogous to, but closer than that of the tubular sponges. These fossils vary from six inches to a foot in diameter. Not having seen any description or name for this fossil, that of *Scyphia digitata* is now proposed, on account of its general resemblance to a grasping hand.

We propose hereafter to give a figure of this interesting fossil, as well as of the other rarer and not less curious form of amorphozoa.

The geological position of these fossils must be not far above the beds furnishing such abundance of *Chœtetes lycoperdon*, in the section above the Arsenal, and below the Cemetery lot at Frankfort, since the same fossils were found near Bright's mill, some distance below where the Scyphia occurs. The position of the Chætetes beds will be seen by consulting the section about to be given of the bluff, on the Kentucky river, near the Arsenal.

There is another interesting fossil locality in this county, on Mr. Crutcher's farm. Fine specimens of a *Cytherina* occur here, in a bed of the blue limestone formation, which has been quarried for building chimneys. This fossil has been referred to the species *C. Baltica*, but it is, no doubt, a distinct and probably undescribed species.

Two hundred feet are well exposed, with some few hidden spaces, in a section on the Kentucky river near the Arsenal at Frankfort, showing the position of the Kentucky river marble, at the base and the Chætetes beds about the middle.

*Feet.*

205. Top of cliff.

195. Bed containing a variety of *Atrypa capax*, of Con; (*increbescens*, of Hall,); *A. Modesta*.

186. Base of bench, composed of thin-bedded limestone, most of which are concretionary or irregularly stratified.

175. Leptæna and orthis bed.

170. Base of upper cliff.

155. Chætetes.

150. Rocks, mostly concealed; a few Chætetes.

140. Chætetes.

130. Chætetes.

126. *Chætetes lycoperdon;* hemispherical form; large and abundant.

125. Thin bedded and irregularly concretionary limestones.

120. Chætetes beds; both branching and dish-shaped forms.

115. Loose specimens of chætetes lycopherdon hemispherical and branching forms.

110. Quarries, in thinly stratified shell-beds, containing Leptæna and other fossil shells.

101. Thin-bedded limestone, charged with very perfect specimens of *Orthis testudinaria.*

Slope rocks concealed.

100. Hydraulic looking layer, containing Conularia (quadrasulcata?) and Avicula.

95. *Orthis testudinaria.*

90. Thin-bedded limestone charged with *Orthis testudinaria.*

85. Semi-crystalline Orthis limestone: near foundation of the Arsenal.

80. Thin earthy layers.

Slope with rocks concealed thirty-one feet.

59. Coarse textured hard limestone, with cherty concretions.

56. Coarse textured hard limestone, less cherty, mottled and silicious.

55. Grey and greenish schistose close textured limestone.

54. Top of Kentucky river marble in beds from eight inches to one and a half feet.

50. Top of Kentucky river marble, close-textured and pure.

48. Greenish close-textured limestone.

41. Green silico-argillaceous schistose limestone.

40. Close textured grey marly limestone.

38. Close textured grey limestone, more schistose, one and a half feet.

*Feet.*
30. Close textured grey limestone, more schistose, one and a half feet.
25. Close textured grey limestone, with some cherty segregations.
20. Kentucky river marble in beds from one foot to eight inches.
15. Kentucky river marble bedded two to four inches.
Space with rocks concealed.
10. Close textured limestone, bedded two to four inches.
7. Close textured limestone, thin bedded.
5. Close textured limestone? concealed.
4. Close textured limestone.
1. Close textured limestone, somewhat schistose.
0. Low water Kentucky river.

The peculiar smooth-textured, dove colored limestone, with disseminated specks and veins of white calcarious spar, which occurs in the base of this section, particularly at twenty and fifty-four feet, has been too hastily referred to the birds-eye limestone of the New York system. It is true that it lies forty feet under the Orthis testudinaria bed, and fifty to sixty feet under the Chætetes lycoperdon bed, at Frankfort; and that the rock has what may perhaps be termed a birds-eye structure, but the disseminated fossil, which is converted into calc. spar has, so far as I have been able to observe its structure, a much greater resemblance to some amorphozoa than to a plant, and the surface markings have not the appearance of figures 1c, 1d, and 1e, plate 8, of Hall's 1st vol., of his New York Palaeontology; but rather linear graphic, or small tubular markings, filled with calc. spar; and some well preserved specimens show, in the tubular branches, a structure analogous to that of some of the sponges. We shall endeavor, hereafter, to give some figures of these fossils.

The beds which occur at one hundred and ten and one hundred and fifteen feet in the above section appear to be the same as those seen below Bright's mill, on Benson, near the locality of the *Scyphia digitata.*

One of the highest beds seen on this part of Benson is charged with *Leptæna,* the shell having been converted into ochre, while the moulds remain, making up almost the entire mass of the rock. Below this, on Mr. Bright's farm, is a bed composed principally of white crystalline calcarious spar. If it lies in a sufficiently solid bed it would make a marble very like that of Trimble county; it is doubtful, however, whether it will be found sufficiently extensive. A good deal of the rock above Bright's mill has the external appearance of hydraulic limestone, specimens of which have been collected for analysis. Some white and pale yellow fluor spar, and sulphuret of zinc, occur in the rocks of Benson, in the vicinity of Bright's mill, but only in small quantities, so far as I have yet seen.

Occasionally, over circumscribed areas, in this and other counties of the limestone regions of the state, small spots of ground may be observed where the grass is quite yellow and sickly looking; such places are known as "sick-spots." Samples of soil have been collected from such places, for chemical analysis, in order to ascertain whether a larger proportion than usual of protoxide of iron, or other peculiarity, could be detected in the soil. My present belief is, however, that analysis will not show any material difference between this and the adjoining fertile ground; because, I think the effect is, in all probability, due to cavernous spaces or fissures in the limestone rock beneath, which evolve carbonic acid in sufficient quantity to displace or exclude atmospheric air and oxygen from the interstices of the soil, without which vegetation cannot flourish; or it may, in part, be caused by the surface-water being drained away into these vacancies. What lends probability to this explanation of the phenomena is the occurrence, near by, of depressions of the surface, like the sink holes of the barrens in miniature.

The reader is referred, for a continuation of the subjects embraced in the preceding chapters, to the second part of this report.

D. D. OWEN, *State Geologist.*

# SECOND CHEMICAL REPORT

OF THE

# ORES, ROCKS, SOILS, COALS,

## MINERAL WATERS, &c.,

## OF KENTUCKY,

BY

ROBERT PETER, M. D.,

CHEMICAL ASSISTANT TO THE GEOLOGICAL SURVEY.

# INTRODUCTORY LETTER.

CHEMICAL LABORATORY OF THE KENTUCKY GEOLOGICAL SURVEY,
*Lexington, Ky., December 8th,* 1856.

D. D. OWEN, M. D.:

*Dear Sir*—In accordance with your instructions I transmit to you my second report of the Chemical Analyses of Kentucky Ores, Soils, Mineral Waters, &c., &c., made at this Laboratory, for the Geological Survey, since the preparation of the first report.

Within about two hundred and twenty-two days, with the occasional aid of an assistant in the minor processes under my immediate supervision, we have succeeded in determining the composition of two hundred and six different objects, and thus, although, as you will discover, the several analyses have been made more minute and accurate; we have increased the amount done, in proportion to the time employed, more than one-sixth over that exhibited in the first report.

The subjects of the analyses reported in the following pages may be summed up as follows:

48 iron ores of the limonite variety.
22 iron ores of the carbonate variety.
43 soils, sub-soils, and marls.
31 limestones.
30 coals.
16 mineral waters and salts.
4 copper and zinc ores and bitumens.
4 iron furnace slags.
4 sandstones.
2 pig iron.
2 shales and slates.
———
206

The greater portion of the large and very interesting collection of soils and sub-soils, made by you during the past summer, amounting to nearly one hundred specimens, sent to this Laboratory for examination, have not yet been analyzed, but the labor will be resumed as soon as possible after the completion of this report.

In regard to soil analysis, a considerable difference of opinion exists in the minds of the agricultural public. When the fact began to be appreciated, that certain organic and mineral substances resident in the soil were essential to its fertility, because they were necessary elements of vegetable and animal tissues, it was natural that the enlightened agriculturist should look to the chemical analysis of the soil, which would give the proportions of these ingredients, as the best index of its value and its adaptedness to his various crops; and full experience, under the proper conditions, will demonstrate that this expectation will not be disappointed. But, when at the demand of the farmer, who perhaps knew little or nothing of the true theory of agriculture and nothing of chemical philosophy, *cheap and superficial analyses* were made, exhibiting only the proportions of the grosser materials of the soil—as of the *sand* and *silica, alumina, oxide of iron, carbonates of lime,* and *magnesia,* and even, perhaps, of the *organic matters,* without showing the amount present of the more valuable and essential ingredients, as the *phosphoric* and *sulphuric acids,* the *potash* and *soda*—this information, purchased by the practical farmer from the scientific man, at however low a price, was found to be dearly bought, and of little real value.

All soils, without exception—the most fertile as well as the most sterile—contain large proportions of *sand* and *silica, alumina* and *oxide of iron,* and they may contain these as well as notable proportions of *lime, magnesia,* and *organic matters,* and yet be sterile to the highest degree; for, although these, with the exception of alumina and sand, enter into vegetable and animal composition, and are essential to their structures, they are of no value in the support of plants, without the aid of the alkalies and the phosphorus and sulphur contained in the phosphoric and sulphuric acids of the soil. These latter ingredients, almost universally found in very small relative proportions in soils, and much more difficult to estimate in a chemical analysis than the preceding, are the elements of the soil, the proportions of which it is most necessary to ascertain, in order to get a proper idea of its value and relationships to the operations of the agriculturist. But these, in consequence of the difficulty of the processes, and the time and care necessary to their estimation, have been generally neglected in ordinary soil analyses. No wonder, therefore, that the practical man, and even some chemists, have begun to doubt whether the so called teachings of science, in this relation, are of any real service.

A full analysis of a soil, giving the correct proportions of all its ingredients, and their various states of combination, is a labor of considerable magnitude, requiring, if the time be devoted to only one soil at once, from ten to fifteen days of work, and demanding in the operator as much special training as to learn to play well on a difficult musical instrument; the farmer, therefore, can never be expected to be able to perform this nice and troublesome operation for himself, any more than he could be expected to make or repair his own watch or time-piece; but he can, by acquiring the necessary elementary knowledge to appreciate the results of chemical analyses, derive great practical advantages from them, and save a great deal of time, labor, and money. He could, it is true, with the aid of his experience, and by the trial of experiments in cropping, ascertain the value of a soil almost as well as it could be set forth by a good chemical analysis; but, in commencing on an unknown specimen, the chemist could, in one week's labor, arrive at results, which could be attained by the practical farmer only at the expense of years of costly agricultural experiments.

The system in which you have collected the specimens of soils, for analysis will aid greatly in giving a practical demonstration of the value of soil analyses. Usually, instead of collecting a single specimen from each locality, you have procured, for *comparative analysis*, specimens of—1. The *Virgin soil;* 2. The *same soil from an old field long in cultivation;* 3. The *sub-soil;* and 4. The *deeper sub-soil,* or *underlying rock stratum.*

By the correct examination of these the following important facts can be ascertained: 1. The change which the soil has undergone under the influence of cropping; and hence the knowledge of what would be necessary to restore it to its original condition, and keep it fertile. 2. What benefit or injury may result from deep sub-soil ploughing or trenching the ground. 3. What influence may be exerted on it by the underlying rock or other *sub-strata.*

By the critical examination of the comparative analyses of soils, &c., &c., already given in this and in the preceding report, it will be observed that chemical analysis is competent, in these respects, to ascertain and report faithfully on changing conditions of the soil in relation to agricultural operations. It will be noticed, in particular, that in every instance where the comparison is made of the proportion of the phosphoric and sulphuric acids, potash, and soda, between the virgin soil, and similar soil which has been long in cultivation, a marked diminution of these most essential ingredients is to be noticed in the old soil. And thus, it is proved, that by careful chemical analysis we can note and estimate the gradual but certain approach to sterility, of soils once very fertile, under the influence of unscientific and thriftless cropping.

The *knowledge* of a defect must naturally precede all efforts for its removal. The full appreciation of the fact, that in yielding its products the soil always gives up a certain amount of its most valuable elements, which are carried off in the crops removed, and which must in some way be restored to it, if it is to be maintained in a fertile condition, is sure to lead, in the end, to an improved system of agriculture, if the education of the people of our state is made to keep pace with the general march of improvement.

The completion of the analyses of the soils of Kentucky, or even of those already collected, ought to exert a beneficial influence on the prosperity of the State. The real agricultural value of the land in its various districts will be to a certain extent demonstrated, and it will be shown more fully, as it is already to some extent exhibited in the analyses given in this and the preceding report, that a great body of lands in the central, eastern, and southern part of the state of Kentucky, held now at prices below, or not much above that of government land in the far west, may be made as valuable as those, to the farmer; whilst, in some localities, they offer superior advantages in the greater proximity of fuel in the form of coal or wood.

These results may, perhaps, help to stimulate our people to endeavor to supply a great necessity of the state, which now operates as an immense incubus on its growth and developement, viz: a chain of great public improvements through the interior, to afford means of communication and channels of commerce, which may bring to the doors of the farmer or manufacturor, who may engage in the business of developing its great mineral and agricultural resources, the markets of the world. The want of these improvements confines the growth of Kentucky, in commerce and the manufactures, mainly to her river banks, and restricts her agriculture to its richest regions, to the neglect of mineral wealth greater than that which has been the basis of the power of England, and a large body of land very susceptible of cultivation. On the other hand, the policy of supplying these public improvements, in the net-work of railroads intersecting the western country, constructed mainly under the patronage of the general government, and with the proceeds of large grants of the public lands, has aided greatly in inviting to its cultivation the hardy yeomanry of the older states, who are tempted to leave their native homes by the inducements of rich soil, at a moderate price, accessible markets for their products, and a prospect of the rapid growth and improvement of the country.

That the reader of this report may be enabled to compare the soil of the fat lands of the western prairies with some of those of Kentucky, usually considered much less valuable to the agriculturist, an analysis of Illinois prairie soil is introduced at the latter end. It will be seen that this prairie soil, now so rich in organic matters, may be considered as the reverse of the heavy *red sub-soil* of some of the southern portions of Kentucky;* in this respect, in particular, in that, from its large proportion of fine *sand and silica,* and small relative amount of *alumina and oxide of iron,* it holds, with a weak affinity, those organic matters derived from the remains of the herbage of thousands of years; and hence gives abundance of rich food to the crops which it supports; until, in the course of time, this deposit is diminished or exhausted. On the other hand, the large proportion of oxide of iron and alumina, of the heavy red sub-soil—which both have a powerful affinity for organic matters—holds them with great tenacity, and thus, under the action of water containing carbonic acid, which is the natural solvent of the mineral and organic matters in the soil employed in vegetable growth, this red sub-soil gives up but a small quantity of solid nutritious matter, especially if there is but a trace of lime or magnesia present. The prairie soil could be rendered more durable, but perhaps less *immediately* fertile, by admixture with clay, containing alumina and oxide of iron, whilst, other things being equal, the heavy

*See Simpson county.

red soil would be made more fertile by the addition of fine sand and lime.

The addition of lime to this heavy red soil, which contains a large proportion of alumina and peroxide of iron, may be beneficial in more than one way: it would not only assist in the solution of the other nutritive elements locked up in the soil, and tend to render it lighter, but from its constant action on the oxygen and nitrogen of the atmosphere, in causing them to combine in the form of nitric acid, soluble nitrates are always present in soil containing much lime or carbonate of lime, which aid in its disintegration, and increase the solubility of its valuable mineral ingredients, besides furnishing a supply of dissolved nitrogen to vegetable roots.

On this principle Leibig has explained the fact, that in the island of Cuba a soil containing a very large proportion of carbonate of lime, can annually produce, without the application of nitrogenous manures, large crops of tobacco—a plant peculiarly rich in nitrogen,—and for the same reason the nitrate of lime, (easily convertible into salt-petre,) is continually formed and effloresces on the porous limestones of the so-called salt-petre caves of Kentucky.

The seventy iron ores which have been analyzed at this Laboratory, since the preparation of the last report, have, with very few exceptions, proved to be rich and valuable, as well those of the *Limonite* variety, composed of hydrated oxide of iron in various states of purity, as the *carbonates of iron*; and afford still further illustration of the great wealth of Kentucky, in ores of this most useful and valuable of metals, and of the fact that a large amount of capital and labor might find room for employment in our state, in the developement of her rich mines, and in the supply of the increasing demand for iron in all its various forms. The analyses of these ores, and of the limestones, &c., which accompany them, will greatly assist the manufacturer in the apportion of his fluxes for the most economical production of the metal.

Amongst these ores are some which doubtless would be found well adapted to the manufacture of steel, and in some localities the association of an easily smelted ore with beds of suitable coal, may induce capitalists to endeavor to supply the very large demand for *cheap* iron for railroads and other purposes.

The thirty kinds of coal which have been examined have been analyzed with more than usual minuteness and labor. Not only have all, not previously analyzed, been submitted to *proximate analysis*, to ascertain their proportions of *moisture*, *volatile matter*, *ashes*, and *coke*, but by separate operations their proportion of *sulphur* and the chemical composition of their *ashes*, have been ascertained; they have also been all submitted to *ultimate* or *orgauic analysis*, to determine their relative proportions of *carbon*, *hydrogen*, *oxygen*, and *nitrogen*, &c., in which analysis, as one of the ingredients—oxygen—is always estimated by the *loss*, or negatively, and therefore, the *control* of the equality of the weight of the sum of the elements found, with the weight of the original compound which was submitted to analysis, being wanting, it was necessary to secure accuracy by a repetition or repetitions of the process; so that the ultimate analysis of these thirty coals required no less than seventy-nine operations of organic analysis. The whole number of analyses of these thirty coals amounted to one hundred and sixty-one. In these various processes several of the most promising of these coals were submitted to destructive distillation, at a heat gradually raised to redness, to ascertain their relative products of *bituminous oils*, *paraffin*, *gas*, &c. In these trials the Breckinridge cannel coal maintained its superiority for this manufacture; but the approach of the Haddock's cannel coal, of the Kentucky river, to it in this respect, encourages the belief that in the course of your investigations amongst the Kentucky coals, especially amongst the cannel coals and bituminous schists, other specimens may be found which may be equally valuable for these products with the Breckinridge coal.

The peculiarity in composition, of the coals which yield the greatest amount of oily and waxey matters on distillation, appears to be the presence, in them, of a larger proportion of hydrogen to the carbon than exists in the *coking coals* or *soft bituminous coals*, which are preferred by the blacksmith and for the manufacture of coke and gas; and of a smaller amount of oxygen than is contained in the *dry coals* or *splint coals*.

It will be seen that the coal fields of Kentucky furnish all these varieties. For the purpose of comparison with the coking varieties of Kentucky coal, an analysis of the Youghiogheny coal of Pennsylvania is given at the end of the report; and to enable the enlightened reader to compare the Breckinridge coal with the celebrated Scotch Bog Head coal, also much used for the production of oils, &c., its organic analysis is stated in connection with that coal, under the head of Hancock county.

The process of *organic analysis* employed may be briefly described. The powdered coal, previously dried at 212° F., was introduced into the hard glass combustion tube, in a small tray of platinum, and submitted to the action of a stream of pure oxygen gas from a gas-holder, dried by passing it through chloride of calcium and dry hydrate of potash; the combustion tube was heated over charcoal, in a common Liebig's furnace; to secure complete combustion of the carbon, the front portion of the tube was filled either with oxide of copper, mixed with copper turnings, or with a tight rolled cylinder of copper gauze which had been previously oxidated at a red-heat in a stream of oxygen. The products of combustion were collected in the usual chloride of calcium tube and potash bulbs; a small tube being interposed to absorb any sulphurous acid, and a *dry* potash tube attached to the bulbs to absorb all the carbonic acid, and prevent the escape of moisture in the stream of dried gas. Thus the proportions of carbon and hydrogen were obtained.

An attempt was made, by collecting the residual gases—mixed nitrogen and oxygen—which passed through this train, and by the removal of the excess of oxygen, by explosion with hydrogen in the *Endiometer*, to estimate, by the same operation, the proportion of nitrogen; but it was soon found that with whatever care the oxygen was procured, the proportion of nitrogen left after the explosion was not constant, and on reflection on the known properties of gases, and the force with which they penetrate each other and porous substances generally, the reason of the failure of this promising process became obvious. The water introduced into the gas-holder to expel the oxygen, contained nitrogen, which gas diffused itself through the atmosphere of oxygen in the gas-holder, and thus, in proportion to the quantity of water forced into it, did the oxygen in it contain more and more nitrogen, as was verified by experiments with the Endiometer. Nor was it found possible, even with the use of a smaller oxygen gas-holder, and of distilled water covered with oil, boiled to expell the gas, wholly to prevent this cause of irregularity, so that the proportion of the nitrogen in the coals was necessarily obtained by a separate process of combustion, by the method of Will and Varrentrapp.

Amongst the limestones and sandstones examined are some quite valuable for building purposes; and others which will be found useful as hydraulic cement, and for agriculture. The magnesian limestone, from Grimes' quarry, and from other neighboring quarries, on the Kentucky river, may be considered one of the best and most durable building stones of the whole country at large, and some others from the Upper Silurian Formation resemble it somewhat closely in composition. The Birds-eye limestone, characterized by its great brittleness, contains but little carbonate of magnesia, and would burn into quite a pure lime; whilst the very fossiliferous limestones of the Blue Limestone Formation, (Lower Silurian,) easily disintegrating and containing, in addition to *lime* and *magnesia*, all the other mineral elements necessary to vegetable nutrition, although they make but poor building stones, are invaluable to the agriculture of the country where they exist, by the enriching influence, on the superincumbent soil, which they exert under the slow solvent action of the natural surface waters, which always contain carbonic acid, and which convey into the soil their valuable ingredients. The waters of such regions are *hard* from this cause, but under their influence the soil is, to a certain extent, constantly renovated.

The sixteen mineral waters, &c., examined, are, mainly, only from one of the Kentucky watering places. The mineral springs of the state are numerous and valuable, and will doubtless repay, in the future, the labor of their exploration.

All of which is respectfully submitted,

ROB. PETER.

A SUMMARY

OF THE

# CHEMICAL ANALYSES

OF

# ORES, ROCKS, SOILS, COALS, MINERAL WATERS, &c.,

OF KENTUCKY,

MOSTLY PROCURED BY DAVID DALE OWEN, M. D., PRINCIPAL GEOLOGIST OF KENTUCKY, AND ANALYZED BY ROBERT PETER, M. D., CHEMICAL ASSISTANT TO THE STATE GEOLOGICAL SURVEY.

ARRANGED IN THE ALPHABETICAL ORDER OF THE COUNTIES IN WHICH THEY WERE OBTAINED.

## ADAIR COUNTY.

No. 233—Soil. *Labeled "Soil from Shaly Geodiferous Limestone, at Clayton Miller's farm, four miles south of Columbia, Adair county, Kentucky."* (*Sub-carboniferous Sandstone, or Knob Formation.*) Growth hickory, sugar tree, white oak, dog-wood, white walnut, and elm.

Color of the dried soil very dark grey. Sifted through a seive, of one hundred and sixty-nine apertures to the inch, it left about one-fifth of its weight of irregular pebbles of ferruginous sandstone. Carefully washed with water it left about fifty-seven per cent. of *sand*, of which 42.3 per cent. is fine enough to pass through fine bolting cloth, of about five thousand apertures to the inch; and 14.7 per cent. is coarser sand, consisting principally of rounded particles of quartz, *hyaline*, and of various shades of yellow, red, and brown, with some few crystalline particles.

One thousand grains of this soil, (air-dried,) digested for one month, in a closely stopped bottle, at a temperature not exceeding 120° F., in water saturated under pressure with carbonic acid gas, gave up to the acidulated water nearly two and a half grains of *solid matter*, which was found to have the following composition, dried at 212° F., viz:

| | *Grains.* |
|---|---|
| Organic and volatile matters, | 1.150 |
| Alumina, oxides of iron and manganese, and trace of phosphates, | .317 |
| Lime, | .447 |
| Magnesia, | .106 |
| Brown oxide of manganese, | .019 |
| Sulphuric acid, | .068 |
| Potash, | .098 |
| Soda, | .024 |
| Silica, | .140 |
| Carbonic acid, chlorine, and loss, | .102 |
| | 2.471 |

The air-dried soil lost 2.50 per cent. of *moisture* when dried at 400° F.

Dried at this temperature its *composition* was found to be as follows, viz:

| | |
|---|---|
| Organic and volatile matters, | 4.440 |
| Alumina and oxides of iron and manganese, | 4.841 |
| Carbonate of lime, | .196 |
| Magnesia, | .046 |
| Phosphoric acid, | .065 |
| Sulphuric acid, | .232 |
| Chlorine, | .005 |
| Potash, | .075 |
| Soda, | .092 |
| Sand and insoluble silicates, | 90.446 |
| | 100.438 |

As explained in the preceding report, the process of digesting the soil for a length of time, in water containing carbonic acid, at a temperature not exceeding that to which it naturally attains under the influence of the sun's heat, is used to ascertain and estimate the proportion contained in it of soluble nutritious matter, *immediately available* for the support of vegetation. In this manner, endeavoring to imitate the usual mode by which these necessary ingredients of organic structures are dissolved out of the soil, and conveyed into the tissues of growing plants in the great operations of nature.

*Pure water* exerts but little solvent action on the carbonates or phosphates of lime or magnesia, but when it is combined with carbonic acid it takes them up in considerable proportions, and especially when aided by the humic acids, so called, which result from the decomposition of vegetable or animal bodies on the soil, and by the small amount of acids of nitrogen which the atmosphere yields under favorable circumstances, it not only brings these and the oxides of iron and manganese and silica to a soluble condition, but also acts gradually on the *insoluble silicates*, to release their lime, magnesia, potash, &c., &c., for vegetable nourishment. These, then, are the solvents which, by their continual action on the soil, and with the aid of frost, slowly disintegrate its hard particles, and gradually dissolve out its available materials. All rain water, and surface water in general, contain more or less carbonic acid, with occasional traces of the acids of nitrogen; and the water acquires in the soil the organic acids which are produced there by the decomposition of vegetable and animal matters.

Although this soil contains a larger proportion than the average of *sand* and *insoluble silicates*—more than ninety per cent.—and less than the usual quantity of phosphoric acid and potash contained in very fertile soils—.075 and .065—it yet contains a pretty large proportion of vegetable nourishment in a *soluble condition*, so that it gave up more than the average quantity of nutritious matter to the carbonated water in which it was digested. Without judicious management—by a course of constant cropping, without returning to it the *essential ingredients* of vegetable nutrition—this soil will more speedily become deteriorated in productiveness, than others which have less sand and less *soluble matters*.

### ANDERSON COUNTY.

No. 484—LIMESTONE. *Labeled "Rock under White Oak Ridge, Mr. Hall's farm, Anderson county, Ky."* (*Lower Silurian Formation.*)

A grey, granular rock, made up of a confused mass of crystalline grains of calcarious spar. No fossils apparent in the specimen sent for analysis.

Composition, dried at 212° F.—

| | | |
|---|---|---|
| Carbonate of lime, - - | 96.65 = | 54.23 *Lime.* |
| Carbonate of magnesia, not estimated. | | |
| Alumina and oxides of iron and manganese, - - - | 1.26 | |
| Phosphoric acid, - - - | .92 | |
| Sulphuric acid, - - - | .25 | |
| Potash, - - - - | .57 | |
| Soda, - - - - - | .39 | |
| Silex and silicates, insoluble in hydrochloric acid, - - | .88 | |
| | 100.92 | |

The air-dried rock lost .30 per cent. of *moisture* at 212°, F.

No. 485—Limestone. *Labeled "Leptæna Limestone," road from Mr. Alexander Julian's to Lawrenceburg, Anderson county, Kentucky.* (*Lower Silurian Formation.*)

A very fossiliferous limestone, of a grey and buff-grey color in the interior; weathered surfaces of a dirty buff colored. Powder light yellowish grey.

Composition, dried at 212° F.—

| | | |
|---|---|---|
| Carbonate of lime, - - - | 83.95 = | 47.11 *Lime.* |
| Carbonate of magnesia, - - | .91 | |
| Alumina, and oxides of iron and manganese, - - - | 2.23 | |
| Phosphoric acid, - - - | .25 | |
| Sulphuric acid, - - - | .34 | |
| Potash, - - - - | .38 | |
| Soda, - - - - - | .47 | |
| Silex and insoluble silicates, - | 11.28 | |
| Loss, - - - - - | .19 | |
| | 100.00 | |

The air-dried rock lost .30 per cent. *moisture* at 212° F.

No. 486—Limestone. *Labeled "Road from A. Julian's to Lawrenceburg, Anderson county, Kentucky."* (*Lower Silurian Formation.*)

A bluish-grey limestone, very full of fossils—Pleurotomaria, Bellerophon, Orthocera, portions of Encrinal stems, &c. Weathered surfaces of a dirty buff color. Powder light grey.

Specific gravity, - - - - - - - 2.653

Composition, dried at 212° F.—

| | | |
|---|---|---|
| Carbonate of lime, - - - | 86.45 | = 48.52 *Lime.* |
| Carbonate of magnesia, - - | 1.57 | |
| Alumina, and oxides of iron and maganese, - - - - | 1.83 | |
| Phosphoric acid, - - - | .12 | |
| Sulphuric acid, a trace. | | |
| Potash, - - - - | .62 | |
| Soda, - - - - - - | .11 | |
| Silex and insoluble silicates, - | 9.57 | |
| | 100.27 | |

The air-dried rock lost .10 per cent. of *moisture* at 212° F.

BALLARD COUNTY.

No. 218—SUB-SOIL. *Labeled "Sub-soil in heavily timbered land, southern part of Ballard county."* (*Quaternary Formation*)

The dried soil is of a light yellowish grey-brown color. Carefully washed with water, one thousand grains of it left about five hundred and ninety-two grains of *sand* of a brownish-grey color, of which only about two grains was too coarse to pass through bolting cloth of five thousand apertures to the inch. The coarser particles were generally rounded, some few angular, consisting of hyaline and milky quartz, with some particles of iron ore.

One thousand grains of this soil, dried at the ordinary temperature, and digested in water containing carbonic acid, for one month, yielded less than *one* grain of *soluble matter.* This dissolved solid extract was found, on analysis, to have the following composition, when dried at 212° F., viz:

| | *Grains.* |
|---|---|
| Organic and volatile matters, - - - - - - - | 0.200 |
| Alumina, oxide of iron, and trace of phosphates, - - - - | ·097 |
| Lime, - - - - - - - - - - - | .064 |
| Magnesia, - - - - - - - - - - | .033 |
| Brown oxide of manganese, - - - - - - - | .047 |
| Potash, - - - - - - - - - - - | .060 |
| Soda, - - - - - - - - - - - | .023 |
| Silica, - - - - - - - - - - - | .180 |
| Sulphuric acid, carbonic acid, and loss, - - - - - | .029 |
| | 0.733 |

The air-dried soil lost 1.80 per cent. of *moisture* when dried at 380° F.

Dried at this temperature, its *composition* was found to be as follows, viz:

| | |
|---|---|
| Organic and volatile matters, | 2.11 |
| Oxide of iron, | 2.24 |
| Alumina, | 2.58 |
| Brown oxide of manganese, | .09 |
| Carbonate of lime, | .15 |
| Magnesia, | .86 |
| Phosphoric acid, | .41 |
| Sulphuric acid, not estimated. | |
| Potash, | .12 |
| Soda, | .02 |
| Sand and insoluble silicates, | 91.72 |
| | 100.30 |

The analysis of this sub-soil may be compared with that of corresponding surface-soil given on pages 259 and 379 of the preceding report, (No. 1.) It will be seen, that whilst it has pretty nearly the same proportions of sand and insoluble silicates, of alumina and oxide of iron, it contains more potash, phosphoric acid, lime, and magnesia, and less of organic and volatile matters, than the surface-soil. It also contains less soluble matter immediately available for the nourishment of vegetables—the surface-soil, No. 1, having yielded 1.53 grains of *solid extract* to water containing carbonic acid, while this sub-soil gave only 0.733 of a grain.

No. 219—SUB-SOIL. *Labeled "Sub-soil from the north-western part of of Ballard county, Ky., from near Col. Gholson's." (Quaternary Formation.)*

Color of the dried soil rather darker than that of the last described, with more of a reddish tinge. Carefully washed with water one thousand grains of it left about 546½ grains of brownish-grey sand, of which all but about eight grains would pass through fine bolting cloth. The coarser particles, under the microscope, appeared to consist principally of rounded fragments of iron ore, mixed with some particles of hyaline and milky and red quartzose mineral.

One thousand grains of this sub-soil, dried at the ordinary temperrative, digested for one month, in water containing carbonic acid, as before described, gave up 1.293 grains of *solid extract,* which, dried at 212° F., was found to have the following *composition,* viz:

| | *Grains.* |
|---|---|
| Organic and volatile matters, | 0.340 |
| Alumina and oxide of iron, | .047 |
| Lime, | .300 |
| Magnesia, | .090 |
| Brown oxide of manganese, | .077 |
| Phosphoric acid, | .011 |
| Sulphuric acid, | .067 |
| Potash, | .110 |
| Soda, | .040 |
| Silica, | .190 |
| Carbonic acid and loss, | .021 |
| | 1.293 |

The air-dried sub-soil lost 2.14 per cent. of *moisture* when dried at 375° F.

Dried at which temperature it was found, on analysis, to have the following *composition,* viz:

| | |
|---|---|
| Organic and volatile matters, | 2.92 |
| Oxide of iron, | 3.39 |
| Alumina, | 2.25 |
| Carbonate of lime, a trace. | |
| Magnesia, | .47 |
| Brown oxide of manganese, | .36 |
| Phosphoric acid, | .18 |
| Sulphnric acid, not estimated. | |
| Potash, | .19 |
| Soda, a trace. | |
| Sand and insoluble silicates, | 90.21 |
| Loss, | .03 |
| | 100.00 |

On comparing this analysis with that of the *surface soil* from the same locality, No. 2, as given on pages 261 and 379 of the preceding report, it will be seen that they present nearly the same differences of composition as were noted in the remarks on the preceding *sub-soil,* (No. 218,) viz: that there is less of *organic and volatile matters,* and less of the nutritious substances soluble in carbonated water, in the *sub-soil,* and rather a larger proportion of phosphoric acid, potash, and magnesia, than in the surface soil.

## BARREN COUNTY.

No. 225—Soil. *Labeled "Soil from Mr. Barlow's farm, Barren county, Kentucky."* (*Sub-carboniferous Limestone Formation.*)

Said to be the best producing soil in the county. Color of the dried soil warm dark grey. On sifting it some cherty fragments were found in it. On carefully washing it with water 45.70 per cent. of *sand*, of a dark brownish grey color, was separated, of which 4.30 per cent. was too coarse to pass through bolting cloth. The coarser sand, examined with the glass, was found to consist of rounded particles of hyaline, milky and red quartz, with some ferruginous mineral.

One thousand grains of the air-dried soil, digested in water containing carbonic acid, as before described, yielded nearly four grains of *solid extract*, dried at 212° F., of which the *composition* is as follows, viz:

| | *Grains.* |
|---|---|
| Organic and volatile matters, | 1.660 |
| Alumina, oxide of iron, and trace of phosphates, | .288 |
| Carbonate of lime, | 1.111 |
| Magnesia, | .046 |
| Brown oxide of manganese, | .019 |
| Sulphuric acid, | .112 |
| Potash, | .144 |
| Soda, | .080 |
| Silica, | .200 |
| Carbonic acid and loss, | .212 |
| | 3.872 |

The air-dried soil lost 2.34 per cent. of *moisture*, when dried at 365° F.; and was found to have the following composition, when thus dried, viz:

| | |
|---|---|
| Organic and volatile matters, | 5.200 |
| Alumina, | 3.460 |
| Oxide of iron, | 2.206 |
| Carbonate of lime, | .366 |
| Magnesia, | .205 |
| Brown oxide of manganese, | .234 |
| Phosphoric acid, | .159 |
| Potash, | .197 |
| Soda, | .090 |
| Sand and insoluble silicates, | 87.686 |
| Sulphuric acid and loss, | .197 |
| | 100.000 |

The cause of the fertility of this soil is obvious, in the large proportion of *soluble matter* which it yields to the water containing carbonic acid, and to the considerable, (although not *large*,) amount of organic and volatile matters, and of phosphoric acid, sulphuric acid, potash, lime, and magnesia, which it is found to contain, in proportion to the sand and insoluble silicates; the alumina and oxide of iron also are in such quantities as to give the proper consistence to the soil.

No. 227—SUB-SOIL. *Labeled "Sub-soil between Big Sink and Bear Wallow, near Mr. Barrow's farm, Barren county, Kentucky."* (*Sub-carboniferous Limestone Formation.*)

Color of the dried soil dull greyish-red, or brick-red. Careful washing with water removed from this soil nearly thirty-nine per cent. of reddish sand, mostly very fine, of which about seven per cent. was coarser sand, containing rounded particles of hyaline and milky quartz, and of some ferruginous mineral.

One thousand grains, dried at the ordinary temperature, and digested for a month in water containing carbonic acid, gave up less than a grain of *solid extract* dried at 212° F., of which the composition was as follows, viz:

| | |
|---|---|
| Organic and volatile matters, | 0.210 |
| Alumina, oxide of iron, and trace of phosphates, | .179 |
| Brown oxide of manganese, | .033 |
| Lime, | .077 |
| Magnesia, | .040 |
| Sulphuric acid, | .075 |
| Potash, | .023 |
| Soda, | .044 |
| Silica, | .139 |
| | 0.820 of a gr. |

The air-dried sub-soil lost 3.90 per cent. of *moisture* at 360° F. The *composition*, thus dried, is as follows, viz:

| | |
|---|---|
| Organic and volatile matters, | 4.730 |
| Alumina, | 10.380 |
| Oxide of iron, | 6.398 |
| Brown oxide of manganese, | .256 |
| Carbonate of lime, | .096 |
| Magnesia, | .522 |
| Phosphoric acid, | .075 |
| Sulphuric acid, | .466 |
| Potash, | .142 |
| Soda, | .082 |
| Sand and insoluble silicates, | 77.067 |
| | 100.214 |

BRECKINRIDGE COUNTY.

No. 487—BITUMEN OR MINERAL PITCH. *Labeled "Bitumen from Mrs. Jackson's spring, one mile east of Tar Springs, Breckinridge county, Kentucky."*

Resembles the bitumen from Tar Springs in Edmonson county. Color dull brownish-black; of the consistence of soft pitch; soft enough to be easily moulded in the fingers; containing some involved sand.

The proximate analysis is as follows, viz:

| | |
|---|---|
| Moisture, | 2.40 |
| Volatile combustible matters, | 36.50 |
| Carbon, in the fixed residue, | 7.30 |
| Ashes, sand, &c., | 53.80 |
| | 100.00 |

No. 312—SHALE. *Labeled "Shale and Marl under the Archimedes Limestone, at Ryan's, four to four and a half miles east of south of the Breckinridge coal mine, Breckinridge county, Kentucky." (Sub-carboniferous Limestone Formation.)*

A dark olive-grey friable shale, containing ferruginous concretions. Rubbed up in a mortar, and washed with water, it left about seventeen per cent. of very fine sand, of which only 0.20 per cent. would not pass through the bolting cloth. These coarser particles, examined with the aid of the glass, were found to be flattened rounded particles of ferruginous sandstone and round particles of hyaline quartz.

One thousand grains, dried at the ordinary temperature, gave up nearly two grains of *solid extract,* when digested for a month in water containing carbonic acid, of which the *composition*, dried at 212°, is as follows, viz:

| | Grains. |
|---|---|
| Organic and volatile matters, - - - - - - - - - | 0.309 |
| Alumina, oxides of iron and manganese, and trace of phosphates, - | .030 |
| Carbonate of lime, - - - - - - - - - | .627 |
| Magnesia, - - - - - - - - - - | .199 |
| Sulphuric acid, - - - - - - - - - - | .287 |
| Potash, - - - - - - - - - - | .062 |
| Soda, - - - - - - - - - - | .051 |
| Silica, - - - - - - - - - - | .210 |
| | 1.775 |

The air-dried shale lost 6.72 per cent. of *moisture* at 400° F.; and when thus dried has the following *composition*, viz:

| | |
|---|---|
| Organic and volatile matters, - - - - - - - - | 7.040 |
| Alumina and oxides of iron and manganese, - - - - | 12.170 |
| Carbonate of lime, - - - - - - - - - | .976 |
| Magnesia, - - - - - - - - - - - | .413 |
| Phosphoric acid, - - - - - - - - - | .101 |
| Sulphuric acid, - - - - - - - - - - | .198 |
| Chlorine, - - - - - - - - - - - | .002 |
| Potash, - - - - - - - - - - - | .556 |
| Soda, - - - - - - - - - - - | .190 |
| Sand and insoluble silicates, - - - - - - - - | 78.680 |
| | 100.326 |

This shale might be usefully applied, as a top dressing, to light and sandy soils, but could not be profitably carried to any great distance. Exposed to the air, water, and frost it would soon be disintegrated into a fertile soil. Its large proportion of potash would make it good for the tobacco or potato crop.

### BULLITT COUNTY.

No. 488—CARBONATE OF IRON. *Labeled "Kidney ore, over the sheet ore," Bellemont Furnace, Bullitt county, Ky. (Sub-carboniferous Sandstone Formation.)*

A dull, dark grey, fine granular mineral; not adhering to the tongue. Exterior surface and fissures reddish and yellowish-brown. The specimen appears to be a portion of a kidney-form mass. Powder yellowish-grey.

| | |
|---|---|
| Specific gravity, - - - - - - - - | 3.446 |

*Composition*, dried at 212° F.—

| | | |
|---|---|---|
| Carbonate of Iron, - - | 67.59 | = 32.62 per cent of *Iron*. |
| Oxide of iron, - - - | 7.77 | |
| Carbonate of lime, - - - | 6.28 | |
| Carbonate of magnesia, - - | 11.76 | |
| Carbonate of manganese, - | 1.32 | |
| Alumina, - - - - - | 1.55 | |
| Phosphoric acid, - - - | .71 | |
| Sulphur, - - - - - | .29 | |
| Potash, - - - - - | .75 | |
| Soda, - - - - - - | .27 | |
| Silica and insoluble silicates, - | 11.18 | |
| Loss, - - - - - | .53 | |
| | 100.00 | |

No. 489—Limonite. *Labeled "Iron ore, in the building stone, not used at Bellemont Furnace, found seventy feet above the black shale, Bullitt county, Ky."* (*Sub-carboniferous Sandstone Formation.*)

Interior of the ore dull reddish and yellowish-brown, glimmering with minute spangles of mica; exterior ochreous; adhering but slightly to the tongue. Powder of a light brown color.

| | |
|---|---|
| Specific gravity, - - - - - - - - | 2.984 |

*Composition*, dried at 212° F.—

| | | |
|---|---|---|
| Oxide of iron, - - - | 62.01 | = 43.46 per cent. of *Iron*. |
| Alumina, - - - - | .68 | |
| Brown oxide of manganese, - | .78 | |
| Carbonate of lime, - - - | .18 | |
| Magnesia, - - - - | 1.02 | |
| Phosphoric acid, - - - | .89 | |
| Sulphur, - - - - | .58 | |
| Potash, - - - - | .36 | |
| Soda, - - - - - | .20 | |
| Silica and insoluble silicates, - | 21.18 | |
| Combined water, - - - | 12.00 | |
| Loss, - - - - - | .12 | |
| | 100.00 | |

The air-dried ore lost 2.00 per cent of *moisture* at 212° F.

This ore is richer in iron, and more silicious than the preceding one, and would require a larger proportion of limestone, in smelting, than that; both contain rather more than is desirable of sulphur and phosphorus; the former, however, can be mainly removed by proper roasting of the ore, and the use of a sufficient amount of limestone; and the latter will not seriously injure the iron, in its ordinary applications in the form of cast-iron.

No. 490—LIMESTONE. *Labeled "Limestone used as a flux at Bellemont Furnace, (in the Black Devonian Shale Formation,") Bullitt county, Ky.*

A fine granular limestone, with bands of bluish and yellowish-grey, containing diffused pyrites, (sulphuret of iron,) and glistening with calcarious spar. Powder white, with a slight greyish tinge.

| | | |
|---|---|---|
| Specific gravity, - - - - - - - - | | 2.766 |
| *Composition*, dried at 212° F.— | | |
| Carbonate of lime, - - | 63.13 | = 35.43 Lime. |
| Carbonate of magnesia, - - | 27,76 | = 13.22 Magnesia. |
| Alumina, and oxides of iron and manganese, - - - | 4.34 | |
| Phosphoric acid, - - - | .19 | |
| Sulphuric acid, - - - | 3.77 | = 1.51 Sulphur. |
| Potash, - - - - - - | .44 | |
| Soda, - - - - - - | .15 | |
| Silica and insoluble silicates, - | 1.63 | |
| | 101.41 | |

The air dried rock lost 0.20 per cent. of *moisture*, at 212° F.

The apparent excess, in the above summary of the analysis, is doubtless due to the oxidation of the sulphur and iron, which were in the form of sulphuret of iron in the mineral, and which are estimated as oxide of iron and sulphuric acid in the analysis. The presence of the sulphur, in notable proportion, in the limestone used as flux, generally exerts an injurious influence upon the iron produced.

No. 491—IRON FURNACE SLAG. *Labeled "Purple Cinder, made when the furnace is producing the best quality of soft grey iron, Bellemont Furnace, (Patterson, Moore & Co.,) Bullitt county, Ky."*

A glassy slag, appearing almost black in the mass; of a dark greyish purple, as seen through the thin edges; containing but few air-bubbles. Before the blow-pipe it fuses pretty readily, with the formation of many minute bubbles.

*Composition*, dried at 212° F.—

| | | | | |
|---|---|---|---|---|
| Silicic acid, - - - - - | 54.60 | Containing oxygen, - | | 28.35 |
| Alumina, - - - - - | 15.90 | " | 7.43 | |
| Lime, - - - - - - | 11.93 | " | 3.39 | |
| Magnesia, - - - - - | 8.09 | " | 3.57 | |
| Protoxide of iron, - - - | 3.29 | " | 1.10 | |
| Protoxide of manganese, - | 1.08 | " | .24 | |
| Potash, - - - - - - | 4.25 | " | .72 | |
| Soda, - - - - - - | 1.31 | " | .33 | |
| | 100.45 | | 16.78 : | 28.35 |
| The oxygen in the bases is to that in the silicic acid, as | | | 1 : | 1.69 |

No. 492—Iron Furnace Slag. *Labeled "Olive-green Cinder, produced when the Furnace is making good forge iron, and yields more, but not so soft iron, as when purple cinder is made, Bellemont Furnace, Bullitt county, Ky."*

An opaque slag, of a dirty olive-green color; full of air bubbles. Before the blow-pipe it behaves like the preceding.

*Composition*, dried at 212° F.—

| | | | | |
|---|---|---|---|---|
| Silicic acid, - - - - - | 53.36 | Containing oxygen, - | | 27.70 |
| Alumina, - - - - - | 17.26 | " | 8.07 | |
| Lime, - - - - - - | 9.74 | " | 2.67 | |
| Magnesia, - - - - - | 8.09 | " | 3.24 | |
| Protoxide of iron, - - - | 6.35 | " | 1.41 | |
| Protoxide of manganese, - - | .89 | " | .20 | |
| Potash, - - - - - - | 4.09 | " | .69 | |
| Soda, - - - - - - | 1.02 | " | .26 | |
| | 100.80 | | 16.54 : | 27.70 |
| The oxygen in the bases is to that in the silicic acid, as | | | 1 : | 1.67 |

In both of these slags there is a considerable amount of oxide of iron, which is so much loss; this might probably be prevented by the use of a purer limestone for the flux. There is a large proportion of *magnesia*, both in the slags and in the limestone employed.

No. 493—Carbonate of Iron. *Labeled "Ironstone, from Button-mould Knob, Bullitt county, Ky."* (*Sub-carboniferous Sandstone Formation.*

A fine-grained, compact, carbonate of iron; interior grey, shading into rust-brown on the exterior; powder dull cinnamon color.

| | | |
|---|---|---|
| Specific gravity, - - - - - - - - | | 3.445 |
| *Composition*, dried at 212° F.— | | |
| Carbonate of iron, - - - | 53.64 | = 31.30 per cent. of *Iron*. |
| Oxide of iron, - - - | 7.71 | |
| Carbonate of lime, - - - | 6.08 | |
| Carbonate of magnesia, - - | 13.99 | |
| Carbonate of manganese, - | 1.94 | |
| Alumina, - - - - - | .55 | |
| Phosphoric acid, - - - | .10 | |
| Sulphuric acid, - - - | 1.37 | = .55 per cent. of *Sulphur*. |
| Potash, - - - - - - | .69 | |
| Soda, - - - - - - | .20 | |
| Silica and insoluble silicates, - | 11.48 | |
| Water and loss, - - - | 2.25 | |
| | 100.00 | |

The air-dried ore lost .50 per cent. of *moisture*, at 212° F.

An ore sufficiently rich for profitable smelting, which could be worked without much additional fluxing materials.

No. 494—LIMESTONE. *Labeled "Magnesian Limestone, on the road from Shepherdsville to Mount Washington, Bullitt county, Kentucky."* (*Lower Silurian Formation.*)

A fine granular rock, of a grey-buff color, rather difficult of fracture; sparkling in spots, with buff-colored calcarious spar; powder of a light grey-buff color.

| | | |
|---|---|---|
| Specific gravity, - - - - - - - | 2.799 | |
| *Composition*, dried at 212°— | | |
| Carbonate of lime, - - - - - - - - | | 63.45 |
| Carbonate of magnesia, - - - - - - - | | 29.64 |
| Alumina and oxide of iron, - - - - - - - | | 3.15 |
| Sulphuric acid, - - - - - - - - - | | .27 |
| Potash, - - - - - - - - - - | | .20 |
| Soda, - - - - - - - - - - | | .21 |
| Silex and insoluble silicates, - - - - - - - | | 2.18 |
| Loss, - - - - - - - - - - - | | .90 |
| | | 100.00 |

The air-dried rock lost 0.20 per cent. of *moisture*, at 212° F.

No. 495—LIMESTONE. *Labeled "Upper Silurian Limestone, Bullitt county, Kentucky, road to Mount Washington."*

A buff-grey, fine granular limestone; not adhering to the tongue.

| | | |
|---|---|---|
| Specific gravity, | 2.765 | |
| *Composition*, dried at 212° F— | | |
| Carbonate of lime, | | 50.25 |
| Carbonate of magnesia, | | 31.05 |
| Alumina and oxides of iron and manganese, | | 5.37 |
| Sulphuric acid, | | 1.46 |
| Phosphoric acid, a trace. | | |
| Potash, | | .59 |
| Soda, | | .20 |
| Silica and insoluble silicates, | | 10.32 |
| Loss, | | .76 |
| | | 100.00 |

The air-dried rock lost .20 per cent. of *moisture*, at 212° F.

No. 496—Sandstone. *Labeled "Building Stone, Knob at Bullitt's Lick, Bullitt county, Kentucky."* (*Sub-carboniferous Formation.*)

A rather soft, fine-grained, buff-grey sandstone; adhering slightly to the tongue; exhibiting, under the lens, minute scales of mica; composed of fine-grained sand, united by an argillaceous cement.

| | | |
|---|---|---|
| Specific gravity, | 2.427 | |
| *Composition*, dried at 212° F— | | |
| Sand and insoluble silicates, | | 93.68 |
| Alumina and oxides of iron and manganese, | | 3.95 |
| Carbonate of magnesia, | | .84 |
| Carbonate of lime, a trace. | | |
| Potash, | | .21 |
| Soda, | | .59 |
| Sulphuric acid and loss, | | .73 |
| | | 100.00 |

The air-dried rock lost .30 per cent. of *moisture*, at 212° F.

No. 497—Sandstone. *Labeled "Building Stone, quarry on the top of Button-mould Knob, Bullitt county, Kentucky."* (*Sub-carboniferous Sandstone Formation.*

A moderately hard, fine-grained sandstone, of grey-buff color; adhering slightly to the tongue; composed of fine grained sand, united by an argillaceous cement.

| | | |
|---|---|---|
| Specific gravity, | 2.415 | |
| *Composition*, dried at 212° F.— | | |
| Sand and insoluble silicates, | | 94.78 |
| Alumina and oxides of iron and manganese, | | 2.85 |
| Carbonate of magnesia, | | 2.29 |
| Carbonate of lime, | | .18 |
| Potash, | | .27 |
| Soda, | | .14 |
| Sulphuric acid, a trace. | | |
| | | 100.51 |

The air-dried rock lost 0.50 per cent. of *moisture*, at 212° F.

No. 498—SANDSTONE. *Labeled "Building Stone, seventy feet above the ——— Shale, Bellemont Furnace, Bullitt county, Ky.." (Sub-carboniferous Sandstone Formation.)*

A dirty buff-colored, fine-grained sandstone; adhering slightly to the tongue; resembling the preceding in structure.

| | | |
|---|---|---|
| Specific gravity, | 2.453 | |
| *Composition*, dried at 212° F.— | | |
| Sand and insoluble silicates, | | 94.75 |
| Alumina, and oxides of iron and manganese, | | 3.48 |
| Lime, | | .16 |
| Magnesia, | | .70 |
| Potash, | | .96 |
| Soda, | | .10 |
| Sulphuric acid, traces. | | |
| | | 100.15 |

The air-dried rock lost 0.30 per cent. of *moisture*, at 212° F.

These three specimens of freestone resemble each other very nearly in composition and structure. They appear to be of uniform texture, sufficiently soft to be easily worked, and yet not so absorbent of water as to be very liable to *scale* under the action of frost. The specimens examined did not contain *pyrites*, (sulphuret of iron,) in any notable quantity; the presence of which, in a sandstone, causes a constant disintegration of the surface, in consequence of the gradual oxidation of the sulphur and iron, and the efflorescence of the sulphate of iron thus produced.

No. 499. *Labeled "Black Devonian Slate, cut of the railroad, Bullitt county, Kentucky."*

A dull slate-colored rock, of an imperfect slatey structure; easily broken into irregular fragments across the layers; some microscopical appearance of pyrites; scarcely adhering to the tongue; powder dark-grey.

| | | |
|---|---|---|
| Specific gravity, | 2.474 | |
| *Composition*, dried at 212° F.— | | |
| Alumina, and oxides of iron and manganese, | | 16.35 |
| Carbonate of lime, | | 2.27 |
| Carbonate of magnesia, | | 3.28 |
| Phosphoric acid, | | .06 |
| Potash, | | 2.49 |
| Soda, | | .18 |
| Bituminous matters, | | 8.80 |
| Silica and insoluble silicates, | | 65.27 |
| Loss, | | 1.30 |
| | | 100.00 |

The air-dried rock lost 1. per cent of *moisture*, at 212° F.

This shale contains a remarkable proportion of *potash*, nearly two and a half per cent. of the dried rock, which may render it useful, in some localities, for the improvement of land which has been exhausted of this alkali by the culture of tobacco, potatoes, &c.

## BUTLER COUNTY.

No. 409—Carbonate of Iron. *Labeled "Carbonate of Iron in the shales of the millstone grit, Woodbury, below the mouth of Barren river, Butler county, Ky."*

A compact, dark-grey, or mouse-colored ore; weathered surfaces and fissures dark reddish-brown; some infiltrations of calcarious matter in the fissures; powder of a dirty buff color.

| | | |
|---|---|---|
| Specific gravity, | | 3.026 |
| *Composition*, dried at 212° F.— | | |
| Carbonate of iron, | 70.20 | = 39.45 per cent. of *Iron*. |
| Oxide of iron, | 9.92 | |
| Carbonate of lime, | 2.55 | |
| Carbonate of magnesia, | 7.04 | |
| Carbonate of manganese, | 1.60 | |
| Alumina, | 1.51 | |

| | |
|---|---|
| Phosphoric acid, - - - | .64 |
| Sulphur, a trace. | |
| Potash, - - - - | .42 |
| Soda, - - - - - | .01 |
| Silica and insoluble silicates, - | 7.65 |
| | 101.54 |

The air-dried ore lost 0.40 per cent. of *moisture*, at 212° F.

A good iron ore.

CARTER COUNTY.

No. 473—LIMONITE. *Labeled "Iron Ore, resting on the sub-carboniferous limestone, Carter county, Ky."*

A dark brown limonite, irregularly cellular; small portions ochreous; powder dirty yellowish-brown.

*Composition*, dried at 212° F.—

| | | |
|---|---|---|
| Oxide of iron, - - - | 78.42 | = 54.93 per cent. of *Iron*. |
| Alumina, - - - - | 1.48 | |
| Brown oxide of manganese, - | 3.17 | |
| Magnesia, - - - - | .30 | |
| Lime, a trace. | | |
| Phosphoric acid, - - - | .73 | |
| Potash, - - - - - | .21 | |
| Soda, - - - - - | .18 | |
| Combined water, - - - | 11.94 | |
| Silica and insoluble silicates, - | 3.77 | |
| | 100.20 | |

The air-dried ore lost 1.20 per cent. of *moisture*, at 212°.

A good iron ore; as rich as it is profitable to smelt in the high furnace; containing more than the usual proportion of oxide of manganese.

CHRISTIAN COUNTY.

No. 216—SUB-SOIL. *Labeled "Sub-soil from the southern part of Christian county, between Dr. Quarles' and Oak Grove, Ky."*

Color of the dried soil light-brownish, with a tinge of dirty orange. Carefully washed with water one thousand grains of this sand left two hundred and ninety-three grains of fine sand, of which only six grains was as coarse as ordinary bar-sand; which was composed generally of small rounded particles of quartz, with a few larger rounded and angular fragments of hyaline and milky quartz, and of a red silicious mineral like carnelian.

One thousand grains of the air-dried soil, digested for one month, in a close bottle, in water charged with carbonic acid, under pressure, gave up nearly a grain of *solid extract*, which, dried at 212°, had the following composition. viz:

| | *Grain.* |
|---|---|
| Organic and volatile matters, | 0.044 |
| Alumina and oxide of iron. | .097 |
| Oxide of manganese, | .157 |
| Lime, | .134 |
| Magnesia, | .033 |
| Phosphoric acid, | .011 |
| Sulphuric acid, | .020 |
| Potash, | .131 |
| Soda, | .015 |
| Silica, | .254 |
| Carbonic acid and loss, | .064 |
| | 0.960 |

The air-dried soil lost 2.24 per cent. of *moisture*, at 300° F.; and thus dried was found to have the following *composition*, viz:

| | |
|---|---|
| Organic and volatile matters, | 2.96 |
| Oxide of iron, | 2.36 |
| Alumina, | 2.39 |
| Phosphoric acid, | .27 |
| Sulphuric acid, not estimated. | |
| Carbonate of lime, | .13 |
| Carbonate of magnesia, | .79 |
| Brown oxide of manganese, | .27 |
| Potash, | .19 |
| Soda, | .04 |
| Sand and insoluble silicates, | 90.26 |
| Loss, | .34 |
| | 100.00 |

The analysis of the surface-soil, (No. 20,) from this locality was given in the preceding report, on pages 272–3, and 379; on reference to which it will be seen, that while the alumina and oxide of iron do not differ much in the soil and sub-soil, there is more *organic matter* in the soil, and a larger proportion of sand and silica in the sub-soil, which also exhibits a somewhat larger amount of phosphoric acid and potash. The quantity of soluble matter, extracted by digestion in water containing carbonic acid, was four times greater from the soil than from the sub-soil.

No. 462—Coal. *Labeled "Woolrich's coal, the most southerly coal in Christian county, Ky."*

A very pure looking glossy, pitch-black coal; not very hard; with no appearance of pyrites or other impurities; breaks in thin layers; having but little fibrous coal between the layers; heated over the spirit-lamp, it does not decrepitate; swells up a good deal, and the fragments agglutinate into a shining, inflated coke. It appears to be a good coking coal.

Specific gravity, - - - - - - - - 1.280

*Proximate analysis.*

| | | | |
|---|---|---|---|
| Moisture, - - - - - | 4 60 | Total volatile matters, | - 39.50 |
| Volatile combustible matters, - | 34.90 | | |
| Carbon in the coke, - - | 58.36 | Moderately dense coke, | - 60.50 |
| Ashes, (dull red,) - - - | 2.14 | | |
| | 100.00 | | 100.00 |

The per centage of *sulphur* was found to be 1.37.

The dull red ashes are composed of about three-fourths alumina and oxide of iron, to about one-third of silica, with traces of lime and magnesia.

Ultimate analysis of this coal gave the following results, dried at 212°, viz:

| | |
|---|---|
| Carbon, - - - - - - - - - - - | 76 636 |
| Hydrogen, - - - - - - - - - - | 4.533 |
| Sulphur, - - - - - - - - - - - | 1.440 |
| Ashes, - - - - - - - - - - - | 2.200 |
| Oxygen, nitrogen, and loss, - - - - - - - | 15.191 |
| | 100.000 |

This coal was not examined as to its yield of oils and gas by destructive distillation, at a low red heat; but its moderate proportion of *hydrogen* to its carbon is unfavorable to the formation of oily products.

CLARKE COUNTY.

No. 500—Soil. *Sent by Dr. S. D. Martin, labeled "Soil from a garden planted in peach-trees, about three years ago; about a foot of the surface-soil well mixed; this ground has been cleared about sixty or seventy years; used as a meadow, and hay cut off of it for many years; then eighteen consecutive crops of hemp were raised on it; in* 1836 *it was sown in grass and small crops of hay cut off; but finally it was taken by the blue grass, and has been used as pasture until three years ago, when it was broken up again and planted with young peach trees, and cultivated ever since as a vegetable garden," Clarke county, Ky.*

Color of the dried soil, dark brownish-grey. Carefully washed with water it left nearly 53. per cent. of very fine sand, containing nearly 7. per cent. of coarser sand, which would not pass through fine bolting cloth; which, examined with the lens, appeared to be, principally, small rounded particles of iron ore, with some milky hyaline, and red and yellow quartz particles, mostly rounded but some angular.

One thousand grains of the air-dried soil, digested for one month in water containing carbonic acid, as above described, gave up more than two grains of light brown *solid extract*, dried at 212°; the composition of which is as follows, viz:

| | |
|---|---|
| Organic and volatile matters, | 0.420 |
| Alumina and oxide of iron and phosphates, | .107 |
| Brown oxide of manganese, | .137 |
| Lime, | .509 |
| Magnesia, | .183 |
| Sulphuric acid, | .030 |
| Potash, | .100 |
| Soda, | .011 |
| Silica, | .178 |
| Carbonic acid and loss, | .418 |
| | 2.093 |

The air-dried soil lost 4.16 per cent. of *moisture*, at 380° F.; and, when thus dried was found to have the following composition, viz:

| | |
|---|---|
| Organic and volatile matters, | 6.10 |
| Oxide of iron, | 4.92 |
| Alumina, | 3.94 |
| Phosphoric acid, | .48 |
| Carbonate of lime, | .47 |
| Magnesia, | .62 |
| Brown oxide of manganese, | .40 |
| Potash, | .32 |
| Soda, | .08 |
| Sand and insoluble silicates, | 82.65 |
| Sulphuric acid not estimated, and loss, | .02 |
| | 100.00 |

Although this soil has been so long in cultivation, it is yet very rich in all the essential elements of vegetable food. It is not stated, on the label accompanying it, whether or not the ground had been manured; from the large proportion of potash and phosphoric acid contained, it is probable that it has been enriched by manure since it has been cultivated as a garden.

No. 501—SUB-SOIL. FROM DR. S. D. MARTIN, CLARKE COUNTY, KY. *Labeled "Sub-soil, eighteen inches below the surface, from the same place as the preceding."*

Color of the dried sub-soil lighter and more yellowish than that of the soil.

Washed with water it yielded nearly 49. per cent. of fine brownish sand, which contained about 7.5 per cent. of coarser sand, composed of rounded particles of a dark color, principally iron ore, but containing some quartzy particles, like the preceding.

One thousand grains of the air-dried sub-soil, digested in the carbonated water, yielded less than a grain and a half of *solid extract,* of a light grey color, containing the following ingredients, viz:

| | Grains. |
|---|---|
| Organic and volatile matters, | 0.080 |
| Alumina and oxides of iron and manganese, | .095 |
| Lime, | .428 |
| Magnesia, | .035 |
| Phosphoric acid, | .030 |
| Potash, | ·059 |
| Soda, | .034 |
| Silica, | .267 |
| Carbonic acid, sulphuric acid, and loss, | .342 |
| | 1.370 |

One thousand grains of the air-dried sub-soil lost 2.96 per cent. of *moisture*, when dried at 370° F.

Its composition, when thus dried, is as follows, viz:

| | |
|---|---|
| Organic and volatile matters, | 4.01 |
| Oxide of iron, | 7.06 |
| Alumina, | 7.71 |
| Phosphoric acid, | .38 |
| Carbonate of lime, | .99 |
| Magnesia, | 1.04 |
| Brown oxide of manganese, | .29 |
| Potash, | .36 |
| Soda, | .03 |
| Sand and insoluble silicates, | 78.03 |
| Sulphuric acid not estimated, and loss, | .10 |
| | 100.00 |

Whilst the sub-soil contains less *organic matter* than the soil above it, and has a smaller proportion of fine sand, the alumina, oxide of iron, lime, and magnesia are found in it in larger proportions.

### CLAY COUNTY.

No. 460—Coal. *Labeled "Col. Garrard's coal, Goose Creek Salt Works, Clay county, Ky."*

A pitch-black, shining, and apparently very pure coal, having some fibrous coal between the layers, but showing no pyrites or other impurities.

Heated over the spirit-lamp it did not decrepitate, but swells up and agglutinates into a very light cellular coke burning with a very smokey reddish-yellow flame. It appears to be a good coking coal.

Specific gravity, - - - - - - - - - 1.259

*Proximate Analysis.*

| | | | |
|---|---|---|---|
| Moisture, | 2.70 | Total volatile matters, | 37.60 |
| Volatile combustible matters, | 34.90 | | |
| Carbon in the coke, | 61.10 | Coke, light and shining, | 62.40 |
| Ashes, (dirty buff,) | 1.30 | | |
| | 100.00 | | 100.00 |

Composition of the ashes—

| | |
|---|---|
| Silica, | 0.49 |
| Alumina and oxide of iron, | .69 |
| Lime, | .05 |
| Magnesia, | .07 |
| | 1.30 |

Submitted to *ultimate analysis*, this coal was found to consist of the following ingredients, dried at 212°:

| | |
|---|---|
| Carbon, | 80.619 |
| Hydrogen, | 5.444 |
| Sulphur, | .575 |
| Nitrogen, | 1.457 |
| Oxygen and loss, | 10.305 |
| Ashes, | 1.600 |
| | 100.000 |

A remarkably pure coal, which would no doubt yield abundance of good gas, and is very fine for coking, containing but a small per centage of ashes.

### CLINTON COUNTY.

No. 222—Soil. *Labeled "Soil, Mr. Andrews', Caney Gap, Clinton county, Ky.; large timber—red oak, white oak, chestnut, hickory, beech, and poplar. Red Ferruginous Sub-soil." (Sub-carboniferous Limestone Formation.)*

Color of the dried soil of a warm grey.

Washed with water it gave more than 51. per cent. of fine *sand*, of a dirty buff color, containing about 12. per cent. of coarser sand, like common bar sand.

One thousand grains of the air-dried soil, digested in the carbonated water, gave up less than a grain and a half of solid extract of a brownish color, dried at 212°, which contained—

| | *Grains.* |
|---|---|
| Organic and volatile matters, | 0.830 |
| Alumina, oxide of iron, and trace of phosphates, | .168 |
| Lime, | .073 |
| Magnesia, | .016 |
| Potash, | .042 |
| Soda, | .038 |
| Silica, | .070 |
| Carbonic acid, sulphuric acid, and loss, | .244 |
| | 1.481 |

The air-dried soil lost 1.96 per cent. of *moisture* at 400° F.; dried at which temperature its composition was as follows:

20

| | |
|---|---|
| Organic and volatile matters, - - - - - - - | 3.970 |
| Alumina, - - - - - - - - - - - - | 1.776 |
| Oxide of iron, - - - - - - - - - - - | 2.466 |
| Brown oxide of manganese, - - - - - - - | .076 |
| Carbonate of lime, - - - - - - - - - | .076 |
| Magnesia, - - - - - - - - - - - - | .131 |
| Phosphoric acid, - - - - - - - - - | .090 |
| Potash, - - - - - - - - - - - - | .085 |
| Soda, - - - - - - - - - - - - | .099 |
| Sand and insoluble silicates, - - - - - - - | 90.720 |
| Sulphuric acid, (not estimated,) and loss, - - - - - | .521 |
| | 100.000 |

No. 412—LIMONITE. *Labeled "Iron Ore, ridge between Wolf river and Spring creek, five miles west of Albany, Clinton county, Ky."*

A dense, dark colored limonite; structure compact and compact fibrous; layers incrusted with yellow ochreous ore; powder dark yellowish-brown.

Specific gravity, - - - - - - - 3.503

Composition, dried at 212° F.—

| | | |
|---|---|---|
| Oxide of iron, - - - | 74.30 | = 52.03 per cent. of *Iron*. |
| Alumina, - - - - | 1.48 | |
| Brown oxide of manganese, - | 1.68 | |
| Phosphoric acid, - - - | .18 | |
| Sulphur, a trace. | | |
| Magnesia, - - - - | .35 | |
| Alkalies, not estimated. | | |
| Silica and insoluble silicates, - | 9.95 | |
| Combined water, - - - | 12.24 | |
| | 100.18 | |

The air-dried ore lost 1.20 per cent. of *moisture* at 212°, F.

A very good iron ore.

CRITTENDEN COUNTY.

No. 25—(*See former report*)—COAL. *From Sneed's mines, on Tradewater river, Crittenden county, Ky.*

This coal, of which the proximate analysis is given on pages 275 and 276 of the former report, has recently been submitted to ultimate analysis, with the following results, viz:

*Composition*, dried at 212° F.—

| | |
|---|---|
| Carbon, | 78.500 |
| Hydrogen, | 5.333 |
| Sulphur, | 1.040 |
| Ashes, | 3.800 |
| Nitrogen, | 1.344 |
| Oxygen and loss, | 9.983 |
| | 100.000 |

CUMBERLAND COUNTY.

No. 232—Soil. *Labeled "Soil, bottom land, between the forks of Sulphur creek, Jacob Speers' land, Cumberland county, Ky."* (*Sub-carboniferous sandstone, or Knob Formation, immediately above, over-lying the Devonian Black Slate.*)

Color of the dried soil very dark grey, nearly slate colored; it contained some fragments of ferruginous sandstone, some of which were rounded at the angles. On careful washing with water, this soil left a considerable proportion of fine sand, and about 10. per cent. of coarser sand, which would not pass through fine bolting cloth, which consisted of rounded particles of quartz and ferruginous sandstone.

One thousand grains, digested in carbonated water, as previously described, gave up more than five grains of *solid extract*, of the following composition, viz:

| | *Grains.* |
|---|---|
| Organic and volatile matters, | 1.530 |
| Alumina, oxides of iron and manganese, and trace of phosphates, | 1.333 |
| Carbonate of lime, | 1.538 |
| Carbonate of magnesia, | .303 |
| Sulphuric acid, | .065 |
| Potash, | .228 |
| Soda, | .045 |
| Silica, | .080 |
| | 5.122 |

The air-dried soil lost 2.40 per cent. of *moisture* at 375° F.; and and was found to contain the following ingredients, viz:

| | |
|---|---|
| Organic and volatile matters, | 5.770 |
| Alumina, | 1.230 |
| Oxide of iron, | 3.140 |
| Carbonate of lime, | .336 |
| Magnesia, | .438 |
| Brown oxide of manganese, | .076 |
| Phosphoric acid, | .127 |
| Sulphuric acid, | .734 |
| Chlorine, | .006 |
| Potash, | .220 |
| Soda, | .029 |
| Sand and insoluble silicates, | 87.110 |
| Loss, | .784 |
| | 100.000 |

This soil is remarkable in the large proportion of soluble matter which it yields to water containing carbonic acid. It is of more than the average fertility.

### DAVIESS COUNTY.

No. 230—Soil. *Labeled "Soil, Daviess county, Ky.; large growth of tobacco; native growth white oak, poplar, hickory, &c.; on the Owensboro' and Henderson road, 1½ miles from Green river."* (*Coal measures, but the soil mostly from the overlying quaternary.*)

Color of the dried soil, brownish-grey. By carefully washing it with water this soil left about 74. per cent. of *fine sand*, of a dirty buff color, of which 24. per cent. was as coarse as bar sand, composed of rounded quartz grains, clear, yellow, and reddish.

One thousand grains of the air-dried soil gave up, when digested in carbonated water for a month, about three and a half grains of brown *solid extract*, dried at 212°, which has the following composition, viz:

| | *Grains.* |
|---|---|
| Organic and volatile matters, | 2.100 |
| Alumina, oxide of iron and phosphates, | .480 |
| Lime, with some oxide of manganese, | .616 |
| Magnesia, | .056 |
| Sulphuric acid, | .041 |
| Potash, | .057 |
| Soda, | .058 |
| Silica, | .184 |
| | 3.592 |

The air-dried soil lost only 1.62 per cent. of *moisture* at 365° F.; and dried at this temperature gave, by analysis, the following ingredients, viz:

| | |
|---|---|
| Organic and volatile matters, | 3.350 |
| Alumina, | 2.026 |
| Oxide of iron, | 2.146 |
| Brown oxide of manganese, | .126 |
| Carbonate of lime, | .176 |
| Magnesia, | .258 |
| Phosphoric acid, | .088 |
| Sulphuric acid, not estimated. | |
| Potash, | .096 |
| Soda, | .053 |
| Silica, | 91.920 |
| | 100.239 |

This soil, which contains so large a proportion of *silicious matter*, and but a moderate quantity of *organic matters*, *potash* and *phosphoric acid*, supported a very luxuriant growth of tobacco, probably because so much of its nutritious ingredients are in the *soluble condition*; as is proved by the large relative proportion of *solid extract* given by it on digestion in the water containing carbonic acid. This circumstance, however, while it increases its present fertility, will hasten the process of exhaustion, under the drain of large herbaceous crops carried off the ground, without any return being made to it in the form of manures.

The rapidity with which the tobacco plant robs the soil of its richness is explained by the fact, that about one-fourth of the weight of the dried plant is composed of the mineral matters essential to vegetable growth, especially potash, lime, magnesia, soda, sulphuric acid, phosphoric acid, &c., as may be seen by reference to table 8, at the end of this report.

No. 189—Coal. *Labeled "Wolf Hill coal, Daviess county, Ky."*

A remarkably pure looking coal; deep black and glossy; with some fibrous coal between the layers, but no appearance of pyrites or other impurities, except some incrustation of sulphate of lime in the joints. Heated over the spirit lamp it swelled up somewhat, but did not agglutinate. Specific gravity 1.275.

This coal, the *proximate* analysis of which was given by Dr. Owen in his first report, page 44, was submitted to *ultimate* organic analysis, and an examination for its proportion of sulphur; the summary of the analysis is as follows:

| | |
|---|---|
| Carbon, | 77.891 |
| Hydrogen, | 5.422 |
| Sulphur, | .300 |
| Nitrogen, | 1.821 |
| Oxygen and loss, | 12.566 |
| Ashes, buff grey, | 2.000 |
| | 100.000 |

This coal has not yet been tried as to its relative yield of illuminating gas, or bituminous oils and paraffine, but its ultimate composition is unfavorable to the production of rich gas, or much oily matter. Coals having less oxygen and nitrogen in their composition are better for illuminating gas; and a larger proportion of hydrogen than that exhibited in this coal is found in those kinds which yield much oil on distillation.

No. 502—Coal. *Labeled "Twenty-four inch coal, on the Triplett place, four miles south-east of Owensboro', Daviess county, Ky."*

A glossy, pitch-black coal, pretty firm, and seemingly pretty free from pyrites; a little sulphate of lime in the joints; not much fibrous coal between the layers. Over the spirit-lamp it softens, swells up, and agglutinates; burns with a smoky flame, and leaves a bright cellular coke. Probably a coking coal.

Specific gravity, - - - - - - - 1.328

*Proximate Analysis.*

| | | | |
|---|---|---|---|
| Moisture, | 6.70 | Total volatile matters, | 42.70 |
| Volatile combustible matters, | 36.00 | | |
| Carbon in the coke, | 51.30 | Moderately light coke, | 57.30 |
| Ashes, (purple-grey,) | 6.00 | | |
| | 100.00 | | 100.00 |

The composition of the ashes is as follows:

| | |
|---|---|
| Silica, | 2.00 |
| Alumina and oxide of iron, | 3.18 |
| Lime, | .27 |
| Magnesia, | .25 |
| Loss, | .30 |
| | 6.00 |

*Ultimate Analysis*, (dried at 212°.)

| | |
|---|---|
| Carbon, | 71.019 |
| Hydrogen, | 5.022 |
| Sulphur, | 2.090 |
| Oxygen, nitrogen, and loss, | 15.069 |
| Ashes, | 6.800 |
| | 100.000 |

## EDMONSON COUNTY.

No. 414—LIMONITE. *Labeled "Iron Ore, from the Nolin Ore Bank, Edmonson county, Ky."*

Composed of hard, dark brown, layers enclosing softer, yellow, and brownish-yellow ore. Powder of a yellow color.

Composition, dried at 212° F.—

| | | |
|---|---|---|
| Oxide of iron, | 60.90 | = 42.64 per cent. of *Iron*. |
| Alumina, | .65 | |
| Brown oxide of manganese, | .75 | |
| Lime, a trace. | | |
| Magnesia, | 1.15 | |
| Phosphoric acid, | .57 | |
| Potash, | .36 | |
| Soda, | .32 | |
| Silica and insoluble silicates, | 23.68 | |
| Combined water, | 11.15 | |
| Loss, | .47 | |
| | 100.00 | |

The air-dried ore lost 1.50 per cent. of *moisture* at 212° F.

A very good silicious limonite.

No. 415—LIMONITE. *Labeled "Iron Ore, in the shales above the coal, Nolin Iron Works, Edmonson county, Ky."*

A dense, dark brown ore, of an irregular cellular structure; powder light yellowish-brown.

*Composition*, dried at 212° F.—

| | | |
|---|---|---|
| Oxide of iron, - - - | 74.70 | = 52.31 per cent. of *Iron*. |
| Alumina, - - - - - | .45 | |
| Brown oxide of manganese, - | .35 | |
| Phosphoric acid, - - - | .55 | |
| Magnesia, - - - - - | .15 | |
| Lime, a trace. | | |
| Silica and insoluble silicates, - | 12.65 | |
| Alkalies, not estimated. | | |
| Combined water, - - - | 11.19 | |
| | 100.04 | |

The air-dried ore lost 0.80 per cent. of *moisture* at 212° F.

Very nearly resembling the preceding, but containing a larger per centage of oxide of iron, and less silica.

No. 416—Carbonate of Iron. *Labeled "Carbonate of Iron, in the shale above the sandstone, Nolin Iron Works, Edmonson county, Kentucky."*

A dense, very fine-grained, dark grey ore; weathered surfaces reddish-brown; powder of a grey color.

Specific gravity, - - - - - - - 3.507

*Composition*, dried at 212° F.—

| | | |
|---|---|---|
| Carbonate of iron, - - - | 65.13 | = 37.04 per cent. of *Iron*. (brace: carbonate and oxide of iron) |
| Oxide of iron, - - - | 7.98 | |
| Carbonate of lime, - - | 1.95 | |
| Carbonate of magnesia, - - | 8.45 | |
| Carbonate of manganese, - | 1.83 | |
| Alumina, - - - - | .95 | |
| Phosphoric acid, - - - | .36 | |
| Sulphuric acid, - - - | .67 | = .10 *Sulphur*. |
| Potash, - - - - - | .57 | |
| Soda, - - - - - | .05 | |
| Silex and insoluble silicates, - | 9.17 | |
| Organic matter, moisture & loss, | 2.89 | |
| | 100.00 | |

The air-dried ore lost 0.50 per cent. of *moisture* at 212°.

No. 419—Limonite. *Labeled "Iron Ore, Mr. W. B. Morris' stock farm, Edmonson county, Ky. (Above his coal.")*

A dull yellowish-brown earthy looking ore; portions of it ochreous, yellow; friable; adhering somewhat to the tongue; powder light yellowish-brown, becoming black when calcined in a covered crucible.

*Composition*, dried at 212° F.—

| | | |
|---|---|---|
| Oxide of iron, - - - | 62.12 | = 43.50 per cent. of *Iron*. |
| Alumina, - - - - | 2.45 | |
| Brown oxide of manganese, - | .05 | |
| Lime, a trace. | | |
| Magnesia, - - - - - | .29 | |
| Phosphoric acid, - - - | .43 | |
| Sulphuric acid, - - - | .06 | |
| Potash, - - - - - | .38 | |
| Soda, - - - - - - | .42 | |
| Silex and insoluble silicates, - | 20.55 | |
| Organic matter, water, and loss, | 13.25 | |
| | 100.00 | |

The air-dried ore lost 2.70 per cent. of *moisture* at 212°.

This ore contains *organic matter*, somewhat similar to that which exists in soils, which causes it to become black when it is heated in a closed vessel. This organic matter can be dissolved out of the ore by alkaline solutions, but was found not to contain either *Crenic* or *Apocrenic* acids. The ore is a good rich mineral of a *silicious* character.

No. 472—Bitumen. (*Mineral pitch.*) *Labeled "From the Tar Spring, near the Nolin Iron Works, Edmonson county, Ky."*

A dull brownish black bitumen of the consistence of pitch, containing involved sand, and portions of vegetable remains; not elastic.

When heated it melts, gives off combustible vapors, and leaves a cellular coke, burning with a smoky yellowish flame. It is soluble in ether, oil of turpentine, naptha, &c., but insoluble in water and alcohol. Acted on by strong nitric acid, sulphuric acid, and caustic potash solution.

*Proximate Analysis.*

| | | | |
|---|---|---|---|
| Moisture, - - - - | 1.80 | Total volatile matters, - | 53.00 |
| Volatile combustible matters, - | 51.20 | | |
| Carbon in the coke, - - | 13.70 | Fixed residium, - - | 47.00 |
| Ashes and sand, - - - | 33.30 | | |
| | 100.00 | | 100.00 |

Doubtless containing the ordinary ingredients of petroleum.

ESTILL COUNTY.

No. 503—COPPER ORE. *Brought from near Irvine, Estill county, Ky., by O. C. Winburn.*

Exterior of the lumps ochreous, brownish-yellow; interior partly of the same character, and partly of a reddish-brown color, with diffused portions of yellow pyrites, and a light greenish substance, (carbonate of copper;) easily broken under the hammer; powder of a dirty olive color.

*Composition*, dried at 212° F.—

| | |
|---|---|
| Copper, | 21.13 |
| Sulphur, | 9.28 |
| Peroxide of iron, | 35.55 |
| Alumina, | .38 |
| Carbonate of lime, | 14.05 |
| Magnesia, | 1.68 |
| Silicious residue, | 19·57 |
| | 101.64 |

The locality of this mineral has not yet been visited by Dr. Owen. Should it be found in sufficient abundance, to warrant the erection of proper furnaces to smelt the ore, it is rich enough to prove a profitable ore.

This is believed to be the first instance of the discovery of copper ore in Kentucky.

FAYETTE COUNTY.

No. 504—SOIL. *Labeled "Virgin soil from a Beech ridge, on Robert Wickliffe's farm, two and a half miles from Lexington, on the Richmond turnpike; much less productive than the neighboring blue limestone soil; Fayette county, Ky."*

Color of the dried soil grey-buff. It contains irregular lumps of soft iron ore, varying in color from nearly black to dark yellow.

One thousand grains carefully washed with water left 489. grains of *pure sand*, of which 113. grains would not pass through fine bolting-cloth, and, examined with the lens, was found to consist of rounded particles of ferruginous mineral, varying from yellowish-brown, to almost black, mostly easily crushed in the fingers; with a few grains of milky quartz.

One thousand grains of the air-dried soil, digested for two months in water containing carbonic acid, gave up more than two and a half grains of dark *brown extract*, of which more than one half was carbonate of lime. Its *composition*, dried at 212°, was as follows:

| | *Grains.* |
|---|---|
| Organic and volatile matters, | 0.680 |
| Alumina, oxide of iron, and phosphates, | .498 |
| Carbonate of lime, | 1.518 |
| Magnesia, | .056 |
| Sulphuric acid, | .036 |
| Potash, | .072 |
| Soda, | .012 |
| Silica, | .199 |
| Oxide of manganese and loss, | .449 |
| | 3.520 |

The air-dried soil lost 4.12 per cent. of *moisture* at 400° F.

Its *composition*, thus dried, is as follows:

| | |
|---|---|
| Organic and volatile matters, | 4.881 |
| Alumina, and oxides of iron and manganese, | 10.306 |
| Carbonate of lime, | .276 |
| Magnesia, | .133 |
| Phosphoric acid, | .254 |
| Sulphuric acid, | .109 |
| Potash, | .139 |
| Soda, | .047 |
| Sand and insoluble silicates, | 83.834 |
| Loss, | .021 |
| | 100.000 |

A comparison between this and the richer *blue limestone* soil of Fayette county can be made by turning to pages 276 and 379 of the preceding report; and its inferiority to that will be seen to depend on its larger proportion of sand and silicious matters, and its smaller proportions of *phosphoric* acid and the *alkalies*, as well as of lime and magnesia, alumina and oxide of iron.

This soil, which in this rich region of country is called a poor soil, by comparison, would be considered quite a good soil in some parts of Kentucky.*

*Compare this with the analysis of Jefferson county soil, O'Bannon's station.

No. 505—SILICIOUS ROCK. *Labeled "Buff silicious rock, underlying the beech ridge soil, Robert Wickliffe's farm, two and a half miles from Lexington, cut of Lexington and Big Sandy railroad, near the Richmond turnpike, Fayette county, Ky."*

A dull, fine-granular rock, of a dirty buff color; adhering slightly to the tongue; quite friable; powder grey-buff.

*Composition*, dried at 212° F—

| | |
|---|---|
| Silica and fine sand, | 87.83 |
| Alumina, oxides of iron and manganese, | 8.65 |
| Carbonate of lime, only a trace. | |
| Carbonate of magnesia, | 1.40 |
| Phosphoric acid, | .25 |
| Sulphuric acid, | .22 |
| Potash, | .27 |
| Soda, | .14 |
| Water and loss, | 1.24 |
| | 100.00 |

This rock, ground-up, might make pretty good *fire-bricks.*

No. 506—SILICIOUS SHALE, *alternating with the preceding buff-colored rock, in the cut of the Lexington and Big Sandy railroad, through the beech ridge on Mr. Robt. Wickliffe's farm, (same locality as the two preceding.) Fayette county, Ky.*

A soft grey-buff clay shale, showing darker discolorations with oxides of iron and manganese; adheres strongly to the tongue; easily disintegrates into clay on exposure to the air; powder grey-buff color.

*Composition*, dried at 212°—

| | |
|---|---|
| Sand and insoluble silicates, | *83.45 |
| Alumina and oxide of iron and manganese, | 10.25 |
| Carbonate of lime, | 1.79 |
| Carbonate of magnesia, | 2.30 |
| Phosphoric acid, | .50 |
| Sulphuric acid, | .92 |
| Potash, | .41 |
| Soda, | .01 |
| Water and loss, | .37 |
| | 100.00 |

*The 83.45 grains of *sand and insoluble silicates* were found, on analysis, to consist of 70 grains of *silica*, and the remainder principally *alumina*, with traces of oxide of iron, lime, and magnesia.

These rocks, which form the sub-strata of this remarkable beech ridge, in this limestone region, are very different in composition from the prevailing rock stratum in Fayette county.

The two varieties of the blue limestone, next to be described, are such as are generally found in this vicinity underlying the soil.

No. 507—LIMESTONE. *Labeled "Upper shelly layer, from Van Akin's quarry, just below Lexington, on the Elkhorn branch, Fayette county, Ky."* (*Blue limestone, of Lower Silurian Formation.*)

A bluish-grey, coarse granular limestone, glimmering with small confused crystals of calcarious spar, and containing many fossil remains, as of small *Encrinal* stems, Atrypa, Modiola, Leptæna, Orthis, Pleurotomaria, &c., &c. Weathered surfaces of a dirty-buff color; powder very light yellowish-grey.

| | | |
|---|---|---|
| Specific gravity, - - - - - - - | | 2.660 |
| *Composition*, dried at 212° F.— | | |
| Carbonate of lime, - - | 92.73 | = 52.03 *Lime.* |
| Carbonate of magnesia, - - | .63 | |
| Alumina, and oxides of iron and manganese, - - - | 2.42 | |
| Phosphoric acid, - - - | .86 | |
| Sulphuric acid, - - - | .34 | |
| Chlorine, - - - - | .05 | |
| Potash, - - - - - | .23 | |
| Soda, - - - - - | .28 | |
| Silica and insoluble silicates, - | 2.18 | |
| Loss, - - - - - | .28 | |
| | 100.00 | |

The air-dried rock lost 0.30 per cent. of *moisture* at 212° F.

No. 508—LIMESTONE. *Labeled "Limestone used for curb-stones, &c., &c., Van Akin's quarry, Fayette county, Ky."*

Underlying the preceding; in thicker layers, and of a darker color and finer grained than that; glimmering with calcarious spar, and containing the usual fossils of the *Trenton limestone*, or blue limestone of the Lower Silurian Formation.

| | | |
|---|---|---|
| Specific gravity, - - - - - - - - | | 2.711 |
| *Composition*, dried at 212° F.— | | |
| Carbonate of lime, - - - | 77.63 = | 43.56 *Lime*. |
| Carbonate of magnesia, - - | 10.00 | |
| Alumina, and oxides of iron and manganese, - - - | 3.23 | |
| Phosphoric acid, - - - | .70 | |
| Sulphuric acid, - - - | 3.12 | |
| Chlorine, not estimated. | | |
| Potash, - - - - - - | .32 | |
| Soda, - - - - - - | .15 | |
| Silica and insoluble silicates, - | 4.98 | |
| | 100.13 | |

The air-dried rock lost 0.20 per cent. of *moisture*, at 212° F.

No. 509—SUB-SOIL. *Labeled "Red clay, under the sub-soil, eastern part of Fayette county, Ky."*

Dried earth of a dirty reddish-brown color.

One thousand grains, washed carefully with water, left 664 grains of *reddish-brown sand*, of which 75 grains was too coarse to pass through the finest bolting cloth, and was composed of rounded particles of soft iron ore, with a few rounded quartzose grains.

One thousand grains of the air-dried soil, digested for two months in water containing carbonic acid, gave up more than four grains of *nearly white extract*, dried at 212°, having the following composition:

| | *Grains.* |
|---|---|
| Organic and volatile matters, - - - - - - - - | 0.350 |
| Alumina, oxides of iron and manganese, and phosphates, - - | .018 |
| Carbonate of lime, - - - - - - - - - - | 3.497 |
| Magnesia, - - - - - - - - - - | .253 |
| Sulphuric acid, - - - - - - - - - - | .055 |
| Potash, - - - - - - - - - - - | .038 |
| Soda, not estimated. | |
| Silica, - - - - - - - - - - - - | .139 |
| | 4.350 |

The air-dried sub-soil lost 7.30 per cent. of *moisture*, at 400°, dried at which temperature its *composition* was:

| | |
|---|---|
| Organic and volatile matters, | 5.242 |
| Alumina, and oxides of iron and manganese, | 19.206 |
| Carbonate of lime, | 1.196 |
| Magnesia, | .426 |
| Phosphoric acid, | .434 |
| Sulphuric acid, | .054 |
| Potash, | .308 |
| Soda, | .086 |
| Sand and insoluble silicates, | 72.994 |
| Loss, | .054 |
| | 100.000 |

No. 510—Sub-soil. *Labeled "Ferruginous clay, under the sub-soil, at Megowan's quarry, terminus of the Big Sandy railroad at Lexington, Fayette county, Ky."*

Dried sub-soil of a greyish-reddish-brown, containing irregular nodules of chert, partly decomposed and porous.

Washed with water one thousand grains left 514 grains or *reddish sand*, of which 160 grains would not go through the finest bolting-cloth, and consisted mainly of rounded particles of soft dark colored iron ore, which could be crushed in the fingers; with a quartzose grains.

One thousand grains of the air-dried sub-soil, digested for two months in water containing carbonic acid, gave up only a little more than one grain of *olive-grey extract*, dried at 212°; the composition of which was as follows:

| | *Grains.* |
|---|---|
| Organic and volatile matters, | 0.280 |
| Alumina, oxides of iron and manganese, and phosphates, | .249 |
| Carbonate of lime, | .278 |
| Magnesia, | .046 |
| Sulphuric acid, | .102 |
| Potash, | .052 |
| Soda, | .026 |
| Silica, | .079 |
| | 1.112 |

The air-dried sub-soil lost 6.38 per cent. of *moisture* at 420° F.; dried at which temperature its *composition* is as follows:

| | Grains. |
|---|---|
| Organic and volatile matters, | 4.913 |
| Alumina and oxides of iron and manganese, | 20.300 |
| Carbonate of lime, | .116 |
| Magnesia, | .034 |
| Phosphoric acid, | .383 |
| Sulphuric acid, | .082 |
| Potash, | .309 |
| Soda, | .159 |
| Sand and insoluble silicates, | 73.874 |
| | 100.170 |

In these two specimens of the red clay, which extensively underlies the upper sub-soil in the *blue grass region*, we find considerable similarity of composition, especially in the proportions of phosphoric acid and alkalies, which are comparatively large. The alumina and oxide of iron, nearly in like quantity in these two Fayette county specimens, is much greater in that brought from Woodford county, near Versailles, (which see;) and in them all the proportion of carbonate of lime is variable. In all of them, a portion of what is stated as *organic and volatile matters*—representing the loss of weight observed on the complete calcination at a red heat, of the well dried soil—must be considered only *combined water*.

Although containing as much as twenty per cent. of *alumina and oxide of iron*, this *red clay* of Fayette county allows water freely to pass through it, so that it does not prevent the drainage of the soil; which is favored by the cavernous nature of the limestone beneath. Whether or not the red clay of Woodford county, which contains more than thirty-three per cent. of these ingredients, causes the surface water to stagnate, the writer is not advised; but it is probable, from its appearance, that it does not act injuriously in this respect.

No. 511—LIMESTONE. *Labeled "Magnesian Limestone, upper layer five inches to a foot thick; not used for building purposes; a bed in the Bird's Eye Limestone of the Lower Silurian Formation, Grimes' Quarry, Horse Shoe Point, Grimes' mill, about one and a quarter miles from the Richmond turnpike near Kentucky river, Fayette county, Kentucky."*

A greyish-buff, fine granular rock, pretty uniform in structure, except for some small cavities lined with light colored ochreous matter; no fossils or pyrites; adhering very slightly to the tongue.

Specific gravity, - - - - - - - - 2.716

*Composition*, dried at 212° F—

| | |
|---|---|
| Carbonate of lime, | 51.57 |
| Carbonate of magnesia, | 29.33 |
| Alumina, and oxides of iron and manganese, | 3.57 |
| Phosphoric acid, | .37 |
| Sulphuric acid, | .34 |
| Potash, | .71 |
| Soda, | .82 |
| Silex and insoluble silicates, | 11.58 |
| Loss, | 1.71 |
| | 100.00 |

The air-dried rock lost 0.10 per cent. of *moisture*, at 212° F.

No. 512—Limestone. *Labeled "Building Stone, from Grimes' Quarry, Fayette county, Ky."*

Some of the layer immediately under the above described, about five feet thick; much used for building purposes.

A light yellowish-grey, fine granular limestone, quite homogeneous in its structure, with no appearance of fossils or pyritous matter. Under the lens appears to be made up of pure crystalline grains, aggregated together without cement; powder nearly white.

Specific gravity, - - - - - - - - 2.703

*Composition*, dried at 212° F.—

| | | |
|---|---|---|
| Carbonate of lime, | 55.54 | = 31.16 *Lime*. |
| Carbonate of magnesia, | 40.80 | = 19.68 *Magnesia*. |
| Alumina, oxide of iron, &c., | .96 | |
| Sulphuric acid, | .02 | |
| Potash, | .36 | |
| Soda, | .22 | |
| Silex and insoluble silicates, | 2.79 | |
| | 100.69 | |

The air-dried rock lost 0.30 per cent of *moisture*, at 212° F.

No. 513—LIMESTONE. *Labeled "Portion of one of the boundary stones of the city of Lexington; originally from Grimes' quarry; locality as above; appearance much the same as that of the preceding; adheres slightly to the tongue."*

| | | |
|---|---|---|
| Specific gravity, | 2.615 | |
| *Composition*, dried at 212° F.— | | |
| Carbonate of lime, | | 55.99 |
| Carbonate of magnesia, | | 37.33 |
| Alumina, oxides of iron, &c., | | .72 |
| Phosphoric acid, | | .25 |
| Sulphuric acid, | | .33 |
| Potash, | | 2.35 |
| Soda, | | .25 |
| Silex and insoluble silicates, | | 3.38 |
| | | 100.60 |

The air-dried rock lost 0.10 per cent. of *moisture*, at 212° F.

The proportion of *potash* in the above specimen is remarkable. The portion analyzed had been broken from the old boundary stone, just at the *surface of the soil*, in order to exhibit the power of this stone to resist the decomposing atmospheric influences, under the most unfavorable circumstances; whether the prolonged contact of the rock with the soil had made any change in its proportion of potash, by interpenetration, or whether there was an error in the determination, would be a subject for further investigation.

This building stone, which has recently been selected by the building committee of the Kentucky Clay Monument Association, for the material of their proposed monument, commends itself, in many respects, as one of the best materials which could be chosen for their purposes.

Its homogeneous structure and purity of composition; its considerable proportion of magnesia, with the absence of fossils, pyrites, or flinty matter; are all favorable to great durability and facility of shaping it with the chissel; and its light warm-grey color is more pleasant to the eyes of most persons than the pure white of statuary marble.

In the city of Lexington the door-steps of some of the oldest houses, made of this rock, exhibit very little sign of disintegration; and, according to the experience of architects in general, a pure homogeneous, magnesian limestone may be classed amongst the most durable of building rocks.

It was of this rock that the block was selected which was sent by the state of Kentucky to the Washington monument, at the capital of the United States.

It will be seen by comparison that the composition of this stone is remarkably similar to that of the Dolomitic limestones of this and other countries.

No. 556—MINERAL WATER. *Water from the bored well at the Lunatic Asylum, Lexington, Ky.*

The water of the large spring, formerly used at this extensive establishment, having become contaminated by the leakage of some of the large sewers, an attempt was made to procure a supply of water by boring; and, after penetrating one hundred and six feet, of which eighty-six feet were through the solid blue limestone rock, abundance of water was obtained. It was found to be a weak saline sulphur water, containing *sulphuretted hydrogen* and *carbonic acid gases*, and left, on evaporation to dryness at the temperature of 212°, about one grain and six-tenths of a grain from the one thousand grains of water, or more than eleven grains of saline matter to the pint.

This saline matter was found to consist of

Carbonate of lime;

Carbonate of magnesia;

Carbonate of iron, a trace;

Chloride of sodium, (common salt,) considerable proportion;

Sulphate of lime;

Sulphate of magnesia;

Silica and *probably* sulphates of soda and potash, with traces of iodine and bromine—one or both.

A full *quantitative* analysis not having been made, as yet, the presence of these minuter ingredients cannot be positively asserted.

This fine well has proved a great boon to this public establishment. It is employed for all the domestic purposes—for washing, drinking, cooking, &c., and since its use the medical superintendent, Prof. W. C. Chipley, thinks the general health of the inmates has been improved: in particular, *endemic diarrhea*, which was formerly a very frequent scourge, has been almost entirely removed. The first influence on the bowels, resulting from its free use, was somewhat constringent, followed by some relaxation, after which their action became natural; it is observed to habitually increase the action of the kidneys.

### FRANKLIN COUNTY.

No. 514—LIMESTONE. *Labeled "Hydraulic? limestone, main Benson, near Bright's mill, Franklin county, Ky."*

A pretty dense, grey, fine granular rock; generally dull, but glimmering in spots with particles of calcarious spar; powder light bluish-grey.

| | | |
|---|---|---|
| Specific gravity, | 2.699 | |
| *Composition*, dried at 212° F.— | | |
| Lime, | | 50.19 |
| Magnesia, | | .66 |
| Alumina and oxide of iron, | | 1.24 |
| Carbonic acid, | | 40.15 |
| Phosphoric acid, | | .44 |
| Sulphuric acid, | | .68 |
| Potash, | | .23 |
| Soda, | | .29 |
| Silex and insoluble silicates, | | 6.94 |
| | | 100.82 |

The air-dried rock lost 0.30 per cent. of *moisture*, at 212° F.

This limestone does not contain enough silica, alumina, &c., to constitute it a good water-lime.

No. 515—LIMESTONE. *Labeled "Near Bridgeport, Franklin county, Ky."*

A fine grained dark bluish-grey rock. Weathered surfaces brownish-buff; no fossils, except what might be the cast of a small *fucoid* body, and certain other similar appearances of small stems traversing the rock, and of a dirty-buff color, very apparent on the generally dark-grey surface; powder light grey.

| | | |
|---|---|---|
| Specific gravity, | 2.700 | |
| *Composition*, dried at 212° F.— | | |
| Carbonate of lime, | | 76.75 |
| Carbonate of magnesia, | | .19 |
| Alumina, oxides of iron, &c., | | 2.25 |
| Phosphoric acid, | | .09 |
| Sulphuric acid, | | .85 |

| | |
|---|---|
| Potash, - - - - - - - - - - - | .48 |
| Soda, - - - - - - - - - - - | .44 |
| Silex and insoluble silicates, - - - - - - | 18.86 |
| Loss, - - - - - - - - - - - | .09 |
| | 100.00 |

The air-dried rock lost 0.20 per cent. of *moisture*, at 212° F.

The proportion of silex in this limestone is sufficient to constitute it a water-lime, provided it is in such a state of aggregation as to unite readily with the lime, which can be ascertained by a practical trial.

No. 516—Limestone. *Labeled "Encrinital limestone from near Bridgeport, Franklin county, Ky."*

On the recent fracture this rock appears to be made up of coarse confused crystalline grains of calcarious spar, colored dark grey and brownish by ferruginous admixture; but on the weathered surfaces, which are of a dirty buff color, innumerable joints and portions of *small* encrinal stems appear.

*Composition*, dried at 212° F.—

| | | |
|---|---|---|
| Carbonate of lime, - - - | 92.65 | = 51.99 per cent. of *Lime*. |
| Carbonate of magnesia, - - | 1.54 | |
| Alumina, oxide of iron, &c., - | 1.19 | |
| Phosphoric acid, - - - | .09 | |
| Sulphuric acid, - - - | 1.27 | |
| Potash, - - - - - | .30 | |
| Soda, - - - - - - | .13 | |
| Silica and insoluble silicates, - | 3.68 | |
| | 100.85 | |

The air-dried rock lost 0.20 per cent. of *moisture*, at 212° F.

No. 517—Soil. *Labeled "Virgin upland soil, from the waters of Benson creek, near Hardinsville, Franklin county, Ky., farm of John J. Julian."*

Color of the dried soil dark, dirty buff-grey.

One thousand grains washed with water left 677 grains of *fine sand*, of which about 90. grains was too coarse to pass through the finest bolting cloth; this consisted mainly of rounded particles of soft iron ore, with a few quartzose grains.

This soil was found to be mixed with fragments of charcoal, which increased its apparent amount of *organic and volatile matters*.

One thousand grains of the air-dried soil, digested for two months in water containing carbonic acid, gave up more than three grains and a half of *dark brown extract*, dried at 212°, which was composed of

| | *Grains.* |
|---|---|
| Organic and volatile matters, | 1.430 |
| Alumina, oxides of iron and manganese, and phosphates, | .758 |
| Carbonate of lime, | .917 |
| Magnesia, | .056 |
| Sulphuric acid, | .037 |
| Potash, | .096 |
| Soda, | .047 |
| Silica, | .339 |
| | 3.680 |

Dried at 400° the air-dried soil lost 5.18 per cent. of moisture; dried at which temperature its *composition* was found to be as follows:

| | |
|---|---|
| Organic and volatile matters, | 9.133 |
| Alumina, and oxide of iron and manganese, | 8.100 |
| Carbonate of lime, | .316 |
| Carbonate of magnesia, | .517 |
| Phosphoric acid, | .243 |
| Sulphuric acid, | .068 |
| Potash, | .173 |
| Soda, | .049 |
| Sand and insoluble silicates, | 80.754 |
| Loss, | .647 |
| | 100.000 |

No. 518—Soil. *Labeled "Same kind of soil and growth as the preceding; has been twelve years in cultivation, in corn and oats chiefly. Waters of Benson creek, near Hardinsville, farm of John J. Julian, Franklin county, Ky.*"

Dried soil a little lighter colored than the preceding.

One thousand grains washed with water left 705 grains of fine greyish sand of which only about 30 grains was too coarse to pass through fine bolting cloth; consisting mainly of rounded and angular fragments of ferruginous and quartzose minerals.

One thousand grains of the air-dried soil, digested for two months in water containing carbonic acid, gave up more than two grains and a half of *yellowish-brown extract*, dried at 212°, of the following composition:

| | Grains. |
|---|---|
| Organic and volatile matters, - - - - - - - | 0.570 |
| Alumina, oxide of iron, and phosphates, - - - - - | .277 |
| Brown oxide of manganese, - - - - - - - | .338 |
| Carbonate of lime, - - - - - - - - - | .857 |
| Magnesia, - - - - - - - - - - - | .100 |
| Sulphuric acid, - - - - - - - - - - | .295 |
| Potash, - - - - - - - - - - - | .050 |
| Soda, - - - - - - - - - - - | .031 |
| Silica, - - - - - - - - - - - | .119 |
| | 2.637 |

Dried at 370° this soil lost 1.98 per cent. of *moisture*, and its composition was found to be as follows:

| | |
|---|---|
| Organic and volatile matters, - - - - - - - | 3.790 |
| Alumina and oxides of iron and manganese, - - - - | 4.589 |
| Carbonate of lime, - - - - - - - - - | .196 |
| Magnesia, - - - - - - - - - - - | .066 |
| Phosphoric acid, - - - - - - - - - | .151 |
| Sulphuric acid, - - - - - - - - - - | .054 |
| Potash, - - - - - - - - - - - | .135 |
| Soda, - - - - - - - - - - - | .026 |
| Sand and insoluble silicates, - - - - - - - | 90.734 |
| Loss, - - - - - - - - - - - | .259 |
| | 100.000 |

The proportions of all the *essential* elements of this soil are smaller than in the preceding virgin soil of the same locality.

No. 518 (*A*)—Soil. *Same kind of soil and growth as the preceding; from a field that has been from forty to fifty years in cultivation; waters of Benson, Franklin county, near Hardinsville, farm of Mr. John J. Julian.*

Dried soil of a grey-buff color.

One thousand grains of the air-dried soil, washed carefully with water, left 720 grains of *fine sand*, of which 21.70 grains would not pass through fine bolting-cloth. This latter portion consisted, principally, of small rounded ferruginous particles.

One thousand grains of the air-dried soil, digested in the usual manner, for a month, in water containing carbonic acid, gave up more than two and a third grains of brownish extract, dried at 212° which exhibited the following composition, viz:

| | *Grains.* |
|---|---|
| Organic and volatile matters, | 0.470 |
| Alumina, oxides of iron and manganese, and phosphates, | .287 |
| Carbonate of lime, | .913 |
| Magnesia, | .091 |
| Sulphuric acid, | .081 |
| Potash, | .086 |
| Soda, | .017 |
| Silica, | .200 |
| Loss, | .222 |
| | 2.366 |

Dried at the temperature of 400° the air-dried soil lost 2.525 per cent. of moisture. Its *composition*, thus dried, is as follows:

| | |
|---|---|
| Organic and volatile matters, | 4.206 |
| Alumina, | 2.120 |
| Oxide of iron, | 2.915 |
| Carbonate of lime, | .173 |
| Magnesia, | .233 |
| Brown oxide of manganese, | .004 |
| Sulphuric acid, | .043 |
| Phosphoric acid, | .128 |
| Potash, | .130 |
| Soda, | .051 |
| Sand and insoluble silicates, | 90.170 |
| | 1.173 |

By comparison with the preceding soil, it will be seen that the soil of this field, which has been from forty to fifty years in cultivation, contains a smaller relative proportion of phosphoric and sulphuric acids, of potash, and of carbonate of lime, than the virgin soil, or the soil from the field which has been but twelve years in cultivation; and that it yielded a smaller quantity of nutritious extract to the carbonated water than those soils.

No. 518 (*B*)—Soil. *Labeled "Sub-soil from a field on John J. Julian's farm, waters of Benson, Franklin county, Ky."*

Dried soil of a dark grey-buff color.

One thousand grains of this sub-soil gave, on washing with water, 630.7 grains of fine sand, of which all but 18 grains passed through fine bolting-cloth. This latter portion consisted of round particles, of a ferruginous mineral, with a few quartzose grains.

One thousand grains of the air-dried soil, digested for a month in the carbonated water, gave up less than a grain of nearly white extract, dried at 212°, composed as follows, viz:

| | Grain. |
|---|---|
| Organic and volatile matters, | 0.217 |
| Alumina, oxides of iron and manganese, and phosphates, | .063 |
| Carbonate of lime, | .181 |
| Magnesia, | .030 |
| Sulphuric acid, | .034 |
| Potash, | .046 |
| Soda, | .038 |
| Silica, | .200 |
| Loss, | .006 |
| | 0.830 |

The air-dried soil lost 3.30 per cent. of *moisture* at 400°. Its composition is as follows:

| | |
|---|---|
| Organic and volatile matters, | 3.179 |
| Alumina, | 4.470 |
| Oxide of iron, | 4.825 |
| Carbonate of lime, | .082 |
| Magnesia, | .312 |
| Brown oxide of manganese, | .005 |
| Sulphuric acid, | .033 |
| Phosphoric acid, | .148 |
| Potash, | .282 |
| Soda, | .002 |
| Sand and insoluble silicates, | 86.380 |
| Loss, | .282 |
| | 100.000 |

## GREENUP COUNTY.

No. 307—LIMONITE. "*Hydrated oxide of iron, in the form of pot ore, associated with the limestone ore, Bellefonte Furnace, Greenup county, Kentucky.*"

A concretionary mass of limonite, with a large irregular cavity lined with an almost black layer; exterior surface, and between the layers, soft and brown; powder brownish-yellow; when calcined, of a handsome spanish-brown color.

Composition, dried at 212° F.—

| | | |
|---|---|---|
| Oxide of iron, | 80.30 | = 56.23 per cent. of *Iron.* |
| Alumina, not estimated. | | |
| Brown oxide of manganese, | .35 | |
| Magnesia, | .40 | |
| Potash, | .34 | |
| Soda, | .01 | |
| Phosphoric acid, | .60 | |
| Silica and insoluble silicates, | 6.55 | |
| Combined water, | 12.12 | |
| | 100.67 | |

The air-dried ore lost 1.00 per cent. of *moisture* at 212° F.

A pure *limonite,* containing only traces of *lime* and *alumina,* and not sufficient silicious matter to form *cinder* enough in the furnace to protect the reduced iron from the action of the oxygen of the blast. It can be smelted successfully by admixture with poorer ores and limestone.

No. 481—LIMESTONE. *Labeled "Limestone used as a flux at the Buffalo Furnace; lies near the level of the Clay creek branch of Little Sandy river, Greenup county, Ky."*

A compact, fine granular, greenish-grey limestone; uniform in texture and appearance.

Specific gravity, 2.691

*Composition,* dried at 212° F.—

| | |
|---|---|
| Carbonate of lime, | 73.90 |
| Carbonate of magnesia, | 2.08 |
| Alumina and oxide of iron, | 1.19 |
| Phosphoric acid, | .46 |
| Potash, | .27 |
| Soda, | .05 |
| Silex and insoluble silicates, | 21.67 |
| Loss, | .38 |
| | 100.00 |

The air-dried rock lost 0.20 per cent. of *moisture* at 212° F.

No. 482—CARBONATE OF IRON. *Labeled "Centre part of the Kidney Ore, which lies over the main block ore, tops of hills, with impure (bastard) limestone under it, Buffalo Furnace, Greenup county, Kentucky."*

Portion of a nodular mass; dull, fine-grained; of which the exterior portion is of a dark brown color, separating in concentric layers; the central part is of a dark grey color, passing, on its exterior, into the yellowish and brown layers, which make up the outside of the mass. (The analysis of the *exterior* portion was given in the previous report.) Powder of the interior grey part of a yellowish-grey color.

Composition, dried at 212° F.—

| | | |
|---|---|---|
| Carbonate of iron, - - - | 70.27 | = 40.70 per cent. of *Iron*. |
| Oxide of iron, - - - | 10.16 | |
| Alumina, - - - - - | .15 | |
| Phosphoric acid, - - - | .73 | |
| Carbonate of lime, - - - | 2.45 | |
| Carbonate of magnesia, - - | 5.52 | |
| Carbonate of manganese, - | 1.46 | |
| Potash, - - - - - - | .40 | |
| Soda, - - - - - - | .09 | |
| Silex and insoluble silicates, - | 8.15 | |
| Loss, - - - - - - | .62 | |
| | 100.00 | |

The air-dried ore lost 0.50 per cent. of *moisture*, at 212°.

No. 474—LIMONITE. *Labeled "Clay Iron Stone, Giger's Hill, Cattlettsburg, Greenup county, Ky."*

Portion of a concretionary mass, irregular in form, with a cavity in the interior, and some concentric layers around it; compact; adhering slightly to the tongue; of a dirty reddish-brown color; powder brownish-ochreous.

*Composition,* dried at 212° F.—

| | | |
|---|---|---|
| Oxide of iron, - - - | 68.30 | = 47.83 per cent. of *Iron*. |
| Alumina, - - - - - | 3.65 | |
| Carbonate of lime, - - | .28 | |
| Magnesia, - - - - - | 2.64 | |
| Potash, - - - - - - | .27 | |
| Soda, - - - - - - | .22 | |
| Silex and insoluble silicates, - | 12.28 | |
| Combined water, - - - | 12.09 | |
| Phosphoric acid and loss, - | .27 | |
| | 100.00 | |

The air-dried ore lost 1.60 per cent. of *moisture*, at 212°.

No. 475—IMPURE CARBONATE OF IRON. *Labeled "Ferrug'nous limestone under the limestone ore, Greenup county, Ky. (How much iron and lime?)"*

A fine granular rock, of a dull aspect; containing small spangles of mica; not adhering to the tongue. Interior of a dark olive-grey color; exterior, to the depth of more than half an inch, dull reddish-brown, shading into dirty yellowish-brown on the outside surface; powder (of an average portion,) of a grey-buff color.

Specific gravity, - - - - - - - - 3.155

*Composition*, dried at 212° F.—

| | | |
|---|---|---|
| Carbonate of iron, - - - | 28.01 | = 23.62 per cent. of *Iron.* |
| Oxide of iron, - - - | 14.42 | |
| Carbonate of lime, - - | 29.37 | |
| Carbonate of magnesia, - - | 5.57 | |
| Carbonate of manganese, - | .18 | |
| Alumina, - - - - - | 1.38 | |
| Phosphoric acid, - - - | .29 | |
| Potash, - - - - - - | .42 | |
| Soda, - - - - - - | .33 | |
| Silex and insoluble silicates, - | 19.98 | |
| Organic matters and loss, - | .05 | |
| | 100.00 | |

The air-dried rock lost 0.60 per cent. of *moisture* at 212° F.

Although this mineral contains rather too small a proportion of iron to be considered a good *ore* of that metal, it yet will answer a profitable purpose when it is mixed, in proper proportion, with some of those limonites of Greenup county which are refractory in the furnace, in consequence of their very large per centage of oxide of iron. The considerable proportion of lime and magnesia, contained in this rock, renders it an appropriate fluxing material for those very rich iron ores which are of a silicious character.

No. 476—LIMONITE. *Labeled "Limestone ore, over the limestone, Pennsylvania Furnace, Greenup county, Ky."*

Exterior of the ore of a dirty yellowish-grey color. On one edge the fracture presented a compact layer of dark brown limonite, which gradually passes into a granular mass, composed of small brownish-red grains, cemented by a whitish and yellowish matter, of which mixture the ore is principally composed, giving it a fine oolitic appearance; powder light brownish-red.

*Composition*, dried at 212° F.—

| | | |
|---|---|---|
| Oxide of iron, - - - | 72.80 | = 50.98 per cent. of *Iron*. |
| Alumina, - - - - - | 2.17 | |
| Brown oxide of manganese, - | .45 | |
| Carbonate of lime, - - | .18 | |
| Magnesia, - - - - | 1.19 | |
| Potash, - - - - - | .48 | |
| Soda, - - - - - - | .02 | |
| Silex and insoluble silicates, - | 10.57 | |
| Combined water, - - - | 11.20 | |
| Loss, - - - - - - | .94 | |
| | 100.00 | |

The air-dried ore lost 2.70 per cent. of *moisture* at 212° F.

No. 316—Limonite. *Labeled "Kidney ore, above the block ore and under the main limestone, Pennsylvania Furnace, Greenup county, Ky."*

A dark, purplish-brown, limonite; compact; adhering slightly to the tongue; containing minute spangles of mica; some of the fissures coated with glimmering dark colored, minute crystals; powder of a spanish-brown color.

Composition, dried at 212° F.—

| | | |
|---|---|---|
| Oxide of iron, - - - | 76 90 | = 53.85 per cent. of *Iron*. |
| Alumina, - - - - | 1.21 | |
| Brown oxide of manganese, - | .25 | |
| Phosphoric acid, - - - | .64 | |
| Magnesia, - - - - - | .28 | |
| Potash, - - - - - | .23 | |
| Soda, - - - - - - | .16 | |
| Silex and insoluble silicates, - | 11.77 | |
| Combined water, - - - | 9.09 | |
| | 100.53 | |

The air-dried ore lost 0.50 per cent. of *moisture* at 212° F.

No. 317—Limonite. *Labeled "Block ore, below the hearth-stone, average seven to eight inches, Pennsylvania Furnace, Greenup county, Ky."*

A dense, compact, limonite of a dark purple-brown color; presenting some cavities lined with ochreous ore; adhering slightly to the tongue; powder of a brownish-red color.

Specific gravity, - - - - - - - 3.292

*Composition*, dried at 212° F.—

| | | |
|---|---|---|
| Oxide of iron, - - - | 68.20 | = 47.76 per cent. of *Iron*. |
| Alumina, - - - - | 2.98 | |
| Brown oxide of manganese, - | .25 | |
| Phosphoric acid, - - - | .99 | |
| Lime, a trace. | | |
| Magnesia, - - - - | 1.02 | |
| Silex and insoluble silicates, - | 17.17 | |
| Combined water, - - - | 8.57 | |
| Alkalies, not estimated, & loss, | .82 | |
| | 100.00 | |

The air-dried ore lost 2.20 per cent. of *moisture*, at 212° F.

No. 318—LIMONITE. *Labeled "Limestone ore, incrusted with ochreous oxide of iron, Pennsylvania Furnace, Greenup county, Ky."*

A friable and porous ore, composed of irregular portions of dark brown hæmatite, imbedded in yellowish, (ochreous) soft matter, of different shades of color; powder brownish-yellow.

*Composition*, dried at 212° F.—

| | | |
|---|---|---|
| Oxide of iron, - - - | 61.10 | = 42.78 per cent of *Iron*. |
| Alumina, - - - - | .85 | |
| Carbonate of lime, - - - | .45 | |
| Magnesia, - - - - | 1.09 | |
| Brown oxide of manganese, - | .95 | |
| Phosphoric acid, a trace. | | |
| Potash, - - - - - | .38 | |
| Soda, - - - - - | .10 | |
| Silica and insoluble silicates, - | 23.85 | |
| Combined water, - - - | 11.67 | |
| | 100.44 | |

The air-dried ore lost 1.50 per cent. of *moisture*, at 212° F.

No. 477—LIMESTONE. *Labeled "Limestone, under the limestone ore, used as a flux, Pennsylvania Furnace, Greenup county, Ky."*

A dark grey, fine grained, compact limestone.

*Composition*, dried at 212° F.—

| | |
|---|---|
| Carbonate of lime, | 91.47 |
| Carbonate of magnesia, | 2.75 |
| Oxide of iron, | 1.82 |
| Brown oxide of manganese, | .05 |
| Alumina, | .48 |
| Potash, | .13 |
| Soda, | .10 |
| Silex and insoluble silicates, | 3.38 |
| | 100.18 |

The air-dried rock lost 0.50 per cent. of *moisture*, at 212° F.

No. 478—LIMONITE. *Labeled "Lower kidney ore, over the one foot sandstone, Raccoon ore banks, Greenup county, Ky."*

A dense dark-colored ore; reddish and purplish brown; with irregular cavities, and portions of soft yellowish and red ochreous mineral.

Specific gravity, - - - - - - - - 3.083

*Composition*, dried at 212° F.—

| | | |
|---|---|---|
| Oxide of iron, | 58.30 | = 40.82 per cent. of *Iron*. |
| Alumina, | 1.05 | |
| Brown oxide of manganese, | .65 | |
| Phosphoric acid, | 1.25 | |
| Carbonate of lime, | .15 | |
| Magnesia, | .77 | |
| Potash, | .40 | |
| Soda, | .08 | |
| Silex and insoluble silicates, | 29.77 | |
| Combined water, | 8.31 | |
| | 100.73 | |

The air-dried ore lost 1.30 per cent. of *moisture*, at 212 F.

No. 289—LIMONITE. *Labeled "Lower six inch black ore, Raccoon Furnace, Greenup county, Ky."*

A dull looking mineral, in irregular hard layers of a dark brown color, coated and separated by soft dirty ochreous ore; powder dull yellow ochre color.

*Composition*, dried at 212° F.—

| | |
|---|---|
| Oxide of iron, - - - | 24.70 = 17.29 per cent. of *Iron*. |
| Alumina, - - - - | 3.75 |
| Brown oxide of manganese - | .05 |
| Phosphoric acid, a trace. | |
| Magnesia, - - - - | .67 |
| Potash, - - - - - | .32 |
| Soda, - - - - - | .01 |
| Silex and insoluble silicates, - | 64.42 |
| Combined water, - - - | 5.66 |
| Loss, - - - - - | .42 |
| | 100.00 |

The air-dried mineral lost 1.20 per cent. of *moisture*, at 212° F.

Rather too poor in iron to be valuable, except for mixture with very rich calcarious ores, to produce cinder.

No. 309—Limonite. *Labeled "Main Kidney Ore, above the limestone ore, Greenup Furnace, Greenup county, Ky."*

A dull looking ore; dirty ochreous on the exterior; dull reddish and yellowish-brown in the interior; apparently a portion of a nodular mass; scarcely adhering to the tongue; powder dirty ochreous.

Specific gravity, - - - - - - 2.770

*Composition*, dried at 212° F.—

| | |
|---|---|
| Oxide of iron, - - - | 41.40 = 28.99 per cent. of *Iron*. |
| Alumina, - - - - | 3.36 |
| Brown oxide of manganese, - | .75 |
| Phosphoric acid, - - - | .54 |
| Carbonate of lime, - - - | 1.15 |
| Magnesia, - - - - | 1.50 |
| Potash, - - - - | .23 |
| Soda, - - - - - | .01 |
| Silica and insoluble silicates, - | 41.47 |
| Combined water, - - - | 10.54 |
| | 100.95 |

The air-dried ore lost 2.30 per cent. of *moisture*, at 212°.

Rather a poor ore, containing a large proportion of silex, which may be made profitable in judicious mixture with other ores.

No. 479—CARBONATE OF IRON. *Labeled "Carbonate of Iron, lowest bed, middle part, Greenup Furnace, Greenup county, Ky."*

Exterior olive-yellow; friable; soiling the fingers; interior dull dark grey, of fine granular, dense structure; powder light grey.

Specific gravity, - - - - - - - 3.497

*Composition*, dried at 212° F.—

| | | |
|---|---|---|
| Carbonate of iron, | 67.84 | 37.46 per cent. of *Iron.* |
| Oxide of iron, | 5.89 | |
| Carbonate of lime, | 3.25 | |
| Carbonate of magnesia, | 4.88 | |
| Carbonate of manganese, | 1.97 | |
| Alumina, | 1.45 | |
| Phosphoric acid, | .60 | |
| Potash, | .50 | |
| Soda, | .09 | |
| Silex and insoluble silicates, | 13.78 | |
| | 100.25 | |

The air-dried ore lost 0.50 per cent. of *moisture* at 212° F.

A valuable ore, which contains within itself nearly enough, or perhaps quite enough, fluxing materials to form its own cinder.

No. 312—CARBONATE OF IRON. *Labeled "Carbonate of Iron, lowest ore obtained at Greenup ore banks, Greenup county, Ky."*

A dull, dark brown, fine granular mineral, with a few minute scales of mica; exterior dirty ochreous; powder dirty orange-brown.

*Composition*, dried at 212° F.—

| | | |
|---|---|---|
| Carbonate of iron, | 56.92 | 37.10 *Iron.* |
| Oxide of iron, | 14.14 | |
| Carbonate of lime, | 1.25 | |
| Carbonate of magnesia, | 5.28 | |
| Carbonate of manganese, | 2.04 | |
| Alumina, | 1.05 | |
| Phosphoric acid, | .99 | |
| Potash, | .61 | |
| Soda, | .01 | |
| Organic matters, | .80 | |
| Silex and insoluble silicates, | 16.15 | |
| Water and loss, | .76 | |
| | 100.00 | |

The air-dried ore lost 0.50 per cent. of *moisture* at 212° F.

This ore nearly resembles the preceding in composition.

No. 311—Carbonate of Iron. *Labeled "Carbonate of Iron, lowest bed of ore, lower part of the bed, Greenup Furnace, Greenup county, Ky."*

A dark-grey, fine granular ore; powder yellowish-grey.

*Composition*, dried at 212° F.—

| | | |
|---|---|---|
| Carbonate of iron, - - - | 60.49 | 32.57 per cent. of *Iron*. |
| Oxide of iron, - - - - | 5.25 | |
| Carbonate of lime, - - - | 3.15 | |
| Carbonate of magnesia, - - | 6.52 | |
| Carbonate of manganese, - | .83 | |
| Alumina, - - - - - | .41 | |
| Phosphoric acid, a trace. | | |
| Potash, - - - - - | .34 | |
| Soda, - - - - - - | .29 | |
| Silex and insoluble silicates, - | 21.82 | |
| Water and loss, - - - | .90 | |
| | 100.00 | |

The air-dried ore lost 0.40 per cent. of *moisture*, at 212°.

Rather less rich than the two preceding ores, but yet a valuable ore of the same general character.

No. 330—Iron Furnace Slag. *Labeled "Pea-green cinder, Buena Vista Furnace, Greenup county, Ky."*

A greyish-green, blebby cinder, containing small nodules of cast iron, with iron rust on the weathered surfaces. Before the blow-pipe it melts, without intumescence, into a clear light bottle-green glass.

*Composition*—

| | | | | |
|---|---|---|---|---|
| Silica, - - - - - - | 58.00 | Containing of oxygen, | | 28.884 |
| Alumina, - - - - - | 20.50 | " | 9.582 | |
| Lime, - - - - - - | 12.06 | " | 3.554 | |
| Magnesia, - - - - - | 2.19 | " | .876 | |
| Protoxide of iron, - - - | 3.51 | " | .778 | |
| Protoxide of manganese, - | 1.21 | " | .272 | |
| Potash, - - - - - | 2.12 | " | .349 | |
| Soda, - - - - - - | .55 | " | .140 | |
| | 100.14 | | 15.551 : | 28.884 |
| The oxygen in the bases to that in the silica is as - - | | | 1 : | 1.78 |

Contains a little more silica, and a little less lime and magnesia, than slag No. 47, from the same furnace, (see former report, page 290.) This contains, also, more protoxide of iron and manganese. From the involved little nodules of iron it is inferred that this was of rather more pasty consistence than that. In this, as well as in No. 47, the bases, especially the alumina, are a little in excess of the proportion to produce the most fusible cinder.

No. 293—LIMONITE. *Labeled "Kidney ore, with sulphate of lime, Birk ore bank, overlaid by sandstone, Laurel Furnace, Greenup county, Ky."*

A dense, dark colored limonite, with many fissures coated with sulphate of lime; powder of a dull spanish-brown color.

Specific gravity, - - - - - - - 3.026

*Composition*, dried at 212° F.—

| | | |
|---|---|---|
| Oxide of iron, - - - | 77.50 | = 54.25 per cent. of *Iron*. |
| Alumina, - - - - | 1.23 | |
| Brown oxide of manganese, - | 1.03 | |
| Phosphoric acid, - - - | .40 | |
| Lime, - - - - - - | .76 | |
| Magnesia, - - - - | .79 | |
| Sulphuric acid, - - - | 1.56 | = .63 *Sulphur*. |
| Potash, - - - - | .20 | |
| Soda, - - - - - - | .14 | |
| Silica and insoluble silicates, - | 7.77 | |
| Combined water, - - - | 9.62 | |
| | 101.00 | |

The air-dried ore lost 3.40 per cent. of *moisture*, at 212°.

The gypsum and sulphate of lime contained in the fissures of this ore is very likely to contaminate the product with sulphur, to a greater or less degree.

No. 433—LIMESTONE. *Labeled "Limestone used as a flux at Laurel Furnace, from Tygert's creek, Greenup county, Ky."*

A compact, light grey limestone; sparkling with small crystals of calcarious spar.

Specific gravity, - - - - - - - 2.699

*Composition*, dried at 212° F.—

| | | |
|---|---|---|
| Carbonate of lime, - - - | 97.90 | = 54.93 per cent. of *Lime*. |
| Carbonate of magnesia, - - | .74 | |
| Alumina, oxide of iron ,and phosphates, - - - - | .53 | |
| Potash, - - - - - - | .28 | |
| Soda, - - - - - - | .08 | |
| Silex and insoluble silicates, - | 1.27 | |
| | 100.80 | |

The air-dried rock lost 0.30 per cent. of *moisture*, at 212° F.

No. 432—FERRUGINOUS LIMESTONE. *Labeled "Ferruginous Limestone, under the limestone ore, near the tops of the hills, waters of Old-town creek, Laurel Furnace, Greenup county, Ky."*

A dark grey, fine granular rock; portions dull brownish and greenish; exterior surface ochreous; not adhering to the tongue; powder of a light grey color.

| | | |
|---|---|---|
| Specific gravity, - - - - - - - - | 2.731 | |
| *Composition*, dried at 212° F.— | | |
| Carbonate of iron, - - | 22.19 | = 11.82 per cent. of *Iron*. |
| Oxide of iron, - - - | 1.49 | |
| Carbonate of lime, - - - | 50.33 | |
| Carbonate of magnesia, - - | 1.83 | |
| Carbonate of manganese, - | .47 | |
| Alumina, - - - - | .77 | |
| Phosphoric acid, - - - | .77 | |
| Sulphur, - - - - - | .26 | |
| Potash, - - - - - - | .38 | |
| Soda, - - - - - | .20 | |
| Silex and insoluble silicates, - | 21.43 | |
| | 100.12 | |

The air-dried rock lost 0.30 per cent. of *moisture*, at 212°.

If it were not for the phosphoric acid and the sulphur present in this limestone, it might advantageously replace the preceding limestone as a flux in the high furnace. It is more fusible and contains a considerable per centage of iron.

No. 294—MIXED CARBONATE AND OXIDE OF IRON. *Labeled "Baker Bank Kidney Ore, near top of hills, Old-town creek, Laurel Furnace, Greenup county, Ky."*

Nodule in the interior of the mass dark grey carbonate of iron; scarcely adhering to the tongue; exterior irregular layers yellowish-brown and dark reddish-brown; adhering to the tongue; powder of the mixed specimen, dirty brownish-yellow.

*Composition* of the mixed mass, dried at 212° F—

| | | |
|---|---|---|
| Carbonate of iron, - - - | 54.42 | = 47.51 per cent. of *Iron*. |
| Oxide of iron, - - - | 30.24 | |
| Carbonate of lime, - - - | .45 | |
| Carbonate of magnesia, - - | .83 | |
| Carbonate of manganese, - | 1.29 | |
| Alumina, - - - - - | 1.86 | |
| Phosphoric acid, - - - | .43 | |
| Sulphur, - - - - - | .35 | |
| Potash, - - - - - | .38 | |
| Soda, - - - - - - | .20 | |
| Silex and insoluble silicates, - | 6.97 | |
| Bituminous matter, water, and loss, - - - - - | 2.58 | |
| | 100.00 | |

The air-dried ore lost 0.60 per cent. of *moisture* at 212°.

This ore is as rich as is desirable for profitable smelting, requiring the addition of lime, and probably of some more silicious ore, to produce a proper amount of cinder.

No. 431—Limonite. *Labeled "Ore, partly roasted, from Laurel Furnace ore banks, Greenup county, Ky." (What is the white mineral?)*

A dark reddish-brown mineral, incrusted on the exterior and in the fissures with a whitish substance, which appears to be principally carbonate of lime; adhering strongly to the tongue; powder of chocolate-brown color; contains no protoxide of iron.

*Composition*, dried at 222° F.—

| | | |
|---|---|---|
| Oxide of iron, - - - | 74.50 | = 52.17 per cent. of *Iron*. |
| Alumina, - - - - | 1.00 | |
| Brown oxide of manganese, - | 2.43 | |
| Carbonate of lime, - - | .77 | |
| Magnesia, - - - - - | 1.81 | |
| Phosphoric acid, - - - | .33 | |
| Sulphur, - - - - - | .57 | |
| Potash, - - - - - | .15 | |
| Soda, - - - - - - | .13 | |
| Combined water, - - - | 3.86 | |
| Silex and insoluble silicates, - | 14.93 | |
| | 100.00 | |

The air-dried ore lost 1.70 per cent. of *moisture* at 212°.

No. 430—LIMONITE. *Labeled "Lower bed of ore used at Laurel Furnace, Greenup county, Ky."*

A concretionary limonite, with irregular cavities, varying, in layers, from dark-brown and compact to yellow and reddish soft mineral; powder of a dirty yellowish ochre color.

*Composition*, dried at 212° F.—

| | | |
|---|---|---|
| Oxide of iron, - - - | 38.38 | = 26.87 per cent. of *Iron*. |
| Alumina, - - - - | 3.54 | |
| Brown oxide of manganese, - | 1.23 | |
| Phosphoric acid, - - - | 1.01 | |
| Sulphur, - - - - | .05 | |
| Lime, a trace. | | |
| Magnesia, - - - - | .60 | |
| Potash, - - - - | .28 | |
| Soda, - - - - - - | .18 | |
| Silex and insoluble silicates, - | 46.83 | |
| Combined water, - - - | 8.12 | |
| | 100.22 | |

The air-dried ore lost 1.50 per cent. of *moisture* at 212°.

The only drawback to the use of this highly silicious ore is in the considerable amount of phosphoric acid which it contains—rather more than one per cent.—which, if it passed mainly into the iron in smelting, as it generally does, unless an excess of lime is used in the flux, would contaminate it with nearly 1.76 per cent. of phosphorus, an ingredient which is always injurious to the strength of the iron, even in as small proportion as the half of one per cent.

Were it not for the phosphoric acid contained in it, this highly silicious ore might be very advantageously used in mixture with the richer ores of Laurel Furnace; but when pure and very tough iron is required such ores as this must be avoided, although the metal which they yield is yet applicable to many common uses.

No. 290—LIMONITE. (*Roasted.*) *Labeled "Kidney ore, showing a prismatic structure only after thorough roasting, Laurel Furnace, Greenup county, Ky."*

Powder dull brownish-red color; structure somewhat like that of starch; in irregular curved prisms; color chocolate-brown; adhering strongly to the tongue.

*Composition*, dried at 212° F.—

| | | |
|---|---|---|
| Oxide of iron, - - - - | 80.03 | = 56.02 per cent. of *Iron*. |
| Alumina, - - - - - | 1.44 | |
| Brown oxide of manganese, - | 2.03 | |
| Phosphoric acid, - - - | .66 | |
| Lime, - - - - - - | .64 | |
| Magnesia, - - - - | 2.87 | |
| Potash, - - - - - | .25 | |
| Soda. - - - - - - | .16 | |
| Silex and insoluble silicates, - | 9.93 | |
| Combined water, - - - | 2.01 | |
| | 100.02 | |

The air-dried ore lost 0.80 per cent. of *moisture* at 212° F.

No sulphur was present in this specimen of roasted ore.

No. 291—LIMONITE. "*Labeled "Main Block Ore, near tops of hills, Old-town creek, Laurel Furnace, Greenup county, Ky.*"

A dark, reddish-brown ore, nearly black in parts; adheres slightly to the tongue; powder brownish-red.

Specific gravity, - - - - - - - - 3.018

*Composition*, dried at 212° F.—

| | | |
|---|---|---|
| Oxide of iron, - - - | 73.90 | = 51.75 per cent. of *Iron*. |
| Alumina, - - - - | 1.71 | |
| Brown oxide of manganese, - | 1.13 | |
| Phosphoric acid, - - - | .62 | |
| Sulphur, - - - - - | .09 | |
| Lime, a trace. | | |
| Magnesia, - - - - - | .39 | |
| Potash, - - - - | .19 | |
| Soda, - - - - - - | .05 | |
| Silex and insoluble silicates, - | 10.43 | |
| Combined water, - - - | 11.51 | |
| | 100.02 | |

The air-dried ore lost 1.90 per cent. of *moisture* at 212°.

No. 292—*Labeled "Kidney Ore, over the Ferruginous Limestone in the hills, Old-town creek, Laurel Furnace, Greenup county, Ky.*"

Color yellowish, reddish, and reddish-brown; containing nodules, irregular cavities, and layers of different degrees of hardness; adhering to the tongue; powder of a dull red color, or spanish-brown color.

| | | |
|---|---|---|
| Specific gravity, | | 3.406 |
| *Composition*, dried at 212° F.— | | |
| Oxide of iron, | 81.40 | = 57. per cent. of *Iron*. |
| Alumina, | .77 | |
| Brown oxide of manganese, | 1.63 | |
| Phosphoric acid, | .24 | |
| Sulphur, | .07 | |
| Lime, a trace. | | |
| Magnesia, | .35 | |
| Potash, | .26 | |
| Soda, | .22 | |
| Silex and insoluble silicates, | 8.33 | |
| Combined water, | 6.72 | |
| Loss, | .01 | |
| | 100.00 | |

The air-dried ore lost 1.90 per cent. of *moisture* at 212°.

This appears to be the *purest* ore used at Laurel Furnace. It contains rather too small a proportion of the materials for the formation of cinder to be profitably smelted with lime alone. The addition of the ferruginous limestone, No. 432, would exactly supply this desideratum; but would, also, render the iron less pure, in consequence of the phosphorus and sulphur which it contains. The use of as large an excess of lime as can be worked, without making the cinder too pastey, is the best means of obviating this disadvantage.

No. 435—Pig-iron. *Labeled "Medium textured Pig-iron, produced frequently at Laurel Furnace when pumice-form slag is formed, Greenup county, Ky. (Does it contain much sulphur?")*

A moderately fine-grained, grey, pig-iron, with brilliant grains; it flattens somewhat under the hammer, but soon crushes to powder; yields easily to the file.

| | | |
|---|---|---|
| Specific gravity, | | 7.009 |
| *Composition*— | | |
| Iron, | 90.00 | |
| Graphite, | 1.77 | Total carbon, 2.67 per cent. |
| Combined carbon, | .90 | |
| Silicon, | 4.28 | |
| Slag, | 1.15 | |
| Aluminium, | .13 | |
| Calcium, | .14 | |
| Magnesium, | .21 | |

| | |
|---|---|
| Potassium, - - - - - | .17 |
| Sodium, - - - - - | .14 |
| Phosphorus, - - - - - | .61 |
| Sulphur, - - - - - | .12 |
| Manganese, - - - - - | .33 |
| Loss, - - - - - - | .05 |
| | 100.00 |

No. 434—PIG-IRON. *Labeled "Soft, but not very strong tough pig-iron, produced at Laurel Furnace when making chiefly dark purple slag, Greenup county, Ky."* (*Does it contain much sulphur?*)

Somewhat coarser-grained, and a little lighter colored, than the preceding (No. 435;) breaks and crushes to powder quite easily under the hammer; yields readily to the file.

Specific gravity, - - - - - - - - 6.886

*Composition*—

| | | |
|---|---|---|
| Iron, - - - - - - | 89.54 | |
| Graphite, - - - - - | 1.87 | Total carbon, 2.03 per cent. |
| Combined carbon, - - - - | .16 | |
| Silicon, - - - - - - | 5.57 | |
| Slag, - - - - - - | 1.25 | |
| Aluminum, - - - - - | .13 | |
| Calcium, - - - - - | .19 | |
| Magnesium, - - - - - | .20 | |
| Potassium, - - - - - | .17 | |
| Sodium, - - - - - | .11 | |
| Phosphorus, - - - - - | .46 | |
| Sulphur, - - - - - | .10 | |
| Manganese, - - - - - | .54 | |
| | 100.29 | |

These specimens of iron do not contain enough sulphur to cause any serious injury to the quality of the metal; the phosphorus, it is true, rather exceeds that proportion, but the principal cause of the want of strength observed in this product is in the large quantity of *silicon* which is found in it, especially in pig-iron No. 434, which appears to have been produced at a higher temperature in the furnace than No. 435. Whether this contamination, which results from the silicious nature of the ores used at Laurel Furnace, or from a too high temperature in the melting, may be prevented by the use of more limestone in the flux, cannot be positively stated, as none of the cinder produced at the furnace was sent to the laboratory for analysis. But it is probable that more limestone could be advantageously added. The admixture of some *aluminous* ores, also, would doubtless improve the quality of the iron.

Difference of opinion has existed amongst writers on iron as to the influence exerted upon it by silicon. Whilst Berzelins and Stromeyer did not find it materially to injure the qualities of the iron, in their experiments, other observers, as Boussingault, Mushet, and Karsten, are positive in the assertion that its presence in considerable proportion—less than that in the above specimens from Laurel Furnace—makes the iron cold-short, or, in other words, diminishes its toughness at the ordinary temperature, whilst it also diminishes its specific gravity. Below the proportion of 0.40 per cent. it is believed to increase the firmness of the iron in the same manner as carbon, but above that proportion it acts on the qualities of this metal in the manner of phosphorus. Indeed, Mushet, who was a practical iron man, who experimented extensively on this metal, believes that the *cold-short* property of iron is generally owing to the presence of an excess of *silicon.*

No. 440—Carbonate of Iron. *Labeled "Grey ore, above the red ore, and next to the top-hill ore, Mount Savage Furnace, Greenup county, Ky."*

A light-grey, granular ore; on the exterior changed to yellowish and reddish-brown; powder (of mixed portions of the interior and exterior,) of a light cinnamon color.

*Composition*, dried at 212° F.—

| | | |
|---|---|---|
| Carbonate of iron, - - - | 43.90 | } = 35.02 per cent. of *Iron*. |
| Oxide of iron, - - - | 23.06 | |
| Carbonate of lime, - - - | 3.87 | |
| Carbonate of magnesia, - - | 3.28 | |
| Carbonate of manganese, - | .65 | |
| Alumina, - - - - - | .33 | |
| Phosphoric acid, - - - | .23 | |
| Sulphur, - - - - | .18 | |
| Potash, - - - - | .23 | |
| Soda, - - - - - | .23 | |
| Silex and insoluble silicates, - | 22.15 | |
| Combined water, - - - | 2.60 | |
| | 100.71 | |

The air-dried ore lost 0.70 per cent. of *moisture*, at 212° F.

No. 441—LIMONITE. *Labeled "Silicious? ore, Mount Savage Furnace, Greenup county, Ky."*

A dull, granular limonite; generally of a dark, brownish-red color; portions ochreous; containing a few minute scales of mica; adheres to the tongue; powder dull brownish-red.

*Composition*, dried at 212° F.—

| | | |
|---|---|---|
| Oxide of iron, - - - | 51.10 | = 35.78 per cent. of *Iron*. |
| Alumina, - - - - | 1.07 | |
| Brown oxide of manganese, - | 1.83 | |
| Phosphoric acid, - - - | .76 | |
| Sulphur, - - - - - | .32 | |
| Lime, a trace. | | |
| Magnesia, - - - - - | .68 | |
| Potash, - - - - - | .38 | |
| Soda, - - - - - - | .10 | |
| Silex and insoluble silicates, - | 35.93 | |
| Combined water, - - - | 8.13 | |
| | 100.30 | |

The air-dried ore lost 1.60 per cent. of *moisture*, at 212° F.

No. 442—LIMONITE. *Labeled "Limestone ore, Mount Savage Furnace, Greenup county, Ky."*

A very dark-brown ore; made up of dense irregular layers, inclosing irregular cavities of various sizes; sometimes coated with ochreous; scarcely adhering to the tongue; powder yellowish-brown.

*Composition*, dried at 212° F.—

| | | |
|---|---|---|
| Oxide of iron, - - - | 83.83 | = 58.70 per cent. of *Iron*. |
| Alumina, - - - - - | .43 | |
| Brown oxide of manganese, - | 1.73 | |
| Phosphoric acid, - - - | .94 | |
| Sulphur, - - - - | .21 | |
| Lime, only a trace. | | |
| Magnesia, - - - - - | .32 | |
| Potash, - - - - - | .30 | |
| Soda, - - - - - - | .11 | |
| Silex and insoluble silicates, - | .83 | |
| Combined water, - - - | 11.30 | |
| | 100.00 | |

The air-dried ore lost 0.70 per cent of *moisture*, at 212° F.

A remarkably pure limonite, containing scarcely anything but hydrated per-oxide of iron, although called limestone ore at the Furnace. As it contains scarcely any of the materials for the formation of cinder, it must be smelted together with other ores containing a larger proportion of earthy matters.

No. 443—CARBONATE OF IRON. *Labeled "Blue Block Ore, Mount Savage Furnace, Greenup county, Ky. (Lies lowest in the hills.")*

A dull dark-grey, fine granular rock, with a few specks of calcareous spar; scarcely adhering to the tongue; powder light-grey.

Specific gravity, - - - - - - - 3.360

*Composition*, dried at 212° F.—

| | | |
|---|---|---|
| Carbonate of iron, - - - | 67.50 | = 33.12 per cent. of *Iron*. |
| Oxide of iron, - - - | 1.28 | |
| Carbonate of lime, - - | 2.15 | |
| Carbonate of magnesia, - - | 4.57 | |
| Carbonate of manganese, - | 1.18 | |
| Alumina, - - - - | .35 | |
| Phosphoric acid, - - - | .36 | |
| Sulphur, - - - - | .17 | |
| Potash, - - - - - | .29 | |
| Soda, - - - - - | .09 | |
| Silica and insoluble silicates, - | 21.45 | |
| Loss, - - - - - | .61 | |
| | 100.00 | |

No. 444—MIXED LIMONITE. *Labeled "Kidney ore, top of the rough block ore, Mount Savage Furnace, Greenup county, Ky."*

A dull-grey, friable nucleus, enclosed in hard layers of blackish-brown limonite. Powder of the mixture of a yellowish-brown, or scotch-snuff color.

*Composition*, dried at 212° F.—

| | | |
|---|---|---|
| Oxide of iron, - - - | 53.44 | = 49.39 per cent. of *Iron*. |
| Carbonate of iron, - - - | 24.79 | |
| Carbonate of lime, - - - | .87 | |
| Carbonate of magnesia, - - | .62 | |
| Carbonate of manganese, - | 1.44 | |
| Alumina, - - - - | .09 | |
| Phosphoric acid, - - - | 1.26 | |
| Sulphuric acid, - - - | .11 | |
| Potash, - - - - | .84 | |

| | |
|---|---|
| Soda, - - - - - - | .08 |
| Silex and insoluble silicates, - | 9.93 |
| Combined water, - - - - | 6.89 |
| Loss, - - - - - - | .14 |
| | 100.00 |

The air-dried ore lost 1.00 per cent. of *moisture*, at 212° F.

Contains rather a larger proportion of phosphoric acid than is desirable, but otherwise, a very good ore.

No. 445—IMPURE CARBONATE OF IRON. *Labeled "Blue Limestone ore, deep in the bed, (with sulphur?,) Mount Savage Furnace, Greenup county, Ky."*

A dull, granular mineral; general color brownish-grey, with a greenish tint in portions, and in others presenting the appearance of pyrites; powder dark-greenish-grey.

Specific gravity, - - - - - - - - 3.567

*Composition*, dried at 212° F.—

| | | |
|---|---|---|
| Carbonate of iron, - - - - | 47.84 | = 41.63 per cent. of *Iron*. |
| Sulphuret of iron, - - - - | 31.60 | = 11.51 per cent. of *Sulphur*. |
| Carbonate of lime, - - - - | 3.25 | |
| Carbonate of magnesia, - - | 3.65 | |
| Carbonate of manganese, - | 6.00 | |
| Alumina, - - - - - | .55 | |
| Phosphoric acid, only a trace. | | |
| Potash, - - - - - - | .34 | |
| Soda, - - - - - - | .08 | |
| Silica and insoluble silicates, - | 4.75 | |
| Organic matter, water, and loss, | 1.94 | |
| | 100.00 | |

The air-dried ore lost 0.30 per cent. of *moisture* at 212° F.

This ore contains entirely too much sulphur. A considerable proportion of it may, however, be removed by thorough roasting.

No. 446. LIMONITE. *Labeled "Best quality of 'rough block ore,' under the 'Kidney ore,' Mount Savage Furnace, Greenup county, Kentucky."*

A dense, very dark-brown limonite; not adhering to the tongue; exhibiting small cavities and minute spangles of mica; the curved layers are covered with brownish-ochreous, soft, mineral; powder brownish-yellow.

*Composition*, dried at 212° F.—

| | | |
|---|---|---|
| Oxide of iron, - - - | 66.76 | = 46.75 per cent. of *Iron*. |
| Alumina, - - - - - | 1.00 | |
| Brown oxide of manganese, - | 1.23 | |
| Phosphoric acid, - - - | 1.41 | |
| Sulphur, a trace. | | |
| Lime, a trace. | | |
| Magnesia, - - - - - | .26 | |
| Potash, - - - - - - | .34 | |
| Soda, a trace. | | |
| Silex and insoluble silicates, - | 17.87 | |
| Combined water, - - - | 11.59 | |
| | 100.46 | |

The air-dried ore lost 1.60 per cent. of *moisture*, at 212°.

Its proportion of phosphoric acid is considerable.

No. 422—LIMONITE. *Labeled "Roasted Kidney ore, rather sandy, Caroline Furnace, Greenup county, Ky."*

Composed of dark reddish-brown layers, enclosing a friable light reddish colored nucleus; adhering to the tongue; powder handsome spanish-brown color.

*Composition*, dried at 212° F.

| | | |
|---|---|---|
| Oxide of iron, - - - | 66.03 | = 46.24 per cent of *Iron*. |
| Alumina, - - - - - | 4.15 | |
| Brown oxide of manganese, - | .55 | |
| Lime, a trace. | | |
| Magnesia, - - - - - | .76 | |
| Phosphoric acid, - - - | .67 | |
| Sulphur, - - - - - | .06 | |
| Potash, - - - - - | .46 | |
| Soda, - - - - - - | .11 | |
| Silex and insoluble silicates, - | 27.15 | |
| Combined water, - - - | .71 | |
| | 100.65 | |

The air-dried ore lost 0.70 per cent. of *moisture* at 212°.

No. 423—IRON FURNACE SLAG. *Labeled "What is the heavy bluish granular material in this slag, from Caroline Furnace, Greenup county, Ky?"*

The granular, nearly opake portion is of steel bluish-grey and pinkish colors, contained in the purple glassy slag. Before the blow-pipe both kinds readily melt into a blebby white glass.

*Composition—*

| | *Granular.* | *Oxygen.* | |
|---|---|---|---|
| Silica, | 48.80 | | 25.338 |
| Lime, | 33.27 | 9.461 | |
| Alumina, | 12.50 | 4.843 | |
| Magnesia, | 1.24 | .495 | |
| Protoxide of iron, | 1.19 | .265 | |
| Protoxide of manganese, | .51 | .115 | |
| Potash, | 1.62 | .275 | |
| Soda, | .18 | .046 | |
| | 99.13 | 15.499 : | 25.338 |

Oxygen in the bases to that in the silica as - - - 1 : 1.63

*Composition—*

| | *Glassy.* | *Oxygen.* | |
|---|---|---|---|
| Silica, | 48.86 | | 25.369 |
| Lime, | 33.05 | 9.398 | |
| Alumina, | 12.86 | 5.011 | |
| Magnesia, | 2.74 | 1.095 | |
| Protoxide of iron, | 1.13 | .251 | |
| Protoxide of manganese, | .51 | .115 | |
| Potash, | 1.54 | .262 | |
| Soda, | .15 | .038 | |
| | 100.85 | 16.169 : | 25.369 |

Oxygen in the bases to that in the silica as - - - 1 : 1.57

No marked difference of composition can be perceived by the analyses of these two varieties of cinder. The granular appearance and change of color were occasioned probably by some irregularity in the cooling of the slag. This cinder contains a larger proportion of lime than is necessary to form a bi-silicate—at least one-third more than is usually present in the Greenup Furnace slags. This *excess* of lime may exert a purifying influence on the iron produced from ores containing much sulphur, but does not increase the fusibility of the cinder.

No. 424—Limonite. *Labeled "Limestone Kidney ore, also associated with the four-feet Limestone, Caroline Furnace, Greenup county, Kentucky."*

Composed of dark brown curved layers, incrusted with dirty yellowish and whitish friable matter; powder of a brownish-buff color.

*Composition*, dried at 212° F.—

| | | |
|---|---|---|
| Oxide of iron, | 63.60 | = 44.54 per cent of *Iron*. |
| Alumina, | .25 | |
| Brown oxide of manganese, | .55 | |
| Phosphoric acid, | .70 | |
| Sulphur, | .06 | |
| Lime, a trace. | | |
| Magnesia, | .99 | |
| Potash, | .25 | |
| Soda, | .05 | |
| Silica and insoluble silicates, | 23.23 | |
| Combined water, and loss, | 10.77 | |
| | 100.45 | |

The air-dried ore lost 1.30 per cent. of *moisture*, at 212° F.

No. 425—Limonite. *Labeled "Hydrated variety of Limestone ore, over the four-feet Limestone, Caroline Furnace, Greenup county, Kentucky."*

A dark-brown limonite, in dense layers, irregularly disposed, involving some small irregular cavities, and covered, in some parts, with a yellow-ochreous soft mineral; powder of a yellowish umber color; when calcined of a purplish-brown color.

*Composition*, dried at 212° F.—

| | | |
|---|---|---|
| Oxide of iron, | 85.91 | = 60.16 per cent. of *Iron*. |
| Alumina, | 1.25 | |
| Brown oxide of manganese, | 2.17 | |
| Phosphoric acid, | .09 | |
| Carbonate of lime, | .17 | |
| Magnesia, | .85 | |
| Potash, | .23 | |
| Soda, | .18 | |
| Silex and insoluble silicates, | 1.25 | |
| Combined water, | 7.90 | |
| | 100.00 | |

The air-dried ore lost 1.60 per cent of *moisture* at 212°.

A very pure iron ore, containing more than the usual proportion of oxide of manganese, and which must be mixed with poorer ores in order to be profitably fluxed in the furnace.

No. 426—FERRUGINOUS LIMESTONE. *Labeled "Bottom portion of Limestone Ore; not considered as good as the red; Caroline Furnace, Greenup county, Ky."*

Irregular portions of compact tawny-brown ferruginous limestone, showing some glimmering crystals of calcareous spar, with friable yellowish and whitish incrusting and included ochreous matter; powder of a grey-buff color; when calcined of a light umber color.

Composition, dried at 212° F.—

| | | |
|---|---|---|
| Oxide of iron, | 25.80 | = 18.06 per cent. of *Iron.* |
| Carbonate of lime, | 65.13 | = 36.55 per cent. of *Lime.* |
| Magnesia, | 1.41 | |
| Brown oxide of manganese, | .17 | |
| Alumina, | .13 | |
| Phosphoric acid, | .17 | |
| Potash, | .11 | |
| Soda, | .06 | |
| Silex and insoluble silicates, | 1.27 | |
| Carbonic acid & combined water, | 5.75 | |
| | 100.00 | |

The air-dried rock lost .60 per cent. of *moisture* at 212°.

This mineral may be profitably mixed with the richer silicious ores of this locality, for smelting, instead of the limestone generally used as a flux.

No. 427—FERRUGINOUS LIMESTONE. *Labeled "Four feet Limestone, under the Limestone Ore, Caroline Furnace, Greenup county, Kentucky."*

A fine grained limestone, glimmering with small plates of calcareous spar, and containing fossil remains; grey, with a portion of a light-grey buff color; powder light yellowish-grey.

Specific gravity, - - - - - - - 2.729

*Composition*, dried at 212° F.—

| | |
|---|---|
| Carbonate of lime, | 84.47 |
| Sulphate of lime, | .71 |
| Carbonate of magnesia, | 3.47 |
| Carbonate of manganese, | .26 |
| Carbonate of iron, | 7.73 |
| Oxide of iron, | 1.77 |
| Alumina, | .25 |
| Phosphoric acid, | .62 |
| Potash, | .32 |
| Soda, | .14 |
| Silex and insoluble silicates, | .55 |
| | 100.29 |

The air-dried rock lost 0.30 per cent. of *moisture* at 212° F.

No. 428—Limonite. *Labeled "Roasted Limestone Ore, Caroline Furnace, Greenup county, Ky."*

Interior of a purplish-brown color; exterior (incrustation) of a dirty light-red, including to pink; friable; adhering firmly to the tongue; powder of a handsome maroon color.

*Composition*, dried at 212° F.—

| | | |
|---|---|---|
| Oxide of iron, | 84.45 | = 59.14 per cent. of *Iron*. |
| Alumina, | 1.20 | |
| Brown oxide of manganese, | .09 | |
| Phosphoric acid, | .38 | |
| Sulphur, | .06 | |
| Magnesia, | 1.43 | |
| Potash, | .44 | |
| Soda, | .10 | |
| Silica and insoluble silicates, | 9.05 | |
| Combined water and loss, | 2.80 | |
| | 100.00 | |

The ore lost .90 per cent. of *moisture*, at 212°.

This specimen contains no appreciable quantity of lime.

No. 429—Limonite. *Labeled "Top-hill Kidney Ore, Caroline Furnace, Greenup county, Ky."*

Formed of irregular curved layers, inclosing cavities; interior of the layers dense and dark reddish-brown; exterior coating friable and yellow (ochreous;) powder of a rich brownish-yellow color; when calcined of a spanish-brown color.

*Composition*, dried at 212° F.—

| | | |
|---|---|---|
| Oxide of iron, | 69.60 | = 48.74 per cent. of *Iron*. |
| Alumina, | .55 | |
| Brown oxide of manganese, | .75 | |
| Phosphoric acid, | .42 | |
| Sulphur, | .07 | |
| Lime, a trace. | | |
| Magnesia, | .35 | |
| Potash, | .42 | |
| Soda, | .01 | |
| Silex and insoluble silicates, | 15.65 | |
| Combined water and loss, | 12.18 | |
| | 100.00 | |

The air-dried ore lost 0.50 per cent. of *moisture*, at 212°.

A very good ore, requiring no other flux than limestone.

No. 299—Limonite. *Labeled "Good red-brown 'Limestone Ore,' under the four feet Limestone, Caroline Furnace, Greenup county, Kentucky."*

A dark, reddish-brown, dull, fine granular ore; glimmering with minute facets of spar; adhering to the tongue; powder of a light spanish-brown color; when calcined of a dark snuff color.

Composition, dried at 212° F.—

| | | |
|---|---|---|
| Oxide of iron, | 53.46 | = 37.44 per cent. of *Iron*. |
| Alumina, a trace. | | |
| Brown oxide of manganese, | .85 | |
| Phosphoric acid, | .87 | |
| Sulphur, | .02 | |
| Carbonate of lime, | 33.85 | = 19. per cent. of *Lime*. |
| Magnesia, | 3.15 | |
| Potash, | .23 | |
| Soda, | .07 | |
| Silex and insoluble silicates, | 1.05 | |
| Combined water and loss, | 6.45 | |
| | 100.00 | |

The air-dried ore lost .80 per cent. of *moisture* at 212°.

This ore, which is rich enough in iron for profitable smelting, contains, like No. 426, an excess of lime and a deficiency of silica; this, however, contains twice as much iron as that. This ore could, no doubt, be advantageously used in mixture with the "Limestone Kidney Ore," No. 424, which is rich in *silica*, and contains no appreciable quantity of lime.

No. 300—CARBONATE OF IRON. *Labeled "Blue Limestone Ore" Caroline Furnace, Greenup county, Ky.*

A dark-grey, fine granular mineral; not adhering to the tongue; portions and fissures dirty yellowish and brownish; under the lens exhibits minute crystalline scales and specks of mica; some little white incrustation in the fissures; powder brownish-cinnamon color.

Specific gravity, - - - - - - - 3.566

*Composition*, dried at 212° F.—

| | | |
|---|---|---|
| Carbonate of iron, - - - | 60.40 | } = 43.82 per cent. of *Iron.* |
| Oxide of iron, - - - | 21.38 | } |
| Carbonate of lime, - - - | 3.17 | |
| Carbonate of magnesia, - - | 3.46 | |
| Carbonate of manganese, - | 1.52 | |
| Alumina, - - - - | .65 | |
| Phosphoric acid, - - - | .63 | |
| Sulphur, a trace. | | |
| Potash, - - - - - | .40 | |
| Soda, - - - - - | .13 | |
| Silex and insoluble silicates, - | 6.03 | |
| Combined water and loss, - | 2.23 | |
| | 100.00 | |

The air-dried ore lost 1.00 per cent. of *moisture*, at 212°.

No. 436—LIMONITE. *Labeled "Red ore of Iron, divide between Tygerts and Kinch creeks, Kenton Furnace, Greenup county, Ky."*

A dull, friable, fine granular limonite, of a dark-purple-brown color, (like that of crocus martis); adhering to the tongue; powder of the same color; when calcined nearly black.

*Composition*, dried at 212° F.—

| | | |
|---|---|---|
| Oxide of iron, - - - | 88.51 | = 61.98 per cent. of *Iron.* |
| Alumina, - - - - | .15 | |
| Brown oxide of manganese, - | 1.95 | |
| Phosphoric acid, - - - | .19 | |
| Sulphur, - - - - | .03 | |
| Lime, a trace. | | |
| Magnesia, - - - - | .78 | |
| Potash, - - - - | .09 | |
| Soda, - - - - - | .17 | |
| Silex and insoluble silicates, - | 2.23 | |
| Combined water, - - - | 6.00 | |
| | 100.10 | |

The air-dried ore lost 1.80 per cent. of *moisture*, at 212 F.

A pretty pure hydrated oxide of iron, requiring for smelting an admixture of the materials for the production of cinder.

No. 437—LIMONITE. *Labeled "Limestone Ore, near the head of Grassy creek, Kenton Furnace, Greenup county, Ky."*

A dull, dark-brown mineral, mixed with ochreous matter in the cavities and between the layers; powder of a light-clove-brown color.

*Composition*, dried at 212° F.—

| | | |
|---|---|---|
| Oxide of iron, - - - | 80.20 | = 56.14 per cent. of *Iron*. |
| Alumina, - - - - - | .47 | |
| Brown oxide of manganese, - | .05 | |
| Phosphoric acid, - - - | .86 | |
| Magnesia, - - - - | .51 | |
| Potash, - - - - | .48 | |
| Soda, - - - - - - | .02 | |
| Silex and insoluble silicates, - | 6.45 | |
| Combined water, and loss, - | 11.31 | |
| | 100.35 | |

The air-dried ore lost 1. per cent. of *moisture* at 212° F.

Nearly as rich in iron as the preceding, and like that, containing in itself too small a proportion of earthy materials for the formation of a sufficient quantity of slag in the furnace.

No. 438—LIMONITE. *Labeled "Earthy variety of 'Block Ore,' Kenton Furnace, Greenup county, Ky."*

A dull, dark-brown ore, in curved layers, inclosing friable, brownish-yellow ochreous matter; powder dirty light-yellowish-brown.

Composition, dried at 212° F.—

| | | |
|---|---|---|
| Oxide of iron, - - - | 49.90 | = 35.06 per cent. of *Iron*. |
| Alumina, - - - - - | 7.00 | |
| Brown oxide of manganese - | .27 | |
| Phosphoric acid, - - - | 1.45 | |
| Carbonate of lime, - - | 8.05 | |
| Magnesia, - - - - - | 4.19 | |
| Potash, - - - - - - | .41 | |
| Silex and insoluble silicates, - | 19.15 | |
| Combined water, - - - | 9.61 | |
| | 100.03 | |

The air-dried ore lost 1.20 per cent. of *moisture*, at 212° F.

This ore, with the only drawback of the considerable proportion of phosphoric acid which it contains, could be profitably smelted with the addition of a very little more lime, or could be employed to great advantage in mixture with the other richer ores of Kenton Furnace, to furnish the ingredients for the formation of cinder, in which they are deficient.

No. 439—LIMONITE. *Labeled "Black Limestone Ore; resting on the limestone, Kenton Furnace, Greenup county, Ky."*

Dull; almost black, with a slight reddish tint; showing a few minute glimmerings of spar; having a somewhat prismatic structure; adhering slightly to the tongue; powder dark brown, nearly black; calcined powder nearly black.

*Composition*, dried at 212° F.—

| | | |
|---|---|---|
| Oxide of iron, | 73.34 | = 51.36 per cent. of *Iron*. |
| Brown oxide of manganese, | 9.41 | |
| Alumina, | .27 | |
| Phosphoric acid, | .36 | |
| Carbonate of lime, | 1.27 | |
| Magnesia, | .83 | |
| Potash, | .40 | |
| Soda, | .03 | |
| Silex and insoluble silicates, | 4.55 | |
| Combined water and loss, | 9.54 | |
| | 100.00 | |

The air-dried ore lost 2.00 per cent. of *moisture* at 212° F.

This mineral owes its dark color, and its property of becoming darker on calcination, to the presence of a large proportion of oxide of manganese. This ingredient in the ore is generally supposed to cause the production of iron which is the best adapted to the manufacture of steel. The alloy of manganese with iron is believed to give it greater firmness and hardness; and the celebrated Swedish chemist, Berzelius, states that the best varieties of steel owe their good qualities partly to the manganese contained in them. It has been found, however, by the careful experiments of Karsten and others, that although the ores containing manganese are the best for the production of iron for making steel, yet some of the best specimens of cast-steel obtained from ores containing oxide of manganese, are destitute of this metal in any notable quantity.

In the smelting of manganesic iron ores there is a great tendency to the production of hard, brittle, *white* iron; not so much because the metal manganese, by its combination with the iron, communicates to it these qualities, but because the oxide of manganese, forming a very fusible slag with the silica in the high furnace, facilitates the reduction and fusion of the iron at a comparatively low temperature, and thus, incidentally prevents the separation of carbon in the form of graphite, which is necessary to the formation of soft grey iron. It thus favors the production of a *pure* hard metal, fitted for the manufacture of steel. Grey soft iron can, however, be produced from manganesic iron ores, either by increasing the heat in the furnace, or by the addition of earthy materials, to counteract the too great fluxing influence of the oxide of manganese, and make the cinder less fusible.

No. 101. (*See former report.*) *Main Ashland coal, above the Clay parting, Greenup county, Ky.*"

This coal, the proximate analysis of which is given on page 318 of the previous report, has been submitted to ultimate analysis. The result of four several operations is as follows, viz:

*Composition*, dried at 212° F.—

| | |
|---|---|
| Carbon, | 79.091 |
| Hydrogen, | 5.111 |
| Sulphur, | .734 |
| Ashes, | 4.000 |
| Oxygen, nitrogen, and loss, | 11 064 |
| | 100.000 |

GRAYSON COUNTY.

No. 408—Limonite. *Labeled "Iron ore, ascending the table land between Cancy and Little Clifty creeks, Grayson county, Ky.*"

A dull friable mineral; adhering to the tongue; presenting various shades of dull red and yellow, in irregular concentric layers; powder light yellowish-brown.

*Composition*, dried at 212° F.—

| | | |
|---|---|---|
| Oxide of iron, | 63.60 | = 44.54 per cent. of *Iron*. |
| Alumina, | 2.36 | |
| Brown oxide of manganese, | .87 | |
| Phosphoric acid, | .89 | |
| Carbonate of lime, | .27 | |
| Magnesia, | 1.22 | |

| | |
|---|---|
| Potash, - - - - - | .25 |
| Silica and insoluble silicates, - | 19.15 |
| Combined water, - - - | 12.02 |
| | 100.63 |

The air-dried ore lost 1.00 per cent. of *moisture*, at 212° F.

No. 456—MAGNESIAN LIMESTONE. *Labeled "Hydraulic Limestone, two miles west of Grayson Springs. (Used for grave stones.")*

A dull, fine granular, light-grey limestone, with a slight tint of greenish; exhibiting a few minute spangles of mica; adhering slightly to the tongue.

| | | |
|---|---|---|
| Specific gravity, - - - - - - - - | | 2.651 |
| *Composition*, dried at 212° F.— | | |
| Carbonate of lime, - - - | 46.83 | = 26.28 *Lime.* |
| Carbonate of magnesia, - - | 26.84 | = 12.96 *Magnesia.* |
| Carbonate of iron, - - - | 3.44 | |
| Brown oxide of manganese, a trace. | | |
| Alumina, - - - - - | .38 | |
| Phosphoric acid, - - - | .12 | |
| Sulphuric acid, - - - | .33 | |
| Potash, - - - - - - | .50 | |
| Soda, - - - - - - | .37 | |
| Silica and insoluble silicates, - | 20.78 | |
| Loss, - - - - - - | .41 | |
| | 100.00 | |

The air-dried rock lost 0.50 per cent. of *moisture* at 212°.

The hydraulic properties of this limestone were not tried at the laboratory.

## HANCOCK COUNTY.

No. 468—COAL. *Labeled "First bed above the Hawesville main coal, under fossiliferous shale, Hancock county, Ky."*

A jet-black coal; specimen tarnished on the surface as though it had been exposed to the weather; seperates in thin layers, which show some fibrous coal on their surfaces, but no pyrites.

Specific gravity, - - - - - - - - - - 1.282

Heated over the spirit-lamp it did not decrepitate; softened and swelled somewhat, but the fragments did not agglutinate; burnt with a smokey flame, leaving a somewhat cellular coke; a splint coal.

*Proximate Analysis.*

| | | | |
|---|---|---|---|
| Moisture, - - - - - | 6.50 | Total volatile matters, - | 41.40 |
| Volatile combustible matters, - | 34.90 | | |
| Carbon in the coke, - - | 53.20 | Dense coke, - - | 58.60 |
| Ashes, (grey-purple,) - - | 5.40 | | |
| | 100.00 | | 100.00 |

The per centage of *sulphur* in the undried coal is 0.47.

The composition of the ashes is as follows:

| | |
|---|---|
| Silica, - - - - - - - - - - - - - | 1.38 |
| Alumina, oxide of iron, &c., - - - - - - - | 2.78 |
| Lime, - - - - - - - - - - - - - | .38 |
| Magnesia, - - - - - - - - - - - | .17 |
| Loss, - - - - - - - - - - - - - | .69 |
| | 5.40 |

The *ultimate* composition of this coal, dried at 212°, was found to be as follows:

| | |
|---|---|
| Carbon, - - - - - - - - - - | 73.255 |
| Hydrogen, - - - - - - - - - - | 5.155 |
| Sulphur, - - - - - - - - - - | .520 |
| Ashes, - - - - - - - - - - - | 5.600 |
| Oxygen, nitrogen, and loss, - - - - - - - | 15.470 |
| | 100.000 |

Like the splint-coals in general, or the so-called *dry* coals, this contains a considerable proportion of oxygen in its composition. The proportion of the nitrogen was not ascertained, but it rarely exceeds two per cent. in coals.

No. 519—COAL. *Labeled "Thirty-three inch coal, fifteen feet below the surface, in Judge Mayhall's shaft, Hancock county, Ky."*

A compact coal, having somewhat the appearance of jet; breaking with a conchoidal fracture in the direction of the layers; not soiling the fingers; some appearance of pyrites, but no fibrous coal. Heated over the spirit-lamp it softened, swelled up, and agglutinated somewhat, and left a spongey coke.

Specific gravity, - - - - - - - - 1.392

*Proximate Analysis.*

| | | | |
|---|---|---|---|
| Moisture, - - - - - | 3.00 | Total volatile matters, | 42.10 |
| Volatile combustible matters, - | 39.10 | | |

| | | | |
|---|---|---|---|
| Carbon in the coke, | 45.40 | Bright, inflated coke, | 57.90 |
| Ashes, (purple-grey,) | 12.50 | | |
| | 100.00 | | 100.00 |

The ashes were found to consist of

| | |
|---|---|
| Silica, | 2.99 |
| Alumina, oxide of iron, &c., | 9.23 |
| Magnesia, | .24 |
| Trace of lime and loss, | .04 |
| | 12.50 |

As the ashes contain but a very small proportion of magnesia, and only a trace of lime, they will require quite a high temperature to fuse them into clinker.

Submitted to *ultimate analysis* this coal gave the following results, dried at 212° F., viz:

| | |
|---|---|
| Carbon, | 63.436 |
| Hydrogen, | 4.622 |
| Sulphur, | 5.866 |
| Ashes, | 13.600 |
| Oxygen, nitrogen, and loss, | 12.476 |
| | 100.000 |

The large proportion of sulphur and of earthy matter in this coal are serious drawbacks on its value. It is probable, however, that the coal may vary as to both these ingredients in other parts of the bed.

No. 520—Coal. *Labeled "Out-crop of coal on Mr. Pate's land, one and a half miles north-west of the house, on the Hardinsburg road, Hancock county, Ky."*

A dull looking, very friable coal, presenting the appearance of having been much weathered; surfaces and seams covered with ochreous incrustation; some fibrous coal between the layers, but no appearance of pyrites. Over the spirit-lamp it swelled up somewhat, burnt with a smokey flame, but the fragments did not agglutinate; probably not a coking coal.

Specific gravity, - - - - - - - - - 1.266

*Proximate Analysis.*

| | | | |
|---|---|---|---|
| Moisture, | 6.30 | Total volatile matters, | 46.10 |
| Volatile combustible matters, | 39.80 | | |

| | | | |
|---|---|---|---|
| Carbon in the coke, - - | 51.40 | Dense coke, - - | 53.90 |
| Ashes, (red-brown,) - - | 2.50 | | |
| | 100.00 | | 100.00 |

The analysis of the ashes is as follows:

| | |
|---|---|
| Silica, - - - - - - - - - - - | 0.49 |
| Alumina, oxide of iron, &c., - - - - - - - - | 1.70 |
| Lime, - - - - - - - - - - - | .30 |
| Magnesia, - - - - - - - - - - - | .10 |
| | 2.59 |

The ultimate composition of this coal was found to be as follows, dried at 212° F:

| | |
|---|---|
| Carbon, - - - - - - - - - - | 75.328 |
| Hydrogen, - - - - - - - - - - | 5.600 |
| Sulphur, - - - - - - - - - - | .890 |
| Ashes, - - - - - - - - - - - | 2.300 |
| Oxygen, nitrogen, and loss, - - - - - - - - | 15.882 |
| | 100.000 |

No. 243—Coal. *Labeled "Breckinridge Cannel Coal, Hancock county, Ky."*

This interesting coal, of which the results of some proximate analyses by Dr. Owen, are given on page 177 of his former report, has been submitted to new examinations in this laboratory. Dull black, with a satiny lustre on the cross fracture; very tough, breaking with great difficulty; cleaving into thin layers; does not soil the fingers; considerable appearance of fine particles of pyrites, but no fibrous coal between the layers. Over the spirit-lamp burns with a yellow smokey flame; the fragments soften a little, but do not swell, alter their form much, nor agglutinate; powder brownish-black.

Specific gravity, - - - - - - - 1.318

On repeating the proximate analysis of this coal the following results were obtained, viz:

| | | | |
|---|---|---|---|
| Moisture, - - - - - | 1.30 | Total volatile matter, | 55.70 |
| Volatile combustible matters, - | 54.40 | | |
| Carbon in the coke, - - | 32.00 | Scarcely coherent coke, | 44.30 |
| Ashes, (umber colored,) - | 12.30 | | |
| | 100.00 | | 100.00 |

On examining different portions of the mass, a large piece about five inches thick, which had been sent for analysis, a considerable difference as to the proportion of ashes, &c., was found to exist—for example: the proportion of *total volatile matters* was found to vary from 55.70 to as high as 71.70 per cent.; of *coke* from 28.30 to 44.30; and of *ashes* from 7. to 12.30 per cent., in the undried coal.

The per centage of sulphur ascertained on one specimen was 2.443 in the air-dried coal.

The composition of the ashes is as follows:

| | |
|---|---|
| Silica, | 3.49 |
| Alumina and oxide of iron, | 7.78 |
| Lime, | .55 |
| Magnesia, | .39 |
| | 12.21 |

By *ultimate* analysis this coal was found to contain

| | |
|---|---|
| Carbon, | 68.128 |
| Hydrogen, | 6.489 |
| Sulphur, | 2.476 |
| Nitrogen, | 2.274 |
| Oxygen and loss, | 5.833 |
| Ashes, | 14.800 |
| | 100.000 |

It will be seen, on making the comparison, that this coal contains a larger proportion of hydrogen and less oxygen than any other of the Kentucky coals hitherto examined. The only other coal which approaches it in this respect is the cannel coal from Haddock's mine, Owsley county, (which see,) which resembles it also in yielding, by destructive distillation, a notable quantity of oils and wax-like matter. There are few coals in the world, so far as yet reported in the journals and works of science, which equal these coals in these characteristics. One of the most noted of these is the Boghead Cannel Coal of Scotland, used extensively for the production of Benzole, illuminating and lubricating oils, and *Paraffine;* of which, for the sake of comparison, we append the *ultimate* composition, as quoted in Liebig and Kopps *Jahresbericht* for 1851, S. 733, from Russell's analysis:

*Ultimate Composition of the Boghead Coal, of Scotland.*

| | |
|---|---|
| Carbon, | 65.34 |
| Hydrogen, | 9.12 |
| Sulphur, | 0.15 |
| Nitrogen, | 0.71 |
| Oxygen, | 5.46 |
| Ashes, | 18.68 |
| | 99.46 |

While the Boghead coal contains a larger proportion of ashes than the Breckinridge coal, the latter contains a much larger quantity of both *sulphur* and *nitrogen.* The Boghead coal also excels this greatly in its proportion of hydrogen to the carbon, approaching thus more nearly than that to the nature of the bitumens. Indeed, the question has been mooted in Europe whether this and similar minerals are stone coals or real *bitumens;* and Geuther* has decided, from the nature of the products of distillation, and of the ashes of the Boghead coal, as well as by its microscopic analysis, that it is simply a bituminous shale or slate.

Abstracting the ashes and sulphur, the composition of the Breckinridge and Boghead coals compare as follows:

| | *Breckinridge Coal.* | *Boghead Coal.* |
|---|---|---|
| Carbon, | 82.355 | 80.487 |
| Hydrogen, | 7.844 | 11.235 |
| Nitrogen, | 2.749 | .874 |
| Oxygen, | 7.051 | 6.726 |

By means of the tables which will be appended to the end of this report, a comparison of the composition of the Kentucky coals can be made.

Eight different trials were made of the distillation of the Breckinridge coal for the production of oils, &c. The operation was generally performed in an iron retort, gradually heated to a moderate red-heat; the products were collected by means of a series of receivers and a graduated bell glass over water to measure the gas obtained. The first receiver was usually kept at a temperature of from 260° to 285° F., by means of a chloride of calcium bath; the second receiver was kept in boiling water; the third was simply exposed to the ordinary temperature of the room, and the last was enveloped with ice, or a mix-

*Liebig and Kopp's Jahresbericht for 1855. S. 896.

ture of ice and salt; the gas, before it was collected, was generally passed through potash, or wash bottles containing Hydrochloric acid, solution of Arsenious acid, and basic acetate of lead, severally. Under these circumstances it was found that a slow gradual application of the heat favored the production of the oily products, and diminished the relative amount of gas.

The first receiver contained a dark-brown tarry product, which became a soft solid on cooling; the second and third receivers contained thinner oils of a light brown color, floating on a strong ammoniacal water, which contained much sulphuret of ammonium, and some little sulphocyanide of ammonium; whilst the last receiver, which was cooled with ice, condensed a clear light-yellow volatile oil—principally Benzole—and besides ammoniacal water, contained limped crystals of bi-carbonate of ammonia. The arsenical and lead solutions showed the presence of abundance of sulphuretted hydrogen, and considerable carbonic acid; and the gas collected had pretty high illuminating powers.

Examined by Dr. Ellett's process—by the action of Bromine vapor—some of the gas was found to contain as much as 9. per cent. of olefiant gas and hydrocarbon vapors.

The products which were separated by this process of *fractional distillation* were not so pure as to induce us to recommend it to the manufacturer on a large scale, for the reason, probably, that some of the more fixed oils were carried forward into the latter receivers in the series, by the mechanical action of the gas, which was continually passing through them; yet the use of a series of receivers may facilitate the subsequent processes for purification. The clear, bright-yellowish, thin, oily matter which passed into the fourth receiver, became gradually brownish after exposure for a few days to the light, after the manner of imperfectly purified Benzole. These products have not as yet been analyzed to ascertain the relative proportions of *Paraffine, Eupione, Benzole, &c., &c.;* they are, indeed, of a very complex nature, containing, besides several neutral hydrocarbons, a number of organic bases and acids. When collected together in one cooled receiver they appear as a fluid dark-colored "*crude oil.*"

This "crude oil," which is produced at the Breckinridge coal and oil company's works, near Cloverport, in the quantity of about 6,000 gallons per week, is manufactured by distillation and purification into various commercial products: as, the *Benzole*, which, from its volatility and combustibility, is employed, mixed with alcohol, as a burning fluid, or used in the form of vapor as a substitute for common illuminating gas; *Naptha* employed as a solvent for Caoutchouc, Gutta Percha, &c.; *illuminating and lubricating oils* as good for these purposes as spermaceti oil; *Paraffin*, a substance resembling spermaceti, obtained from the coal in the proportion of only about one per cent., used for burning in the form of candles, and for giving a finish to some kind of leather; and a residuary black substance used as asphaltum.

This new manufacture, in view of the increasing scarcity of spermaceti oil, is of very great value to the whole country, and will probably be expanded to a great extent.

It has generally been believed that no other than the Breckinridge Cannel Coal could be profitably used, in this country, for this purpose; but, doubtless, amongst the Cannel Coals and Bituminous Schists of our state, some may be found which may exceed the Haddock's Cannel Coal, and equal the Breckinridge coal in this particular.

To ascertain whether the proportion of the oily products might be increased by the use of sur-heated steam, instead of simple dry-heat applied to the coal, an apparatus for the purpose was constructed. The coal, introduced in a semi-cylindrical tray, into a tubular iron retort, was subjected to the action of steam, which had passed through tubes kept heated in the fire—receivers being attached, as above described. The results obtained did not, however, show any great advantage of this over the simple application of heat to the retort containing this coal. It was, indeed, somewhat difficult to regulate the heat of the tubes, and thus it is probable the steam was used at too high a temperature in the experiments. The results of the eight operations, as far as noted, are as follows, calculated to 1,000 grains of the undried coal:

<table>
<tr><th>Experiments.</th><th>Crude oil.</th><th>Ammoniacal water.</th><th>Coke.</th><th>Gas.</th><th>Weight of gas, and loss.</th></tr>
<tr><td>1st, - - - - -</td><td>334.10</td><td>66.10</td><td>468.7</td><td>465. cubic inches,</td><td>131.1 grains.</td></tr>
<tr><td>2nd, coal dried at 600°,</td><td>290.00</td><td>43 50</td><td>470.</td><td>425. cubic inches,</td><td>196.5</td></tr>
<tr><td>3rd, - - - - -</td><td colspan="2">400.</td><td>460.</td><td>—</td><td>140.</td></tr>
<tr><td>4th, - - - - -</td><td>–</td><td>–</td><td>450.</td><td>—</td><td>–</td></tr>
<tr><td>5th, very slow, - -</td><td>348.80</td><td>37.50</td><td>471.</td><td>—</td><td>142.7</td></tr>
<tr><td>6th, - - - - -</td><td>300.</td><td>61.30</td><td>437.</td><td>—</td><td>201.2</td></tr>
<tr><td>7th, with steam, - -</td><td colspan="2">427.5</td><td>412.</td><td>—</td><td>160.5</td></tr>
<tr><td>8th, - - - - -</td><td>–</td><td>–</td><td>464.</td><td>—</td><td>–</td></tr>
<tr><td>Average, - -</td><td>318.2</td><td>52.10</td><td>455.</td><td>445.</td><td>161.8</td></tr>
</table>

This average yield of crude oil corresponds nearly with that given by Dr. Owen in his former report, as the result of his experiments, and verifies the extraordinary fact that this singular coal, when submitted to slow distillation below a bright-red heat, will give almost one-third of its weight of *oily matters*, besides yielding more than 45. per cent. of *coke*, and good illuminating gas in the proportion of nearly two cubic feet to the avoirdupois pound. It will be sufficiently near the results obtained to sum up the per centage of the products of the Breckinridge coal as follows:

| | | |
|---|---|---|
| Crude oil, - - | 32. | per cent. |
| Ammoniacal water, - | 5.5 | per cent. |
| Coke, - - - | 45.5 | |
| Gas and loss, - | 17. | equal to 2227. cubic inches to the pound avoirdupois. |
| | 100.0 | |

In consequence of the large proportion of nitrogen in this coal the ammoniacal liquor is unusually strong, and might be used to yield ammonia and its salts; it also contains much sulphur, of which the coal has a very large amount. The gas which is produced, therefore, has a large admixture of sulphuretted hydrogen, and, if used for illuminating purposes, must be purified with more than the usual care from this injurious and offensive ingredient. But when the object of the manufacturer is simply to obtain the oily products and the *paraffine*, the gas produced in the operation might be economically used under the retorts, or in the processes of re-distillation and purification of these products.

As stated above, the only Kentucky coal hitherto examined, which resembles the Breckinridge in its composition, (particularly in its proportion of hydrogen,) and its yield of oily products on distillation, is the cannel coal from Haddock's mine, Kentucky river.

The other coals which were submitted to distillation did not yield enough of fluid matter to make them at all valuable for this purpose.

Under this head, for convenience of reference, I will give the comparative results obtained from the several coals examined, including a good specimen of Pennsylvania coal, (Youghiogheny coal,) used by the Lexington gas company as the best adapted to their purposes.

| Coals. | Crude oils. | Ammoniacal water. | Coke. | Gas, (cubic inches.) |
|---|---|---|---|---|
| Breckinridge cannel, - - - | 318.20 | 52.10 | 455. | 445 good. |
| Haddock's cannel, - - - | 248.50 | 54 50 | 589. | 370. very good. |
| Union Company's coal, bottom part, | 148. | 38. | 750. | 465. very good. |
| Mulford's five-foot, or Main coal, - | 136.50 | 64 75 | 684. | 567. very good. |
| Robert's, or Muddy river coal, - | 102.10 | 119.80 | 659.50 | 370. good. |
| Ice house coal, - - - | 108. | 73. | 714. | 465. very good. |
| Youghiogheny coal, - - - | 136. | 52. | 710. | 545. very good. |

These results are calculated to 1,000 grains of each of the coals in the air-dried condition.

The low temperature at which the distillation was carried on is unfavorable to the production of much gas, as is proved by the fact, that in the ordinary course of the manufacture of illuminating gas, from the Youghiogheny coal, fully twice as much is obtained from it as was procured in our slower process. But as all these coals were submitted, as nearly as possible, to the same temperature, in the above described experiments, it is believed that the relative proportions and quality of the gas obtained from them would hold good also under conditions of heat more favorable for the formation of gaseous products. If this be true, the Mulford's main coal, and the Ice-house coal, and Union Company's coal, will prove to be as good, or nearly as good, for gas and coke as the best Pennsylvania bituminous coal; but with the drawback that they contain a larger proportion of sulphur. For the composition of these several coals we refer to their descriptions under their appropriate counties.

## HOPKINS COUNTY.

No. 463—Coal. *Labeled "Hall's coal, Clear creek, Hopkins county, Kentucky."*

A shining pitch-black coal; not very hard; dividing into thin layers separated by fibrous coal, on which there were some microscopical appearance of pyrites. Over the spirit-lamp it softened and swelled up a good deal, and the fragments became agglutinated into a light cellular coke. Probably a coking coal.

Specific gravity, - - - - - - - - - 1.277

*Proximate Analysis.*

| | | | |
|---|---|---|---|
| Moisture, - - - - - | 3.20 | Total volatile matters, - | 38.60 |
| Volatile combustible matters, - | 35.40 | | |
| Carbon in the coke, - - | 57.80 | Inflated coke, - - | 61.40 |
| Ashes, (reddish-grey,) - - | 3.60 | | |
| | 100.00 | | 100.00 |

The composition of the ashes is as follows:

| | |
|---|---|
| Silica, - - - - - - - - - - - - | 1.59 |
| Alumina and oxide of iron, - - - - - - - | 1.58 |
| Lime, a trace. | |
| Magnesia, - - - - - - - - - - | .10 |
| Loss, - - - - - - - - - - - | .33 |
| | 3.60 |

The coal, on *ultimate analysis*, dried at 212°, was found to contain

| | |
|---|---|
| Carbon, - - - - - - - - - - | 75.491 |
| Hydrogen, - - - - - - - - - - | 5.088 |
| Sulphur, - - - - - - - - - - | 1.520 |
| Oxygen, nitrogen, and loss, - - - - - - - | 14.101 |
| Ashes, - - - - - - - - - - - | 3.800 |
| | 100.000 |

This appears to be quite a good coal, with a small proportion of ashes, containing, however, a rather more than an average quantity of sulphur.

No. 465—Coal. *Labeled "Mr. Samuel's coal, two and a half feet thick, Hopkins county, Ky."*

A dull looking coal, with the appearance of having been weathered; separating easily into thin layers; oxide of iron, as from decomposed pyrites, on the surfaces of the seams. Over the spirit-lamp it decrepitates, and burns with a smokey flame; some of the fragments soften and swell a little, but most of them retain their original form; coke easily burnt to ashes; a splint coal.

Specific gravity, - - - - - - - - - 1.422

*Proximate Analysis.*

| | | | |
|---|---|---|---|
| Moisture, - - - - - | 5.00 | Total volatile matters, - | 33.40 |
| Volatile combustible matters, - | 28.40 | | |

| | | | |
|---|---|---|---|
| Carbon in the coke, - - | 53.50 | Coke, (not adherent,) - | 66.60 |
| Ashes, (reddish-grey,) - - | 13.10 | | |
| | 100.00 | | 100.00 |

The ashes were found to be composed of

| | |
|---|---|
| Silica, - - - - - - - - - - - - | 7.19 |
| Alumina and oxide of iron, - - - - - - - | 5.68 |
| Lime, - - - - - - - - - - - - | .05 |
| Magnesia, - - - - - - - - - - - | .06 |
| Loss, - - - - - - - - - - - - | .12 |
| | 13.10 |

The *ultimate composition* of this coal, dried at 212°, is as follows:

| | |
|---|---|
| Carbon, - - - - - - - - - - - | 66.000 |
| Hydrogen, - - - - - - - - - - - | 4.244 |
| Sulphur, - - - - - - - - - - - | .820 |
| Oxygen, nitrogen, and loss, - - - - - - - | 13.436 |
| Ashes, - - - - - - - - - - - - | 15.500 |
| | 100.000 |

The large proportion of earthy matter in this coal considerably diminishes its value; but, as the ashes contain but very small quantities of lime and magnesia, they will not be likely to fuse into clinker, except at an exceedingly high temperature. The specimen examined appeared to have been taken from the *out-crop* of the coal; it is possible that the interior portion may be more pure, although it is likely to contain rather more sulphur.

No. 135—COAL. *Labeled "Wright's Mountain Coal, Townes and Kirkwell, Hopkins county, Ky."*

This coal, of which the description and proximate analysis are given on page 339 of the former report, has been submitted to *ultimate analysis* with the following results, viz:

| | |
|---|---|
| Carbon, - - - - - - - - - - - | 77.400 |
| Hydrogen, - - - - - - - - - - - | 4.999 |
| Sulphur, - - - - - - - - - - - | *1.060 |
| Nitrogen, - - - - - - - - - - - | 1.620 |
| Oxygen and loss, - - - - - - - - - | 12.521 |
| Ashes, - - - - - - - - - - - - | 2.400 |
| | 100.000 |

*Erroneously printed in the former report 0.106.

JEFFERSON COUNTY.

No. 521—HYDRAULIC LIMESTONE (UNBURNT.) "*From the Falls of the Ohio river at Louisville, Jefferson county, Ky.*"

A greenish-grey, dull, fine granular limestone; adheres slightly to the tongue; powder light-grey.

*Composition*, dried at 212° F.—

| | | |
|---|---|---|
| Carbonate of lime, | 50.43 | = 28 29 *Lime.* |
| Carbonate of magnesia, | 18.67 | = 8.89 *Magnesia.* |
| Alumina, and oxides of iron and manganese, | 2.93 | |
| Phosphoric acid, | .06 | |
| Sulphuric acid, | 1.58 | |
| Potash, | .32 | |
| Soda, | .13 | |
| Silica and insoluble silicates, | 25.78 | Silica, 22.58; Alumina colored with oxide of iron, 2.88; Lime, magnesia & loss, .32; = 25.78 |
| Loss, | .10 | |
| | 100.00 | |

The air-dried rock lost .70 per cent. of *moisture*, at 212° F.

The analysis of this well-known water-lime will serve for comparison with that of other limestones supposed to possess *hydraulic* qualities.

No. 522—SOIL. *Labeled "Virgin Soil, from E. B. O'Bannon's farm, O'Bannon's station, overlying cellular magnesian limestone, of the Upper Silurian Formation, twelve miles from Louisville, Jefferson county, Ky.*"

Dried soil of a grey-brown color; some small rounded particles of iron ore noticed in it. As this and the following soils were received just before this report was made up there was not time for digestion in water containing carbonic acid, to ascertain the relative amount of matters soluble in that menstruum; they were therefore submitted to ordinary analysis, dried at 370 F.

The composition of this soil is as follows:

| | |
|---|---|
| Organic and volatile matters, | 7.996 |
| Alumina, and oxides of iron and manganese, | 7.480 |
| Carbonate of lime, | .394 |
| Magnesia, | .240 |
| Phosphoric acid, | .205 |
| Sulphuric acid, | .082 |
| Potash, | .200 |
| Soda, | .043 |
| Sand and insoluble silicates, | 83.134 |
| Loss, | .226 |
| | 100.000 |

The air-dried soil lost 4.42 per cent. of *moisture* at 370°.

No. 523—Soil. *Labeled "Soil from an old field, over cellular magnesian limestone of the Upper Silurian Formation, which lies from six to twelve feet beneath the surface. Has been from twenty-five to thirty years in cultivation; E. B. O'Bannon's farm. (Would it be a good soil for the cultivation of the grape?)"*

Color of the dried soil light greyish-brown; lighter than the preceding.

*Composition*, dried at 400° F.—

| | |
|---|---|
| Organic and volatile matters, | 4.506 |
| Alumina, and oxides of iron and manganese, | 6.240 |
| Carbonate of lime, | .316 |
| Magnesia, | .200 |
| Phosphoric acid, | .191 |
| Sulphuric acid, | .067 |
| Potash, | .158 |
| Soda, | .070 |
| Sand and insoluble silicates, | 88.318 |
| | 100.000 |

The air-dried soil lost 2.80 per cent. of *moisture*, at 400° F.

By comparison of the two preceding analyses it will be seen what the soil, which has been in cultivation for twenty-five to thirty years, has lost of its original value: *First.* It has lost *organic and volatile matters*, which is evinced also in its lighter color, and in the smaller quantity of *moisture* which it is capable of holding at the ordinary temperature, but which was driven off at the heat of 400°. These organic matters absorb and retain moisture with great power. Besides the nourishment which organic matters in the soil give directly to vegetables, by their gradual decomposition and change, these substances also greatly increase the solubility of the earthy and saline ingredients in the soil, which are necessary to vegetable growth. *Second.* It has lost *some of every mineral ingredient of the soil which enters into the vegetable composition:* as lime, magnesia, oxide of iron, phosphoric acid, sulphur, and the alkalies. The only apparent exception to this is in the greater proportion of soda in the old soil than in the virgin soil. This increase may have been occasioned by the ordinary free use of salt on the farm, and its transfer to the cultivated field by the animals feeding on it. It will be seen, in the *third* place, that the propor- of *alumina and oxide of iron* to the *sand and silicates* is smaller in the soil of the old field than in the virgin soil, cultivation having, perhaps, favored the washing down into the sub-soil those ingredients which are the most readily transported by water. To renovate this field to its original state would require the application of ordinary barn-yard manure, which contains all the ingredients which have been removed from it except the *alumina, and oxides of iron and manganese.* To supply these, if it be deemed desirable, the *red sub-soil found on the washed slopes* of the old field, presently to be described, would answer very well, applied as a top dressing; but the immediate sub-soil, next to be described, does not, by its analysis, promise to be of any service in this or in any other respect.

*Would this be a good soil for the cultivation of the grape?* If it has sufficient drainage to prevent the habitual lodgement of water in the sub-soil, there is nothing in the composition of the soil to forbid its use for this purpose. The soil which will produce good indian corn will generally produce the grape. The vine requires for its growth, and the production of its fruit, precisely the same mineral ingredients which are necessary to every other crop which may be produced on the soil, differing in this respect from them only in the proportion of these sevral ingredients. The juice of the grape contains a considerable proportion of *potash,* much of which is deposited in the wine-cask, after fermentation, in the form of tartar, (acid tartrate of potash,) and which must be supplied to the growing vine from the soil to enable it to produce the grape. It has hence been generally believed that vineyard culture tends speedily to exhaust the soil of its alkalies, unless they are habitually re-applied in manures. This is true in regard to every *green* crop which is carried off the ground: as hay, turnips, potatoes, and especially tobacco, and the fruits of the orchard; whilst the indian corn and other grains carry off less of the alkalies they also require and remove them in considerable proportion; and Boussingault, of France, has arrived at the conclusion, from his experiments on his vineyard of 170 acres, in Alsace, that the grape does not remove any more of the valuable mineral substances from the soil, annually, than the ordinary grain and root crops.

The following tabular view of his results, from an equal surface of ground to each crop, is given in Liebig and Kopp's *Jahresbericht fur* 1850.

| Removed by— | | Potash. | Soda. | Lime. | Magnesia | Phosphoric acid. | Sulphuric acid. |
|---|---|---|---|---|---|---|---|
| The vine. | Wine, | 11.53 | 0.13 | 17.48 | 3.91 | 6.66 | 1.02 |
| | Husks, | 12.07 | 0.13 | 3.50 | 0.72 | 3.50 | 1.77 |
| | Small wood, | 4.64 | – | 0.51 | 0.95 | 2.27 | 0.53 |
| Total, | | 28.24 | 0.26 | 21.49 | 5.58 | 12.43 | 3.32 |
| Potatoes, | | 107.1 | | – | – | 23.8 | – |
| Beets, | | 153.0 | | – | – | 20 4 | – |
| Wheat with straw, | | 45.9 | | – | – | 323.0 | – |

The leaves of the vine were not taken into the estimation, because they fall and decay on the soil.

The quantity of grape-juice produced, per acre, is greater in this county than that obtained in the vineyards of France, but in the above figures, if they are to be taken as correct data, there is a wide margin for increase. The great reputation of Boussingault as an accurate analyst and observer must be the guarantee of their correctness. In corroboration of those facts are the more recent analyses, by Berthier, of the fruit and wine, stems, and leaves of the vine, showing that the great demand made upon the soil for alkali is not so much for the *grapes*, but for growth of the *wood* and the *leaves*, so that, if these are not removed, the crop does not prove inordinately exhaustive.

To return to the two comparative soil analyses. The difference between the proportions, of the valuable ingredients of the two above stated, may seem quite unimportant on a superficial examination, but when we apply these differences to the more than *three millions of pounds of soil* which are contained in an acre of ground, calculated only to the depth of one foot, we may see their significance. Thus the

the *potash* in the virgin soil is in proportion of 0.200 per cent., and in the soil of the *old field* in that of 0.158. This proportion gives *six thousand pounds* of potash to the acre of earth, one foot deep, in the *new soil,* and *four thousand seven hundred and forty pounds* only into the *old,* showing, that if the old soil was originally like the neighboring virgin soil, it has lost, amongst other ingredients, as much as one thousand two hundred and sixty pounds of potash from the acre, within one foot of the surface only. To restore to it this amount of alkali alone would require the application of a large amount of ordinary manure.

No. 524—SUB-SOIL. *Labeled "Sub-soil, seven to twelve inches under the surface, old field twenty-five to thirty years in cultivation, over cellular magnesian limestone of the Lower Silurian Formation, E. B. O'Bannon's farm, Jefferson county, Ky."*

Color of the dried soil light greyish-brown.

*Composition,* dried at 400° F.—

| | |
|---|---|
| Organic and volatile matters, | 2.844 |
| Alumina, and oxides of iron and manganese, | 6.335 |
| Carbonate of lime, | .256 |
| Magnesia, | .226 |
| Phosphoric acid, | .099 |
| Sulphuric acid, | .082 |
| Potash, | .181 |
| Soda, | .028 |
| Sand and insoluble silicates, | 89.900 |
| Loss, | .049 |
| | 100.000 |

The air-dried sub-soil lost 2.98 per cent of *moisture* at 400° F.

By the examination of this upper sub-soil it does not appear that any of the valuable ingredients of the surface-soil have lodged in it. It contains, it is true, more *potash,* and has less organic matter, but in other respects does not materially differ from the upper soil. A greater difference may be seen in the *deeper* sub-soil, the analysis of which will next be given.

No. 525—SUB-SOIL. *Labeled "Red sub-soil, on the washed slopes of an old field, found almost universally a few feet under the surface, E. B. O'Bannon's farm, Jefferson county, Ky."*

Color of the dried soil light brick-red; it contains some small nodules of iron ore.

*Composition*, dried at 400° F.—

| | |
|---|---|
| Organic and volatile matters, | 3 112 |
| Alumina, and oxides of iron and manganese, | 17.020 |
| Carbonate of lime, | .194 |
| Magnesia, | .366 |
| Phosphoric acid, | .497 |
| Sulphuric acid, | .088 |
| Potash, | .297 |
| Soda, | ·111 |
| Sand and insoluble silicates, | 77.434 |
| Loss, | .881 |
| | 100.000 |

The air-dried *sub-soil* lost 3.60 per cent. of *moisture*, at 400° F.

No. 526—Soil. *Labeled "Soil from a poor point of old field, where gravel iron ore prevails, E. B. O'Bannon's farm, Jefferson county, Ky."*

Color of the dried soil rather lighter than that of the preceding; soft pebbles of iron ore, very dark in appearance when broken.

*Composition*, dried at 380° F.—

| | |
|---|---|
| Organic and volatile matters, | 4.390 |
| Alumina, and oxides of iron and manganese, | 11.849 |
| Carbonate of lime, | .236 |
| Magnesia, | .216 |
| Phosphoric acid, | .126 |
| Sulphuric acid, | .109 |
| Potash, | .239 |
| Soda, | .043 |
| Sand and insoluble silicates, | 82.694 |
| Loss, | .458 |
| | 100.000 |

The air-dried soil lost 3.91 per cent. of *moisture* at 380° F.

The cause of the unproductiveness of this soil lies more in the *state of aggregation* than the composition, as shown by the chemical analysis. The valuable ingredients necessary to vegetable growth are contained in it in at least as large proportions as in the earth from the other portions of the field; but in this there is doubtless a *larger quantity of them locked up in the pebbles of so-called iron ore,* which the fibres of the vegetable roots cannot penetrate. If, by any means, these were to be disintegrated or pulverised, the soil would doubtless be rendered more fertile. Doubtless if these several soils had been digested in the carbonated water this one would have given up much less of *soluble extract* to that menstruum than the others. The iron gravel, diffused through this soil, has been also submitted to analysis.

No. 527—FERRUGINOUS GRAVEL. *Labeled "Gravel of Iron Ore, disseminated in the sub-soil over cellular magnesian limestone, E. B. O'Bannon's farm, Jefferson county, Ky."*

Irregular tuberculated lumps, from the size of a large hickory nut down to that of a mustard seed; easily broken; fracture showing a general dark appearance, like that of peroxide of manganese; some of the lumps presented some included lighter earthy matter like clay; powder of a snuff-brown color. It dissolved in hydrochloric acid with the escape of chlorine. It contained no protoxide of iron, but much oxide of manganese.

*Composition*, dried at 212° F.—

| | |
|---|---|
| Oxide of iron and alumina, | 33.90 |
| Brown oxide of manganese, | 4.28 |
| Carbonate of lime, | .58 |
| Carbonate of magnesia, | 1.22 |
| Alkalies and acids not estimated. | |
| Silex and insoluble silicates, | 58.18 |
| Combined water, | 8.20 |
| Loss, | 1.64 |
| | 100.00 |

Dried at 212° it lost 2.80 per cent. of *moisture.*

No. 528—LIMESTONE. *Labeled "Cellular (Magnesian?) Limestone, found about six to ten feet under the surface of the ground, where the preceding soils were collected, O'Bannon's farm, Jefferson county, Ky. Upper Silurian Formation."*

A light-grey, friable cellular rock, layers and cavities covered with minute crystals.

*Composition*, dried at 212° F.—

| | | |
|---|---|---|
| Carbonate of lime, - - - | 50.76 | = 28.49 *Lime.* |
| Carbonate of magnesia, - - | 45.00 | |
| Alumina, oxides of iron and manganese, and phosphates, | 1.78 | |
| Sulphuric acid, - - - | .04 | |
| Potash, - - - - - - | .21 | |
| Soda, - - - - - - | .35 | |
| Silex and insoluble silicates, - | 2.48 | |
| | 100.62 | |

The air-dried rock lost 0.20 per cent. of *moisture* at 212°.

No. 529—SOIL. *Labeled "Virgin soil, over compact magnesian building stone of the Upper Silurian Formation, White Oak ridge, at Pleasant Grove meeting house, Wm. Galey's farm, Jefferson county, Ky. (This soil is considered not more than one half as productive as that over the cellular magnesian limestone.")*

Dried soil of a dirty grey-buff color.

*Composition*, dried at 400° F.—

| | |
|---|---|
| Organic and volatile matters, - - - - - - - | 3.761 |
| Alumina and oxides of iron and manganese, - - - | 6.952 |
| Carbonate of lime, - - - - - - - - - | .156 |
| Magnesia, - - - - - - - - - - | .240 |
| Phosphoric acid, - - - - - - - - - | .088 |
| Sulphuric acid, - - - - - - - - - | .340 |
| Potash, - - - - - - - - - - | .177 |
| Soda, - - - - - - - - - - | .031 |
| Silex and insoluble silicates, - - - - - - | 88.294 |
| | 100.039 |

The air-dried soil lost 3.22 per cent. of *moisture* at 400°.

Contains less organic matters, phosphoric acid, and alkalies, and a larger proportion of sand and silicates, than the soil over the cellular magnesian limestone.

No. 530—LIMESTONE. *Labeled "Magnesian Building Stone, found under the preceding soil, Upper Silurian Formation, same locality as the last, Jefferson county, Ky."*

A fine grained, light-grey limestone; weathered surfaces having a buff discoloration, with peroxide of iron; under the lens appears to be made up of a mass of pure crystalline grains.

*Composition*, dried at 212° F.—

| | | |
|---|---|---|
| Carbonate of lime, - - | 56.36 | = 31.62 of *Lime*. |
| Carbonate of magnesia, - - | 37.07 | |
| Alumina, oxides of iron and manganese, and phosphates, | 1.28 | |
| Sulphuric acid, a trace. | | |
| Potash, - - - - - - | .33 | |
| Soda, - - - - - - | .35 | |
| Silex and insoluble silicates, - | 5.68 | |
| | 101.07 | |

The air-dried rock lost 0.10 per cent. of *moisture* at 212°.

This is probably a very durable stone; and, in consequence of its very slow disintegration, can communicate very little soluble material to the soil above it. It resembles a good deal, in composition, the magnesian building stone from Grimes' quarry, in Fayette county, which is remarkable for its great durability amongst the rocks of that region.

## LAUREL COUNTY.

No. 406—Impure Carbonate of Iron. *Labeled "Iron Ore, White Oak, Laurel county, Ky., from General Jackson. (Examine for other metals.")*

A dark-grey, fine granular mineral, showing minute spangles of mica, and some incrustation, in parts, with sulphate of lime; weathered surface of yellowish and reddish-brown color.

Specific gravity, - - - - - - - - 3.126

*Composition*, dried at 212° F.—

| | | |
|---|---|---|
| Carbonate of iron, - - - | 32.29 | } = 19.10 per cent. of *Iron*. |
| Oxide of iron, - - - | 5.01 | } |
| Carbonate of lime, - - - | 2.95 | |
| Carbonate of magnesia, - - | 3.60 | |
| Carbonate of manganese, - | .64 | |
| Alumina, - - - - | 1.55 | |
| Phosphoric acid, - - - | 1.00 | |
| Potash, - - - - | .42 | |
| Soda, - - - - - | .01 | |
| Silex and insoluble silicates, - | 51.55 | |
| Organic matter and loss, - | .98 | |
| | 100.00 | |

The air-dried mineral lost 0.30 per cent. of *moisture*, at 212°.

Contains too small a proportion of *iron* to be profitable smelted alone.

No. 410—CARBONATE OF IRON. *Labeled "Iron Ore, Craig's creek, Laurel county, Ky. (Examine for other metals.")*

A dense, dark-grey, fine granular rock; exhibiting some minute spangles of mica; weathered surface dark-reddish and yellowish-brown; powder grey.

| | | |
|---|---|---|
| Specific gravity, | | 3.395 |
| *Composition*, dried at 212° F.— | | |
| Carbonate of iron, | 68.46 | = 35.45 per cent. of *Iron*. |
| Oxide of iron, | 3.41 | |
| Carbonate of lime, | .75 | |
| Carbonate of magnesia, | 3.73 | |
| Carbonate of manganese, | 1.31 | |
| Alumina, | 1.43 | |
| Phosphoric acid, | .52 | |
| Potash, | .34 | |
| Soda, | .07 | |
| Organic matter, | .79 | |
| Silex and insoluble silicates, | 19.65 | |
| | 100.46 | |

The air-dried ore lost 0.40 per cent. of *moisture* at 212°.

A very good iron ore, which could be readily smelted, after roasting, by the aid of the ordinary flux of limestone.

No. 411—CARBONATE OF IRON. *Labeled "Iron Ore, two and a half miles from Mr. Hargal's, Robinson creek, Laurel county, Ky. (Examine for other metals.")*

A dark-grey, dull, fine granular ore, with a shining mineral resembling zinc blend or brown spar, filling some of the small fissures; not adhering to the tongue; powder yellowish-grey.

| | | |
|---|---|---|
| Specific gravity, | | 3.352 |
| *Composition*, dried at 212° F.— | | |
| Carbonate of iron, | 66.01 | = 33.05 per cent. of *Iron*. |
| Oxide of iron, | 2.67 | |
| Carbonate of lime, | 5.85 | |
| Carbonate of magnesia, | 9.19 | |
| Carbonate of manganese, | .86 | |
| Alumina, | .35 | |
| Phosphoric acid, | .63 | |

| | |
|---|---|
| Potash, - - - - - - | .34 |
| Soda, - - - - - - | .33 |
| Silex and insoluble silicates, - | 12.68 |
| Traces of sulphur, zinc, & loss, | 1.09 |
| | 100.00 |

The air-dried ore lost 0.40 per cent. of *moisture* at 212°.

Resembling the preceding, but containing rather less silica and more carbonates of lime and magnesia. This ore would require little or no limestone to flux it; and would most probably yield its iron with facility, without any addition.

No. 224—Soil. *Labeled "Soil of Laurel county, Kentucky, derived from the argillaceous shale and soft sandstone, near the base of the Coal Measures, above the Conglomerate and near the base of the Muriatiferous groupe."*

Color of the dried soil dark-grey; sifted through a seive, having one hundred and sixty-nine apertures to the inch, it left about one-eighth of its weight of fragments of soft reddish and dark-brown ferruginous sandstone. The finer portion, carefully washed with water, left about 42. per cent. of *sand*, of which all but about 5.5 per cent. was fine enough to pass through the finest bolting-cloth. The coarser portion was found, by the lens, to consist principally of rounded and flat fragments of ferruginous sandstone. One thousand grains of the air-dried soil, digested for a month in water containing carbonic acid, gave up nearly two and a half grains of *soluble extract*, of the following composition, dried at 212°:

| | *Grains.* |
|---|---|
| Organic and volatile matters, - - - - - - - | 0.710 |
| Alumina, oxide of iron, and phosphates, - - - - - - | .287 |
| Lime, - - - - - - - - - - - - | .937 |
| Magnesia, - - - - - - - - - - | .066 |
| Brown oxide of manganese, - - - - - - - - | .029 |
| Sulphuric acid, - - - - - - - - - - | .171 |
| Potash, - - - - - - - - - - - | .067 |
| Soda, - - - - - - - - - - - - | .007 |
| Silica, - - - - - - - - - - - - | .130 |
| | 2.404 |

The air-dried soil lost 3.60 per cent. of *moisture* at 400° F.

Dried at which temperature its *composition* is as follows:

| | |
|---|---|
| Organic and volatile matters, | 6.190 |
| Alumina and oxide of iron, | 8.926 |
| Oxide of manganese, | .078 |
| Carbonate of lime, | .116 |
| Magnesia, | .280 |
| Phosphoric acid, | .139 |
| Sulphuric acid, | .355 |
| Potash, | .239 |
| Soda, | .021 |
| Chlorine, | .009 |
| Sand and insoluble silicates, | 83.626 |
| Loss, | .021 |
| | 100.000 |

LAWRENCE COUNTY.

No. 466—Coal. *Labeled "McHenry's big seven feet coal, branch of Three Mile creek, over sandstone, between Tug and Louisa forks of Big Sandy river, Lawrence county, Ky."*

A moderately soft, glossy-black coal, with some fibrous coal between the layers, and occasionally some pyritous matter, and a little ochreous incrustation resulting from its decomposition. Heated over the spirit-lamp it decrepitated, softened, and swelled up considerably into an inflated coke, burning with a smokey yellow flame. A coking coal.

Specific gravity, - - - - - - - - 1.326

*Proximate Analysis.*

| | | | |
|---|---|---|---|
| Misture, | 3.60 | Total volatile matters, | 39.50 |
| Volatile combustible matters, | 35 90 | | |
| Carbon in the coke, | 53.50 | Coke, cellular, | 61.50 |
| Ashes, (lilac colored,) | 7.00 | | |
| | 100.00 | | 100.00 |

The composition of the *ash* is as follows:

| | |
|---|---|
| Silica, | 2.18 |
| Alumina, and oxide of iron, | 4.23 |
| Lime, | .08 |
| Magnesia, | .28 |
| Loss, | .23 |
| | 7.00 |

Submitted to ultimate analysis it gave the following results:

| | |
|---|---|
| Carbon, | 72.655 |
| Hydrogen, | 5 111 |
| Sulphur, | 1.750 |
| Oxygen, nitrogen, and loss, | 13 084 |
| Ashes, | 7.400 |
| | 100.000 |

No. 469—Coal. *Labeled "Keener's coal, three to four feet thick, three miles above Terman's Ferry, Big Sandy river, Lawrence county, Ky."*

A shining pitch-black coal; not very hard; breaking into thin layers; fibrous coal, with pyritous matter, evident by the lens, between the layers, and some efflorescence of sulphate of iron. Over the spirit-lamp it softens, and the fragments agglutinate and swell into a moderately dense cellular coke; it does not decrepitate much.

Specific gravity, - - - - - - - - 1.358

*Proximate Analysis.*

| | | | |
|---|---|---|---|
| Moisture, | 4 10 | Total volatile matters, | 37.80 |
| Volatile combustible matters, | 33.70 | | |
| Carbon in coke, | 53 00 | Dense coke, | 62.20 |
| Ashes, (light grey,) | 9.20 | | |
| | 100.00 | | 100.00 |

The *composition* of the ash of this coal was found to be—

| | Grains. |
|---|---|
| Silica, | 3 39 |
| Alumina and oxide of iron, | 4.48 |
| Magnesia, | .18 |
| Lime, a trace, and loss, | .35 |
| | 8.40 |

On *ultimate analysis* this coal gave

| | |
|---|---|
| Carbon, | 70.200 |
| Hydrogen, | 4.777 |
| Sulphur, | 1 470 |
| Oxygen, nitrogen, and loss, | 13.953 |
| Ashes, | 9 600 |
| | 100.000 |

Both very good coals, with a little more than the average proportion of ashes.

No. 325—Ferruginous Limestone. *Labeled "Segregations in the creek, seven miles below Millions', (How much Iron?) Big Sandy Railroad, Lawrence county, Ky."*

A dark-grey, fine granular ore, with a dirty-yellowish-brown weathered surface; resembles a dark colored limestone; powder of a light-grey color.

*Composition*, dried at 212° F.—

| | |
|---|---|
| Carbonate of lime, | 50.95 |
| Carbonate of magnesia, | 4.53 |
| Carbonate of iron, | 3.01 |
| Oxide of iron, | 4.62 |
| Alumina, | 1.91 |
| Phosphoric acid, | .36 |
| Potash, | .57 |
| Soda, | .31 |
| Organic matters, | 2.00 |
| Silex and insoluble silicates, | 32.17 |
| | 100.43 |

The air-dried mineral lost 0.70 per cent. of *moisture* at 212°.

LINCOLN COUNTY.

No. 531—Mineral Water. *Labeled "From the Grove Spring, in the yard of the proprietor of the Crab Orchard Springs, Mr. Caldwell, Lincoln county, Ky."*

A chalybeate water, which had deposited a little of its oxide of iron in the bottle in which it was brought.

Evaporated to dryness, at 212°, this water left 0.384 of a grain of solid residuum to the 1,000 grains of the water.

The composition of this mineral water was found to be, in 1,000 grains of the water—

| | | |
|---|---|---|
| Carbonate of iron, | 0.021 | Held in solution in the water, by carbonic *acid*. |
| Carbonate of manganese, | .005 | |
| Carbonate of lime, | .195 | |
| Carbonate of magnesia, | .041 | |
| Sulphate of magnesia, | .056 | |
| Sulphate of potash, | .013 | |

30

| | |
|---|---|
| Chloride of sodium, - - | .013 |
| Silica, - - - - - | .040 |
| Nitric acid, a trace. | |
| | 0.384 |

It contained also free carbonic acid, which was not estimated.

No. 532—MINERAL WATER. *Labeled "From the 'Brown Spring,' half a mile from Crab Orchard, on the Lancaster Turnpike."*

A chalybeate water; it had been partly decomposed, and the oxide of iron separated in the bottle during carriage, but its composition was ascertained after mixing the sediment fully with the water. One thousand grains of the water left 0.442 grains of solid residuum, on evaporation to dryness, at 212°.

The composition of this water was found to be, in one thousand grains—

| | | |
|---|---|---|
| Carbonate of iron, - - - | 0.028 | Held in solution in the water, by free carbonic acid. |
| Carbonate of manganese, - | .005 | |
| Carbonate of lime, - - | .117 | |
| Carbonate of magnesia, - - | .020 | |
| Sulphate of magnesia, - - | .112 | |
| Sulphate of lime, - - - | .015 | |
| Sulphate of potash, - - | .028 | |
| Chloride of sodium, - - | .018 | |
| Silica, - - - - - | .046 | |
| Moisture and loss, - - - | .053 | |
| | 0.442 grs. | |

The free carbonic acid, also present, was not estimated.

No. 533—MINERAL WATER. *"From the 'Field Spring,' on the lot of the proprietor of the Crab Orchard Springs, Mr. John H. Caldwell, Lincoln county, Ky."*

A chalybeate water. A thousand grains of the water contain about 0.446 of a grain of solid matter, dried at 212°.

The *composition* of this mineral water may be thus stated, in one thousand grains of the water—

| | | |
|---|---|---|
| Carbonates of iron & manganese, | 0.015 | Held in solution by carbonic acid. |
| Carbonate of lime, - - - | .139 | |
| Carbonate of magnesia, - - | .131 | |
| Sulphate of magnesia, - - | .066 | |
| Sulphate of soda, - - - | .024 | |
| Sulphate of potash, - - | .022 | |
| Chloride of sodium, - - | .008 | |
| Silica, - - - - - - | .041 | |
| | 0.446 | |

The free carbonic acid present was not estimated.

Whilst these three chalybeate waters each contain about the same amount of saline matters, they present some differences in the proportions of the several ingredients. The "Brown Spring" contains rather the largest quantity of carbonate of iron, and the "Field Spring" the least. The carbonate of magnesia is in larger amount in the Field spring, and the sulphate of magnesia in the Brown spring. The proportion of carbonate of lime is highest in the Field spring.

All these waters are good *saline chalybeates,* and applicable in all cases to which such remedies are appropriate. Whether experience in the use of the waters from these several wells has exhibited any difference in the effects on the system, attributable to the slight differences of composition, the writer is not informed.

By comparing these with the waters from the several chalybeate springs at Bryant's springs, near Crab Orchard, to be described further on, a considerable analogy of composition will also be observed.

No. 534—Mineral Water. *Labeled "From Howard's Sulphur Well, one and a half miles from Crab Orchard, on the Mt. Vernon road, Lincoln county, Ky."*

A *white sulphur* water; but all the sulphuretted hydrogen had been decomposed by carriage.

Specific gravity, - - - - - - - - - 1.00007

One thousand grains of the water contained 0.164 of a grain of solid matter, dried at 212°.

The *composition* of the water is as follows, in 1,000 grains:

| | | |
|---|---|---|
| Carbonate of magnesia, - | 0.065 | Held in solution by *carbonic acid.* |
| Carbonate of lime, - - | .013 | |
| Sulphate of magnesia, - - | .012 | |
| Sulphate of potash, - - | .008 | |
| Alumina, and trace of phosphate, - - - - - | .002 | |
| Chloride of sodium, - - | .017 | |
| Silica, - - - - - - | .022 | |
| Moisture and loss, - - | .025 | |
| | 0.164 | |

It contained also sulphuretted hydrogen and carbonic acid gases—amount of which was not estimated.

The medicinal virtues of the water would depend principally on the sulphuretted hydrogen, and on the depurative influence of the water taken in considerable quantities; whether the saline ingredients, which together amount only to a little more than a grain to the pound avourdupois, are sufficient to exert much sensible action on the system, is somewhat questionable, especially as they are not of a very potent nature. This water is, nevertheless, a good weak sulphur water.

No. 535—Mineral Water. *Labeled "Water from 'Epsom Spring,' (No. 1) one mile from Crab Orchard, on the Lancaster Turnpike, Lincoln county, Ky."*

Specific gravity, - - - - - - - - - - 1.0041

One thousand grains evaporated to dryness at 212° left 5.428 grains of solid saline matter. The water was found, also, to contain a considerable amount of free carbonic acid, which was not estimated.

*Composition*, in 1,000 grains of the water.

| | *Grains.* | |
|---|---|---|
| Carbonate of lime, - - - | 0.673 | Held in solution by carbonic acid. |
| Carbonate of magnesia, - - | .116 | |
| Carbonate of iron, a trace. | | |
| Sulphate of magnesia, - - | 3.454 | |
| Sulphate of lime, - - - | .203 | |
| Sulphate of potash, - - | .067 | |
| Sulphate of soda, - - - | .774 | |
| Chloride of sodium, - - | .081 | |
| Silica, - - - - - | .060 | |
| | 5.428 | |

No. 536—MINERAL WATER. *Labeled "Water from the Epsom Spring at Foley's, half a mile from the centre of Crab Orchard, on the Fall Dick road."*

Specific gravity, - - - - - - - - 1.0068

One thousand grains of the water, evaporated to dryness at 212°, left 6.884 grains of solid saline matter.

The *composition* of this water may be stated as follows, in 1,000 grains of the water:

| | *Grains.* | |
|---|---|---|
| Carbonate of lime, - - - | 0.912 | Held in solution by carbonic acid. |
| Carbonate of manganese, - | .131 | |
| Carbonate of iron, a trace. | | |
| Sulphate of magnesia, - - | 3.520 | |
| Sulphate of lime, - - - | .185 | |
| Sulphate of potash, - - | .170 | |
| Sulphate of soda, - - - | 1.013 | |
| Chloride of sodium, - - | ·304 | |
| Silica, - - - - - - | .056 | |
| Moisture and loss, - - - | .593 | |
| | 6.884 | |

The amount of free carbonic acid present was not estimated.

Although the sulphate of magnesia, (Epsom Salt,) is the principal saline ingredient of these "Epsom Springs" at and near Crab Orchard, the presence of the other saline ingredients, and of the carbonate of iron, modifies greatly the action of that well known salt, so that the medicinal effects, from the use of these waters, is considerably different from that of a pure solution of sulphate of magnesia, and they are applicable to a greater variety of cases.

The medicinal virtues of the saline matter of the Crab Orchard Springs have been so highly appreciated of late that a large quantity of "Crab Orchard Salts," obtained by evaporating the water to dryness in iron kettles, has been sold by our druggists, and it has become an *officinal* article. Some of this salts, as manufactured by Mr. B. H. Sowder, from the water of "Sowder's Spring," near Crab Orchard, presently to be described, was submitted to chemical examination.

No. 537—CRAB ORCHARD SALTS. *Brought by Mr. B. H. Sowder.*

A moist granular powder, with a slight tinge of brownish, like the whitest Havana sugar, in appearance.

Dried at 212° it lost more than twenty per cent. of *moisture.*

*Composition*, dried at 212° F.—

| | |
|---|---|
| Sulphate of magnesia, | 63.19 |
| Sulphate of soda, | 4.20 |
| Sulphate of potash, | 1.80 |
| Sulphate of lime, | 2.54 |
| Chloride of sodium, | 4.77 |
| Carbonate of lime, magnesia, and iron, and silica, | .89 |
| Bromine, a trace. | |
| Water of crystallation and loss, | 22.61 |
| | 100.00 |

This salt had been obtained by the evaporation of the water of the spring next to be described. The water was boiled down in an iron kettle, to a certain density, and then, after allowing it to stand for some time in a wooden vessel, the clear liquid, drawn off from the mixed deposit of carbonates of lime and magnesia and oxide of iron, thrown down by boiling, was evaporated to full dryness.

By some of the manufacturers much attention is paid to this process of "purification" of the salt, so that it is entirely freed from oxide of iron and the precipitated carbonates, and is perfectly white; but whether the removal of these ingredients of the water is not injurious to the full medicinal virtue of the saline matter may well be questioned.

The Crab Orchard salts have been much employed by the physicians of Lexington. They find them less drastic, and more tonic, than pure unmixed Epsom Salts, and more likely to act on the liver, in the manner of calomel, when taken in small doses.

No. 538—Mineral Water. "*Sent by B. H. Sowder from "Sowder's Spring," about a mile and a half from Crab Orchard, on the north of the hill towards Dick's river, near its base, and some* 300 *yards from the river. Spring yields about two hundred gallons a day.*

Specific gravity, - - - - - - - - 1.006

One thousand grains of the water, evaporated to dryness at 212°, left 7.153 grains, of saline matter, in one thousand grains of the water.

*Composition—*

| | *Grains.* | |
|---|---|---|
| Carbonate of lime, | 0.506 | Held in solution by carbonic acid. |
| Carbonate of magnesia, | .375 | |
| Carbonate of iron, a trace. | | |
| Sulphate of magnesia, | 2.989 | |
| Sulphate of lime, | 1.566 | |
| Sulphate of potash, | .298 | |
| Sulphate of soda, | .398 | |
| Chloride of sodium, | 1.000 | |
| Silica, | .021 | |
| Bromine, a trace. | | |
| | 7.153 | |

The amount of free carbonic acid present in this water was not estimated.

No. 539—MINERAL WATER. *From Bryant's Springs, near Crab Orchard. Labeled (No.* 1) *"Chalybeate Fountain in the valley," Lincoln county, Ky.*

The water had deposited a slight brownish sediment in the bottle, and the cork was somewhat blackened; it gave a little brownish-white deposit on boiling; reaction neutral.

It was found to have the following *composition*, in 1,000 *grains:*

| | | |
|---|---|---|
| Carbonate of lime, | 0.118 | Held in solution by carbonic acid. |
| Carbonate of magnesia, | .024 | |
| Carbonate of iron, with trace of manganese, | .007 | |
| Sulphate of magnesia, | .027 | |
| Sulphate of potash, | .010 | |
| Chloride of sodium, | .088 | |
| Silica, | .017 | |
| | 0.291 of a grain. | |

Free carbonic acid not estimated.

No. 540—MINERAL WATER. *From Bryant's Springs, Lincoln county, Ky. Labeled "Chalybeate (No.* 2) *from the Pasture Spring."*

A very slight, dark sediment had formed in the bottle, and the cork was more blackened than by the above described water; tastes more chalybeate than that; gave a slight, brown precipitate on boiling; reaction neutral.

*Composition*, in 1,000 grains—

| | *Grain.* | |
|---|---|---|
| Carbonate of lime, - - - | 0.095 | Held in solution by carbonic acid. |
| Carbonate of magnesia, - - | .037 | |
| Carbonate of iron, - - - | .021 | |
| Sulphate of lime, - - - | .010 | |
| Sulphate of magnesia, - - | .070 | |
| Sulphate of potash, - - | .026 | |
| Chloride of sodium, - - | .015 | |
| Silica, - - - - - | .046 | |
| | 0.320 | |

Free carbonic, not estimated.

A somewhat stronger chalybeate than the "Valley Spring."

No. 541—Mineral Water. *From Bryant's springs, Lincoln county. Labeled "Sulphur Water (No. 3) Valley spring."*

No sediment in the bottle; no discoloration of the cork; a very faint taste and smell of sulphuretted hydrogen; gave no sediment on boiling.

*Composition*, in 1,000 grains—

| | *Grain.* | |
|---|---|---|
| Carbonate of lime, - - - | 0.093 | Held in solution by carbonic acid. |
| Carbonate of magnesia, - - | .048 | |
| Carbonate of iron, a trace. | | |
| Sulphate of lime, a trace. | | |
| Sulphate of magnesia, - - | .006 | |
| Sulphate of potash, - - | .025 | |
| Chloride of sodium, - - | .175 | |
| Chloride of magnesium, - - | .042 | |
| Silica, - - - - - | .015 | |
| | 0.404 | |

The free *carbonic acid* and *sulphuretted hydrogen* present in the water were not estimated; a weak saline *sulphur water*.

No. 542—Mineral Water. *From Bryant's Springs, Lincoln county, Ky. Labeled "Sulphur Water (No. 4) from the Knob Spring."*

A little floculent black precipitate in the bottle; the cork was somewhat blackened; a more decided taste and smell of sulphur than in the last; a slight taste of common salt evident; no sediment formed on boiling.

*Composition*, in 1,000 grains.

| | Grains. |
|---|---|
| Carbonate of iron, a trace. | |
| Chloride of sodium, - - | 0.933 |
| Sulphate of magnesia, - - | .069 |
| Sulphate of lime, - - - | .104 |
| Sulphate of soda, - - - | .205 |
| Sulphate of potash, - - | .016 |
| Silica, - - - - - - | .015 |
| | 1.342 |

The examination of the saline residuum, obtained by the evaporation of some gallons of this water, would doubtless give evidence of the presence of traces of *iodine* and *bromine*. The free carbonic acid and sulphuretted hydrogen were not estimated. A stronger and more active saline sulphur water than the preceding.

No. 543—Mineral Water. *From Bryant's Springs, Lincoln county, Ky. Labeled "Mr. Stone's sulphur water."*

A very little black flocculent sediment in the bottle, and the cork had been somewhat blackened; the odor of sulphuretted hydrogen was scarcely perceptible, and the taste very faint. Gave a light colored sediment on boiling.

*Composition*, in 1,000 grains of the water—

| | |
|---|---|
| Carbonate of lime, - - | 0.058 |
| Carbonate of magnesia, - - | .116 |
| Carbonate of iron, - - - | .026 |
| Sulphate of lime, - - - | .012 |
| Sulphate of magnesia, - - | .023 |
| Sulphate of potash, - - | .007 |
| Chloride of sodium, a trace. | |
| Silica, - - - - - | .030 |
| | 0.272 of a grain. |

The free *carbonic acid* and *sulphuretted hydrogen* were not estimated.

In this, and in the other sulphur waters described, the *sulphuretted hydrogen* had been decomposed during the transportation of the water to the laboratory; and the proportion of this gas, as well as of the carbonic acid, can only be correctly ascertained in water examined a:

the fountain, or introduced there with care into bottles containing the proper re-agents, to bring these gases to a fixed state.

No. 544—Mineral Water. *From Bryant's Springs, Lincoln county, Ky. Labeled "Well in front of Bryant's house."*

A little dark sediment in the bottle; no odor; a slight taste of Epsom salt; deposits a whitish sediment on boiling.

*Composition*, in 1,000 grains—

| | *Grains.* | |
|---|---|---|
| Carbonate of lime, | 0.480 | Held in solution by carbonic acid |
| Carbonate of magnesia, | .013 | |
| Carbonate of iron, | .019 | |
| Sulphate of magnesia, | .904 | |
| Sulphate of lime, | .966 | |
| Sulphate of potash, | .066 | |
| Sulphate of soda, | 0.028 | |
| Chloride of sodium, | .278 | |
| Silica, | .090 | |
| | 2.844 | |

The amount of free carbonic acid in this water was not estimated.

This water resembles, in composition, the Epsom waters of the Crab Orchard springs, but, whilst it contains a smaller proportion of saline matters, it has a rather larger proportion of carbonate of iron and of sulphate of lime.

No. 545—Mineral Water. *Labeled "Stone Spring," from Bryant's Springs, Lincoln county, Ky.*

Presented nothing remarkable in taste and smell. There was a little flocculent precipitate of oxide of iron in the bottle, but the cork was not perceptibly blackened. One thousand grains of the water, evaporated to dryness, left but 0.05 of a grain of saline matter, which consisted of

Sulphate of magnesia;
Sulphate of lime;
Chloride of sodium;
Carbonate of iron;
Carbonate of lime; and
Carbonate of magnesia.

This is a remarkably *pure* water, and slightly chalybeate. Very few spring waters contain so small a proportion of saline matter as this, which has only about a third of a grain to the avoirdupois pound.

No. 407—CARBONATE OF IRON. *Labeled "'Flat Lick Iron Ore,' in burnt shale, near Stanford, on Thomas Holmes' land, Lincoln county, Ky."*

A dense, fine granular, greenish-grey carbonate of iron; powder of a dirty buff color.

Specific gravity, - - - - - - - 3.339

*Composition*, dried at 212° F.—

| | | |
|---|---|---|
| Carbonate of iron, - - - | 47.97 | 30.77 per cent. of *Iron*. |
| Oxide of iron, - - - | 10.66 | |
| Alumina, - - - - | 2.99 | |
| Phosphoric acid, - - - | .36 | |
| Carbonate of lime, - - - | 7.25 | |
| Carbonate of magnesia, - - | 12.13 | |
| Carbonate of manganese, - | 3.03 | |
| Sulphur, - - - - - | .21 | |
| Potash, - - - - - | .57 | |
| Soda, - - - - - - | .24 | |
| Silex and insoluble silicates, - | 13.95 | |
| Water and loss, - - - - | .64 | |
| | 100.00 | |

The air-dried ore lost 0.40 per cent. of *moisture*, at 212° F.

A very good iron ore, which would require very little addition of limestone to flux it.

LIVINGSTON COUNTY.

No. 240—COAL. *Labeled "Union Company's Coal, bottom part, Livingston county, Ky."*

A glossy pitch-black coal; firm, but not very hard; a little fibrous coal, and some pyritous matter between the layers. Over the spirit-lamp it does not decrepitate except in the pyritous portions; softens, swells a good deal, and agglutinates into a light cellular coke; burning with a smokey flame.

Specific gravity, - - - - - - - - - 1.356

The *ultimate analysis* of this coal gave the following results:

| | |
|---|---|
| Carbon, | 78.000 |
| Hydrogen, | 4.977 |
| Sulphur, | .630 |
| Nitrogen, | .628 |
| Oxygen and loss, | 7.165 |
| Ashes, | 8.600 |
| | 100.000 |

Submitted to distillation, at a heat gradually increased to dull redness, one thousand grains of this coal gave

148.00 grains thick dark-colored "crude oil."
38.00 grains ammoniacal water.
750.00 grains light cellular coke; and
465 cubic inches of good gas.

Like the rest of the soft bituminous coals, this does not yield a large quantity of oily or waxy matters on distillation; but it gives a large proportion of good coke, and doubtless would answer well for the production of illuminating gas. Its proportion of sulphur is moderate.

## LOGAN COUNTY.

No. 217—Sub-soil. *Labeled "Sub-soil, southern part of Logan county, ten miles from Franklin, on the road from Keysburg. (Sub-carboniferous Limestone Formation.)"*

Color of the dried sub-soil of a handsome dirty orange, or brownish-reddish-yellow color. Carefully washed with water this soil left 69½ per cent. of greyish-red sand, of which all but about 6½ per cent. was *very fine;* this coarser portion consists of rounded particles of hyaline and milky quartz, with a little admixture of ferruginous mineral.

One thousand grains, dried at the ordinary temperature, were digested for a month in water containing carbonic acid, to which it gave up only about half a grain of solid extract, dried at 212°, the *composition* of which was as follows:

| | |
|---|---|
| Organic and volatile matters, | 0.260 |
| Oxide of iron and alumina, | .036 |
| Brown oxide of manganese, | .027 |
| Phosphoric acid, | .011 |
| Sulphuric acid, not estimated. | |
| Lime, | .037 |
| Magnesia. | .033 |

| | |
|---|---|
| Potash, | .054 |
| Soda, | .017 |
| Silica, | .110 |
| | 0.585 of a gr. |

Dried at 400° F., this sub-soil lost 2.80 per cent. of *moisture*.

Its *composition*, thus dried, was found to be—

| | |
|---|---|
| Organic and volatile matters, | 3.14 |
| Oxide of iron, | 3.66 |
| Alumina, | 4.77 |
| Phosphoric acid, | .14 |
| Sulphuric acid, not estimated. | |
| Carbonate of lime, | .30 |
| Magnesia, | .40 |
| Brown oxide of manganese, | .18 |
| Potash, | .12 |
| Soda, | .03 |
| Silex and insoluble silicates, | 89.27 |
| | 102.01 |

The analysis of this *sub-soil* may be compared with that of its superincumbent surface-soil, No. 141, given on pages 342 and 379, of the preceding report. It will be seen, that whilst that gave a larger proportion of soluble matter to the water containing carbonic acid, and contains a trifle more of organic and volatile constituents, and less phosphoric acid, yet the composition of the two is strikingly alike—making allowance for an evident error in the estimation of the oxide of iron and alumina, in this latter analysis, which causes the apparent excess of about two per cent. in the sum.

### MONROE COUNTY.

No. 418—LIMONITE. *Labeled "Iron Ore, Malone's farm, Cole's fork of Mill creek, Monroe county, Ky."*

A dense limonite, of a dark yellowish-brown color; with irregular cavities; and shining portions of a nearly black color; powder light yellowish-brown.

*Composition*, dried at 212° F.—

| | | |
|---|---|---|
| Oxide of iron, | 76.90 | = 53.85 per cent. of *Iron*. |
| Alumina, | .27 | |
| Brown oxide of manganese, | .95 | |
| Carbonate of lime, | .27 | |
| Magnesia, | .73 | |
| Phosphoric acid, | .30 | |
| Potash, | .20 | |
| Soda, | .08 | |
| Silex and insoluble silicates, | 9.35 | |
| Combined water, | 11.79 | |
| | 100.84 | |

The air-dried ore lost 1.40 per cent. of *moisture*, at 212° F.

Quite a pure hydrated oxide of iron, which, for successful smelting, must be mixed with poorer ores and limestone. As previously stated, experience has proved that iron ores containing more than fifty per cent. of iron cannot be so cheaply smelted in the high furnace as those which contain a larger proportion of earthy ingredients.

No. 228—Soil. *Labeled "Soil from the dividing ridge between Barren and Cumberland rivers, where the broom-sedge grass prevails, Monroe county, Ky. (Sub-carboniferous Sandstone or Knob Formation.)"*

Color of the dried soil dark yellowish-grey. Sifted through a seive, with about 169 apertures to the inch, some cherty fragments were left. Carefully washed in water it gave about 52. per cent. of *sand*, of which all but about 14. per cent. would go through bolting cloth of 5,000 apertures to the inch; this coarser sand consisted of rounded particles of hyaline and milky quartz, and of yellow, red, and brown ferruginous quartz.

One thousand grains, dried at the ordinary temperature, and digested for a month in water containing carbonic acid, gave up nearly three grains of *soluble extract*, of which the *composition* was—

| | |
|---|---|
| Organic and volatile matters, | 0.920 |
| Alumina, oxide of iron and phosphates, | .468 |
| Brown oxide of manganese, not estimated. | |
| Carbonate of lime, | 1.078 |
| Magnesia, | .026 |
| Sulphuric acid, | .119 |
| Potash, | .040 |

| | |
|---|---|
| Soda, | .040 |
| Silica, | .110 |
| Loss, | .052 |
| | 2.853 |

The air-dried soil lost 1.82 per cent of *moisture* at 365° F.; dried at which temperature its *composition* was found to be as follows:

| | |
|---|---|
| Organic and volatile matters, | 4 130 |
| Alumina, | 2.700 |
| Oxide of iron, | 2.120 |
| Carbonate of lime, | .106 |
| Magnesia, | .200 |
| Brown oxide of manganese, | .116 |
| Phosphoric acid, | .075 |
| Sulphuric acid, not estimated. | |
| Potash, | .119 |
| Soda, | .122 |
| Silex and insoluble silicates, | 89.393 |
| Water and loss, | .913 |
| | 100.000 |

No. 454—ZINC ORE. *Labeled "Zinc and Lead Ore," from the rocks under the Devonian Black Slate, Sulphur Lick, Monroe county, Ky."*

A fine granular rock, containing carbonate of lime, with sulphurets of zinc and lead disseminated through it.

*Composition*—

| | | |
|---|---|---|
| Sulphuret of zinc, | 77.33 | = 51.77 per cent. of *Zinc*. |
| Silica, &c., | 17.48 | |
| Carbonates of lime and magnesia, and sulphuret of lead disseminated, | 5.19 | |
| | 100.00 | |

If found in sufficient abundance might be profitably employed in the manufacture of zinc white paint.

MUHLENBURG COUNTY.

No. 464—COAL. *Labeled "Walker's Coal, one and a half miles west of Turners, Muhlenburg county, Ky;"*

A dull looking coal, with the appearance of having been weathered; some signs of decomposed pyrites on its exposed surfaces; separates easily into thin layers, between which are fibrous coal and impressions as of broad reed leaves. Over the spirit-lamp it does not decrepitate; burns with a smokey flame; softens and agglutinates somewhat, and swells into a moderately dense coke.

Specific gravity, - - - - - - - - 1.271

*Proximate Analysis.*

| | | | |
|---|---|---|---|
| Moisture, - - - - - | 3.80 | Total volatile matters, - | 45.30 |
| Volatile combustible matters, - | 41.50 | | |
| Carbon in the coke, - - | 53.60 | Coke, (bright cellular,) | 54.70 |
| Ashes, (dirty buff,) - - | 1.10 | | |
| | 100.00 | | 100.00 |

The composition of the ashes of this coal is as follows:

| | |
|---|---|
| Silica, - - - - - - - - - - - | 0.29 |
| Alumina and oxide of iron, - - - - - - - - | .58 |
| Lime, - - - - - - - - - - - | .10 |
| Magnesia, - - - - - - - - - - | .13 |
| | 1.10 |

The considerable proportion of lime and magnesia in this ash will make it more than usually fusible in a strong fire.

By *ultimate analysis* this coal, dried at 212°, was found to be composed of

| | |
|---|---|
| Carbon, - - - - - - - - - - | 79.577 |
| Hydrogen, - - - - - - - - - - | 5.199 |
| Sulphur, - - - - - - - - - - | .640 |
| Nitrogen, oxygen, and loss, - - - - - - - | 13.384 |
| Ashes, - - - - - - - - - - - | 1.200 |
| | 100.000 |

Quite a pure coal, but its large proportion of oxygen and nitrogen prevents it from being a very good coking coal.

No. 191—Coal. *Labeled "Robert's Main Muddy River Coal, Muhlenburg county, Ky."*

A very pure looking, dark, glossy coal, with scarcely any appearance of fibrous coal between the layers, and only microscopical appearances of pyrites in a few spots; not soiling the fingers; firm, but not very hard. Over the spirit-lamp it decrepitates a little; burns with a smokey flame; softens, swells up, and agglutinates into a moderately dense cellular coke, with botryoidal prominences.

Specific gravity, - - - - - - - - - 1.263

*Proximate Analysis.*

| | | | |
|---|---|---|---|
| Moisture, - - - - - | 5.80 | Total volatile matters, - | 38.30 |
| Volatile combustible matters, - | 32.50 | | |
| Carbon in the coke, - - | 56.70 | Bright coke, - - | 61.70 |
| Ashes, (light-grey,) - - | 5.00 | | |
| | 100.00 | | 100.00 |

The proximate analysis of the lower portion of this coal was given by Dr. Owen in his first report, page 142.

The *composition* of the ashes was found to be—

| | |
|---|---|
| Silica, - - - - - - - - - - - - | 2.99 |
| Alumina, with little oxide of iron, - - - - - - | 1.68 |
| Lime, - - - - - - - - - - - - - | .27 |
| Magnesia, - - - - - - - - - - - | .05 |
| Loss, - - - - - - - - - - - - - | .01 |
| | 5.00 |

Submitted to *ultimate analysis* this coal, dried at 212° F, gave of

| | |
|---|---|
| Carbon, - - - - - - - - - - | 74.455 |
| Hydrogen, - - - - - - - - - - | 4.933 |
| Sulphur, - - - - - - - - - - | .906 |
| Nitrogen, - - - - - - - - - - | 1.030 |
| Oxygen and loss, - - - - - - - - - | 13.076 |
| Ashes, - - - - - - - - - - - | 5.600 |
| | 100.000 |

By destructive distillation, at a moderate heat, one thousand grains of Roberts' coal gave

102.10 grains of thick black tarry matter;
119.80 grains of ammoniacal water of a dark purple color;
659.50 grains of bright coke;
Leaving 118.60 grains for loss and gas.

1000.00

The gas collected measured only 370 cubic inches, and was of moderately good quality.

This coal is not, therefore, very well suited to the manufacture of gas, nor for the production of *Paraffin* and Benzole, &c., by destructive distillation.

Its large proportions of oxygen and nitrogen injure it somewhat for these purposes: the hydrogen being, to an equivalent amount, monopolized in the production of water and ammonia, by union with these gases; but it is a very good coal for domestic and manufacturing purposes generally, and no doubt yields a very good coke.

The dark purple color of the *ammoniacal water*, obtained by its distillation, is due to the presence of sulpho-cyanide of ammonium, which, by action on the iron of the tube of the retort produced the characteristic dark-purple compound, sulpho-cyanide of iron. Besides this compound the ammoniacal water contained hydrosulphate of ammonia and carbonate of ammonia.

No. 156—COAL. *Labeled "(McLean) Airdrie Coal, below the clay parting, six and three-twelfths feet thick, Muhlenburg county, Ky."*

The proximate analysis of this coal was given on page 352, of the former report.

*Ultimate Analysis.*

| | |
|---|---|
| Carbon, | 76.091 |
| Hydrogen, | 5.222 |
| Sulphur, | 1.350 |
| Oxygen, nitrogen, and loss, | 13.937 |
| Ashes, | 3.400 |
| | 100.000 |

Does not differ much in composition from Roberts' coal, but the specimen examined contained rather more sulphur.

No. 157—COAL. *Labeled "Eades Coal, two and a half miles southwest of Greenville, Muhlenburg county, Ky."*

The proximate analysis of this coal is also given on page 352 of the former report.

*Ultimate Analysis.*

| | |
|---|---|
| Carbon, | 76.855 |
| Hydrogen, | 5.244 |
| Sulphur, | .654 |
| Oxygen, nitrogen, and loss, | 13.847 |
| Ashes, | 3.400 |
| | 100.000 |

Closely resembles the two preceding in composition and properties.

## OHIO COUNTY.

No. 405—LIMONITE. *Labeled "Iron Ore? Top of the hill at Mr. French's, seven miles north of Hartford, Ohio county, Ky."*

A porous, yellowish-brown mass, containing a small bi-valve shell; under the lens exhibiting a few minute spangles of mica and grains of sand, united by a ferruginous cement; powder yellowish-brown.

*Composition*, dried at 212° F.—

| | | |
|---|---|---|
| Oxide of iron, - - - | 39 48 | = 27.64 per cent. of *Iron*. |
| Alumina, - - - - | 1.81 | |
| Brown oxide of manganese, - | 1.77 | |
| Phosphoric acid, - - - | .64 | |
| Lime, a trace. | | |
| Magnesia, - - - - | 1.12 | |
| Potash, - - - - | .34 | |
| Soda, - - - - | .06 | |
| Silex and insoluble silicates, - | 47.37 | |
| Combined water, - - - | 8.28 | |
| | 100.87 | |

The air-dried ore lost 1.00 per cent. of *moisture*, at 212°.

Rather a poor silicious ore.

No. 413—LIMONITE. *Labeled "Argillaceous Iron Ore, at Livermore's landing, Ohio county, Ky."*

Portion of a flat nodular mass, formed of concentric layers of brownish-yellow hydrated oxide of iron; dull; adhering strongly to the tongue; powder brownish-yellow; when it has been calcined, of a bright spanish-brown color.

*Composition*, dried at 212° F.—

| | | |
|---|---|---|
| Oxide of iron, - - - | 60.18 | = 42.14 per cent. of *Iron*. |
| Alumina, - - - - | 4.85 | |
| Phosphoric acid, - - - | .60 | |
| Lime, a trace. | | |
| Magnesia, - - - - | .73 | |
| Brown oxide of manganese, - | .27 | |
| Potash, - - - - | .40 | |
| Soda, - - - - - | .08 | |
| Silex and insoluble silicates, - | 19.75 | |
| Combined water, - - - | 13.14 | |
| | 100.00 | |

The air-dried ore lost 1.80 per cent. of *moisture* at 212° F.

No. 455—LIMESTONE. *Labeled "Hydraulic Limestone, five miles north of Hartford, Ohio county, Ky."*

A dark-grey, compact, limestone; glimmering with minute facets of calcareous spar; not adhering to the tongue; powder light-grey.

Specific gravity, - - - - - - - - 2.721

*Composition*, dried at 212° F.—

| | |
|---|---|
| Carbonic acid, | 38.55 |
| Sulphuric acid, | .80 |
| Phosphoric acid, | .12 |
| Lime, | 47.06 |
| Magnesia, | 2.39 |
| Alumina and oxide of iron, | 1.44 |
| Potash, | .29 |
| Soda, | .24 |
| Silex and insoluble silicates, | 9.96 |
| | 100.85 |

The air-dried rock lost only 0.30 per cent. of *moisture* at 212° F.

No. 459—COAL. *Labeled "Pitchener's Coal, Green river, two miles above Livermore, Ohio county, Ky."*

A shining pitch-black coal; not very hard; with some little infiltrated pyrites. Heated over the spirit-lamp did not decrepitate much, the fragments softened, swelled up, and agglutinated, forming an inflated coke.

Specific gravity, - - - - - - - - 1.272

*Proximate Analysis.*

| | | | |
|---|---|---|---|
| Moisture, | 5.50 | Total volatile matters, | 46.70 |
| Volatile combustible matters, | 41.20 | | |
| Carbon in the coke, | 48.90 | Shining cellular coke, | 53.30 |
| Ashes, (yellowish-grey,) | 4.40 | | |
| | 100.00 | | 100.00 |

The *composition* of the ashes was found to be

| | |
|---|---|
| Silica, | 2.18 |
| Alumina and oxide of iron, | 1.98 |
| Lime, a trace. | |
| Magnesia, | .10 |
| Loss, | .14 |
| | 4.40 |

The *ultimate composition* of this coal, dried at 212°, is as follows:

| | |
|---|---|
| Carbon, | 71.618 |
| Hydrogen, | 5.377 |
| Sulphur, | 1.750 |
| Oxygen, nitrogen, and loss, | 16.455 |
| Ashes, | 4.800 |
| | 100.000 |

No. 470—COAL. *Labeled "Barret's Coal, two miles north of Hartford, Ohio county, Ky."*

A shining, pitch-black coal; apparently pure, except from some microscopical appearance of pyrites in the fibrous coal which separates the layers, and some efflorescence of sulphate of iron. Over the spirit-lamp it decrepitates a little; burns with a yellow smokey flame; softens and wells up a good deal, the fragments agglutinating into a light cellular coke.

Specific gravity, - - - - - - - - 1.311

*Proximate Analysis.*

| | | | |
|---|---|---|---|
| Moisture, | 4.70 | Total volatile matters, | 42.60 |
| Volatile combustible matters, | 37 90 | | |
| Carbon in the coke, | 52.02 | Light shining coke, | 57 40 |
| Ashes, (light-chocolate-brown,) | 5 38 | | |
| | 100.00 | | 100.00 |

The composition of the ashes is as follows:

| | |
|---|---|
| Silica, | 1.24 |
| Oxide of iron and alumina, | 3.88 |
| Traces of lime and magnesia, and loss, | .26 |
| | 5.38 |

The *ultimate composition* of this coal, dried at 212°, is as follows:

| | |
|---|---|
| Carbon, | 74.510 |
| Hydrogen, | 5.332 |
| Sulphur, | 3 054 |
| Oxygen, nitrogen, and loss, | 12.504 |
| Ashes, | 4.600 |
| | 100.000 |

The per centage of *sulphur* is quite considerable.

No. 461—Coal. *Labeled "Mr. Jackson's Coal, one mile below Cromwell, Green river, Ohio county, Ky."*

A very dark and glossy coal; easily breaking into cuboidal fragments; fibrous coal between some of the layers, but no appearance of pyrites or other impurities. Over the spirit-lamp it does not decrepitate; burns with a very smokey flame; softens very much; agglutinates, and swells into a cellular shining coke.

Specific gravity, - - - - - - - - 1.272

*Proximate Analysis.*

| | | | |
|---|---|---|---|
| Moisture, - - - - - | 5.60 | Total volatile matter, | 43.90 |
| Volatile combustible matters, - | 38.30 | | |
| Carbon in the coke, - - | 53.60 | Bright cellular coke, - | 56.10 |
| Ashes, (reddish-grey,) - - | 2.50 | | |
| | 100.00 | | 100.00 |

The *composition* of the ashes was found to be—

| | |
|---|---|
| Silica, - - - - - - - - - - - | 1.19 |
| Alumina, and oxide of iron, - - - - - - | 1.28 |
| Lime, - - - - - - - - - - | .10 |
| Magnesia, - - - - - - - - - - | .06 |
| | 2.63 |

Submitted to *ultimate analysis*, dried at 212°, this coal gave

| | |
|---|---|
| Carbon, - - - - - - - - - - | 75.219 |
| Hydrogen, - - - - - - - - - - | 5.177 |
| Sulphur, - - - - - - - - - - | 1.704 |
| Oxygen, nitrogen, and loss, - - - - - - | 14.900 |
| Ashes, - - - - - - - - - - | 3.000 |
| | 100.000 |

No. 223—Soil. *Labeled "Soil, one foot deep, Mr. Harris', Morgantown road, Ohio county, Ky. (Coal Measures.)"*

Color of the dried soil light yellowish-grey. Carefully washed in water it left more than 52. per cent. of sand, of which less than 1. per cent. did not pass through fine bolting cloth. This consisted of flattened rounded particles of ferruginous sandstone.

One thousand grains, dried at the ordinary temperature, and digested for a month in water containing carbonic acid, gave up about two and a third grains of *brown extract*, which had the following composition, dried at 212°:

| | Grains. |
|---|---|
| Organic and volatile matters, | 0.770 |
| Alumina, oxide of iron, and phosphates, | .317 |
| Lime, | .274 |
| Magnesia, | .123 |
| Brown oxide of manganese, | .049 |
| Sulphuric acid, | .067 |
| Potash, | .081 |
| Soda, | .144 |
| Silica, | .230 |
| Carbonic acid and loss, | .275 |
| | 2.330 |

Dried at 400° the air-dried soil lost 1.74 per cent. of *moisture;* and its *composition,* thus dried, was found to be as follows:

| | |
|---|---|
| Organic and volatile matters, | 5.080 |
| Alumina, and oxides of iron and manganese, | 4.349 |
| Carbonate of lime, | .176 |
| Magnesia, | .166 |
| Phosphoric acid, | .101 |
| Sulphuric acid, | .413 |
| Chlorine, | .016 |
| Potash, | .157 |
| Soda, | .015 |
| Sand and insoluble silicates, | 90.166 |
| | 100.639 |

### OWSLEY COUNTY.

No. 160—Coal. *Labeled "Cannel Coal from Haddock's mine, between south and middle forks of Kentucky river, Owsley county, Ky."*

This coal, of which the *proximate analysis* was given in the former report, page 354, has been submitted to ultimate analysis, with the following results, viz:

| | |
|---|---|
| Carbon, | 76.791 |
| Hydrogen, | 6.177 |
| Sulphur, | .241 |
| Oxygen, nitrogen, and loss, | 13.791 |
| Ashes, | 3.000 |
| | 100.000 |

As it contains a larger relative proportion of hydrogen than any other of the coals examined, except the Breckinridge coal, it was submitted to destructive distillation at a heat gradually raised to dull redness, and the quantity of liquid combustible products is second only to that obtained from that coal. One thousand grains of this cannel coal, dried at the ordinary temperature, gave, on distillation,

| | |
|---|---|
| 248.50 | grains of crude oil, (thick and dark colored); |
| 54 50 | grains of ammoniacal water; |
| 589.00 | grains of dense coke. |
| 892.00 | |

Leaving 108.00 grains for loss and gaseous product.

The *gas* collected measured 370 cubic inches, and had a very high illuminating power.

This coal, as well as the cannel coal on Troublesome creek, Breathitt county, described by Dr. Owen in the preceding report, might doubtless be profitably employed in the manufacture of Benzole, lubricating oils, Paraffin, &c.

### PULASKI COUNTY.

No. 452—CARBONATE OF IRON. *Labeled "Headwaters of Indian and Rockhouse creeks, Grassy Gap Survey, Pulaski county, Ky."*

A dense, fine granular, dark-grey carbonate of iron, with a thin exterior layer of hydrated oxide; powder yellowish-grey.

Specific gravity, - - - - - - - - 3.344

*Composition*, dried at 212° F.—

| | | |
|---|---|---|
| Carbonate of iron, - - - | 53 02 | = 35.60 per cent. of *Iron*. |
| Oxide of iron, - - - | 20.13 | |
| Carbonate of lime, - - - | 5.36 | |
| Carbonate of magnesia, - - | 7.48 | |
| Carbonate of manganese, - | .71 | |
| Alumina, - - - - - | 1.95 | |
| Phosphoric acid, - - - | 1.13 | |
| Potash, - - - - - - | .54 | |
| Soda, - - - - - - | .08 | |
| Silex and insoluble silicates, - | 9.45 | |
| Organic matter, trace of sulphur, and loss, - - - - | .16 | |
| | 100.00 | |

The air-dried ore lost 0.50 per cent. of *moisture* at 212°.

This could be very economically smelted, because it contains within itself all the materials for the flux and the formation of cinder.

Ores containing about this per centage of iron are more profitably worked than those which are richer.

No. 467—Coal. *Labeled "Sears' Coal, Pitman hill, waters of Pitman and Buck creeks, Pulaski county, Ky."*

A glossy, pitch-black coal; seemingly pretty pure, with only microscopical appearances of pyritous matter in the fibrous coal, which separates the thin layers, into which it easily cleaves. Over the spirit lamp it softens a little, does not decrepitate, nor swell up much; the fragments agglutinate only at the angles in contact.

Specific gravity, - - - - - - - - 1.274

*Proximate Analysis.*

| | | | |
|---|---|---|---|
| Moisture, - - - - - | 2.20 | Total volatile matters, - | 41.10 |
| Volatile combustible matters, - | 38.90 | | |
| Carbon in the coke, - - | 57.00 | Dense coke, - - | 58.90 |
| Ashes, (light-buff-grey,) - - | 1.90 | | |
| | 100.00 | | 100.00 |

The composition of the ash was found to be—

| | |
|---|---|
| Silica, - - - - - - - - - - - | 0.69 |
| Alumina and oxide of iron, - - - - - - - - | .88 |
| Magnesia, - - - - - - - - - - | .10 |
| Lime, a trace, and loss, - - - - - - - - | .23 |
| | 1.90 |

On *ultimate analysis* this coal, dried at 212°, gave

| | |
|---|---|
| Carbon, - - - - - - - - - - | 78.608 |
| Hydrogen, - - - - - - - - - - | 5.311 |
| Sulphur, - - - - - - - - - - | .380 |
| Oxygen, nitrogen, and loss, - - - - - - - | 13.451 |
| Ashes, - - - - - - - - - - - | 2.250 |
| | 100.000 |

This coal is remarkable for its small proportions of earthy matters and sulphur.

No. 471—Coal. *Labeled "Lower Bed of Coal, sixty feet under the Main Coal, Cumberland Mines, Pulaski county, Ky."*

A pitch-black, pretty hard coal; cleaving into thin layers, which are separated by fibrous coal, in which there is scarcely any appearance of pyrites or other impurities. Over the spirit-lamp it does not decrepitate; swells up very little; burns with a reddish-yellow smoky flame; leaving a pretty dense shining coke.

Specific gravity, - - - - - - - - 1.311

*Proximate Analysis.*

| | | | |
|---|---|---|---|
| Moisture, - - - - | 4.40 | Total volatile matters, - | 38.20 |
| Volatile combustible matters, | 33.80 | | |
| Carbon in the coke, - - | 58.80 | Coke scarcely coherent, | 61.80 |
| Ashes, (nearly white,) - - | 3.00 | | |
| | 100.00 | | 100.00 |

The ashes were found to be composed of

| | |
|---|---|
| Silica, - - - - - - - - - - - - | 1.69 |
| Alumina, with a trace of oxide of iron, - - - - - - | 1.38 |
| Lime, - - - - - - - - - - - - | .10 |
| | 3.17 |

Submitted to *ultimate analysis*, dried at 212°, gave the following results, viz:

| | |
|---|---|
| Carbon, - - - - - - - - - - - | 76.364 |
| Hydrogen, - - - - - - - - - - | 5.200 |
| Sulphur, - - - - - - - - - - | .420 |
| Oxygen, nitrogen, and loss, - - - - - - - | 14.716 |
| Ashes, - - - - - - - - - - - | 3.300 |
| | 100.000 |

In its large proportion of oxygen and nitrogen we probably see the cause why it does not soften and swell up much in burning.

This, like the preceding, which it resembles, is also a remarkably pure coal.

No. 546—IMPURE LIMONITE. *Labeled "Iron Ore, found in masses of tons weight, near the mill of Dr. Graham, Rockcastle river, Pulaski county, Ky."*

A dark-red, mottled with lighter-red, fine granular ore, containing grains of sand, and glimmering under the lens; does not adhere to the tongue.

Specific gravity, - - - - - - - - 2.696

*Composition*, dried at 212° F.—

| | |
|---|---|
| Sand and insoluble silicates, - - - - - - - | 69.18 |
| Oxide of iron, - - - - - - - - - | 27.18 |
| Alumina, - - - - - - - - - - | .70 |
| Combined water, - - - - - - - - - | 3.50 |
| Magnesia and lime, traces. | |
| | 100.56 |

Contains too much sand to be a profitable iron ore, yet it might be used to mix with calcareous ores, or, with limestone added, to assist in fluxing ores which were difficult to smelt in consequence of their too great purity from earthy matters.

RUSSELL COUNTY.

No. 226—Soil. *Labeled "Soil and sub-soil, table land of Russell county, Ky., four miles north of Jamestown. (Sub-carboniferous Sandstone, or Knob Formation.)*

Dry soil of a dark buff-grey color; sifted through a seive, with one hundred and sixty-nine apertures to the inch, some ferruginous and quartz pebbles were removed from it. Carefully washed in water it left 57. per cent. of fine sand, of which all but ahout 5. per cent. passed through fine bolting cloth. These coarser particles appeared, under the lens, as rounded fragments of quartz and ferruginous sandstone, mixed with a few small *entrochites*.

One thousand grains of the air-dried soil, digested for a month in water containing carbonic acid, gave up more than two grains of *soluble brown extract*, which, dried at 212°, had the following composition:

| | *Grains.* |
|---|---|
| Organic and volatile matters, - - - - - - - - - | 0.910 |
| Alumina, oxide of iron, and phosphates, - - - - - - | .287 |
| Lime, - - - - - - - - - - - - - | .369 |
| Magnesia, - - - - - - - - - - - - | .080 |
| Brown oxide of manganese, - - - - - - - - | .059 |
| Potash, - - - - - - - - - - - - | .143 |
| Soda, - - - - - - - - - - - - - | .050 |
| Sulphuric acid, - - - - - - - - - - - | .089 |
| Silica, - - - - - - - - - - - - - | .150 |
| Carbonic acid and loss, - - - - - - - - - | .084 |
| | 2.221 |

The air-dried soil lost 3.44 per cent. of *moisture* at 400° F. Dried at which temperature it was found to contain

| | |
|---|---|
| Organic and volatile matters, - - - - - - - - - | 4.170 |
| Alumina, and oxides of iron and manganese, - - - - - | 4.478 |
| Carbonate of lime, - - - - - - - - - - - | .176 |
| Magnesia, - - - - - - - - - - - - | .066 |
| Phosphoric acid, - - - - - - - - - - - | .083 |
| Sulphuric acid, - - - - - - - - - - - | .227 |

| | |
|---|---|
| Potash, | .063 |
| Soda, | .068 |
| Sand and insoluble silicates, | 90.786 |
| Chlorine, a trace. | |
| | 100.122 |

The large proportion of sand and silicious matters, and the small relative amount of alumina and oxide of iron, and especially of phosphoric acid and the alkalies, explain the poverty of this soil of the Knob Formation. Yet, as the silicious matter is in a state of very fine division, even this, by skillful management in the proper application of manures, may be made and kept quite productive. Whether this could be *profitably* done would depend upon local circumstances.

### SIMPSON COUNTY.

No. 480—Sub-soil *Labeled "Red sub-soil, northern part of Simpson county, three-fourths of a mile from the Warren county line, Ky. (Sub-carboniferous Limestone Formation.)"*

Dried soil of a handsome brick-red, or light orage-red color. Carefully washed in water it left about 45½ per cent. of fine sand, mixed with some larger rounded fragments of quartz mineral, some clear, some milky, others colored light-red with oxide of iron, and about 4 per cent. of coaser sand about as fine as bar sand, composed of rounded particles of the same minerals.

One thousand grains of the soil, dried at the ordinary temperature, digested for a month in water containing carbonic acid, gave up only about two-thirds of a grain of greyish *extract*, which had the following composition:

| | |
|---|---|
| Organic and volatile matters, | 0.260 |
| Oxide of iron, alumina, oxide of manganese, and phosphates, | .047 |
| Lime, | .064 |
| Magnesia, | .033 |
| Potash, | .027 |
| Soda, | .020 |
| Silica, | .157 |
| | 0.608 of a gr. |

The air-dried soil lost 4.14 per cent. of *moisture* at 360°; dried at which temperature its composition was found to be as follows:

| | |
|---|---|
| Organic and volatile matters, | 7.02 |
| Oxide of iron, | 8.82 |
| Alumina, | 11 98 |
| Phosphoric acid, | .24 |
| Carbonate of lime, | .21 |
| Magnesia, | .20 |
| Brown oxide of manganese | .13 |
| Potash, | .19 |
| Soda, | .06 |
| Sand and insoluble silicates, | 71.13 |
| Loss, | .02 |
| | 100.00 |

A portion of the *volatile matter* stated above is no doubt *water* combined with the oxide of iron and alumina, which are present in unusually large proportions in this soil, and to the former of which it owes its fine red color. These ingredients give the soil the property of forming quite a fixed compound with *organic matters*, as is shown by the fact that although this soil contains as much as 7. per cent. of organic and volatile substances, one thousand grains, digested for one month, gave up only about a quarter of a grain to the carbonated water. These substances also have a considerable attraction for ammonia, absorb it with great facility, retain it with such tenacity that water will not remove it, and are always found to contain some of it after exposure to the atmosphere. Some of this red soil examined for *ammonia* was found to yield only 0.025 per cent. of that compound, but this is equal to seven hundred and fifty pounds to the acre, to one foot depth. This amount is probably but a part of that really contained in this soil. According to the recent experiments of Th. Way, of England, all the soils examined exhibited considerable power of absorption of ammonia, from an atmosphere containing it, and will remove it from water which holds it in solution. By the analysis of Dr. Kroker, in the Giessen Laboratory, and of several chemists in the employ of the Royal Prussian College of Husbandry, in Berlin, all the soils submitted to analysis, for the detection of ammonia, were found to yield quite large proportions, amounting, in some of the German soils, to as much as 18,040 pounds to the acre of ground, to twelve inches of depth, and in a remarkable Russian black soil to nearly 50,000 pounds! From these facts Liebig, in his recent publication *On the theory nad practice of agriculture:"* (*"Uber Theorie and Praxis in der Landwirth-*

*schaft.*" *Braunschweig,* 1856,) not yet translated into English, triumphantly contends, that as nitrogen (contained abundantly in ammonia,) is so constantly and plentifnlly supplied by the atmosphere, the *mineral ingredients* of the soil are the only essential elements of vegetable structures which are in danger of exhaustion, and which need be restored to the soil to maintain it in a state of fertility.

By comparing the above analysis of this peculiar *sub-soil,* with that of the *surface-soil* from the same locality, detailed in the preceding report, pages 355–6 and 379, marked differences of composition and properties will be noticed.

The surface soil gave nearly three grains of *extract* to the carbonated water, although containing less organic and volatile matters, but it contains only about one-third as much oxide of iron and alumina as this sub-soil, and considerably more fine sand and silicates. The *sub-soil* contains rather more phosphoric acid and alkalies than the *soil;* and, if *gradually* mixed with the surface soil, by deep ploughing, would give greater tenacity and strength to it, as it became exhausted by cropping. The great affinity of this red sub-soil, for organic matters, might, however, cause too great a mixture of it with the soil to be at first rather injurious than beneficial, but the simultaneous application of lime to the land might be useful.

## TRIGG COUNTY.

No. 420—LIMONITE. *Labeled "Iron Ore,—"honey-comb ore"—Capt. Williams', waters of Little river, Trigg county, Ky.*"

A porous, friable mineral, composed of numerous thin contorted layers of reddish-brown dense limonite, separated by soft ochreous matter; powder light yellowish-brown.

*Composition,* dried at 212° F.—

| | | |
|---|---|---|
| Oxide of iron, - - - | 56.10 | = 39.28 per cent. of *Iron.* |
| Alumina, - - - - | .45 | |
| Phosphoric acid, - - - | .38 | |
| Sulphur, a trace. | | |
| Lime, a trace. | | |
| Magnesia, - - - - | .57 | |
| Brown oxide of manganese, - | 1.05 | |
| Potash, - - - - - | .34 | |
| Soda, - - - - - | .08 | |

| | |
|---|---|
| Silex and insoluble silicates, - | 30.15 |
| Combined water, - - - | 10.70 |
| Loss, - - - - - - | .18 |
| | 100.00 |

The air-dried ore lost 1.00 per cent. of *moisture* at 212°.

A good silicious limonite.

No. 421—LIMONITE. *Labeled "Iron Ore, Hœmatitic variety, Capt. Williams', waters of Little river, Trigg countg, Ky."*

A dense, dark-reddish-brown limonite, with some red and yellow ochreous incrustation, and cavities lined with botryoidal concretions; powder rich brownish-yellow.

Specific gravity, - - - - - - - - 3.778

*Composition*, dried at 212° F.—

| | | |
|---|---|---|
| Oxide of iron, - - - | 79.40 | = 55.60 per cent. of *Iron*. |
| Alumina, - - - - - | .45 | |
| Phosphoric acid, - - - | .87 | |
| Brown oxide of manganese, - | .67 | |
| Magnesia, - - - - | .60 | |
| Potash, - - - - - - | .21 | |
| Soda, - - - - - - | .05 | |
| Silex and insoluble silicates, - | 5.75 | |
| Combined water, - - - | 11.98 | |
| Loss, - - - - - - | .02 | |
| | 100.00 | |

The air-dried ore lost 0.70 per cent. of *moisture*, at 212°.

A very rich limonite, which will not only require the addition of limestone but also of some silicious matter or ore to form a sufficient amount of cinder in the furnace to protect the reduced iron from the direct influence of the oxygen of the blast. The ore described just preceding this would, no doubt, answer this purpose admirably.

No. 457—LIMESTONE. *Labeled "Hydraulic Limestone near Mr. Hendricks', four miles above the mouth of Little river, Trigg county, Ky."*

A dull, fine, granular, grey limestone; not adhering to the tongue; exhibiting a few small specks of of calcareous spar; powder of a light grey color.

| | | |
|---|---|---|
| Specific gravity, | 2.702 | |
| *Composition*, dried at 212° F.— | | |
| Carbonic acid, | | 40.90 |
| Phosphoric acid, | | .06 |
| Lime, | | 43.91 |
| Magnesia, | | 7.00 |
| Alumina, oxide of iron, &c., | | .36 |
| Potash, | | .21 |
| Soda, | | .09 |
| Silex and insoluble silicates, | | 8.36 |
| | | 100.89 |

The air-dried rock lost 0.30 per cent. of *moisture* at 212° F.

No. 458—LIMESTONE. *Labeled "Hydraulic Limestone."*

| | | |
|---|---|---|
| Specific gravity, | 2.596 | |
| *Composition*, dried at 212° F.— | | |
| Carbonic acid, | | 38.85 |
| Phosphoric acid, | | .92 |
| Sulphuric acid, | | .29 |
| Lime, | | 28.61 |
| Magnesia, | | 14.77 |
| Alumina, | | 1.23 |
| Oxide of iron, | | .73 |
| Potash, | | .27 |
| Soda, | | .30 |
| Silica and insoluble silicates, | | 13.68 |
| Loss, | | .35 |
| | | 100.00 |

The air-dried rock lost 0.20 per cent. of *moisture* at 212°.

UNION COUNTY.

No. 237—SOIL. *Labeled "Soil, taken ten inches below the surface on Pond creek bottom, eight miles north-east of Caseyville, Union county, Ky; called there "Black Bottom;" land of Esquire Gains, (No. 1.) (Coal Measures.)*

Dried soil of a mouse color. Washed carefully with water it left about 28. per cent. of mouse-grey fine sand, which contained about 3. per cent. of coarser quartz grains, mixed with rounded particles of a ferruginous mineral.

One thousand grains, digested in water containing carbonic acid for a month, gave up about two and a quarter grains of *brown extract*, which, dried at 212°, has the following composition, viz:

| | |
|---|---|
| Organic and volatile matters, - - - - - - - - - - | 1.190 |
| Alumina, oxides of iron and manganese, and phosphates, - - - | .039 |
| Lime, colored with oxide of manganese, - - - - - - - | .386 |
| Magnesia, - - - - - - - - - - - - - | .083 |
| Sulphuric acid, - - - - - - - - - - - - | .188 |
| Potash, - - - - - - - - - - - - - | .046 |
| Soda, - - - - - - - - - - - - - - | .187 |
| Silica, - - - - - - - - - - - - - - | .161 |
| | 2.280 |

The air-dried soil lost 2.76 per cent. of *moisture*, at 365° F.

Its composition, thus dried, was found to be—

| | |
|---|---|
| Organic and volatile matters, - - - - - - - - - | 4.580 |
| Alumina, - - - - - - - - - - - - - | 2.986 |
| Oxide of iron, - - - - - - - - - - - - | 2.666 |
| Carbonate of lime, - - - - - - - - - - | .396 |
| Magnesia, - - - - - - - - - - - - - | .390 |
| Brown oxide of manganese, - - - - - - - - - | .056 |
| Phosphoric acid, - - - - - - - - - - - | .115 |
| Sulphuric acid and chlorine, not estimated. | |
| Potash, - - - - - - - - - - - - - | .139 |
| Soda, - - - - - - - - - - - - - - | .116 |
| Sand and insoluble silicates, - - - - - - - - - | 88.426 |
| Loss, - - - - - - - - - - - - - - | .130 |
| | 100.000 |

No. 235—Sub-soil. *Labeled "Sub-soil from the land of Esquire Gains, on the points making to Pond creek; taken a quarter of a mile distant from No. 1, (the preceding,) Union county, Ky."*

Color of the dried soil buff-grey; when calcined of a brick-red. Washed carefully with water it left a considerable proportion of fine sand, (weight lost,) all of which passed through the finest bolting cloth.

One thousand grains of the air-dried soil, digested for a month in water containing carbonic acid, gave up nearly a grain and a half of light-brown extract, composed of—

| | Grains. |
|---|---|
| Organic and volatile matters, | 0.700 |
| Alumina, oxides of iron and manganese, and phosphates, | .048 |
| Carbonate of lime, | .128 |
| Carbonate of magnesia, | .105 |
| Sulphuric acid, | .051 |
| Potash, | .058 |
| Soda, | .030 |
| Silica, | .360 |
| | 1.480 |

The air-dried sub-soil lost 3.16 per cent. of *moisture*, at 400° F.

Its composition, when thus dried, is as follows:

| | |
|---|---|
| Organic and volatile matters, | 2.740 |
| Alumina, and oxides of iron and manganese, | 9.530 |
| Carbonate of lime, | .276 |
| Magnesia. | .287 |
| Phosphoric acid, | .147 |
| Sulphuric acid, | .288 |
| Chlorine, | .003 |
| Potash, | .185 |
| Soda, | .056 |
| Sand and insoluble silicates, | 86.130 |
| Loss, | .358 |
| | 100.000 |

No. 236—Soil. *Labeled "Soil, derived from the shaley rock above the Anvil Rock, forming remarkable flat Post Oak glades, Shawneetown road, two and a quarter miles north-east of Mulford Page's land, Union county, Ky.* (*Coal Measures.*")

Color of the dried soil light-grey. Carefully washed with water one thousand grains of the air-dried soil left five hundred and seventeen grains of *fine sand*, of which one hundred and eighteen grains were too coarse to go through fine bolting cloth, and consisted principally of nearly spherical particles of ferruginous sandstone and iron ore, with rounded grains of quartz—hyaline, milky, yellow, and red.

One thousand grains of the soil, digested in water containing carbonic acid for a month, gave up about two grains of light clove-brown colored *extract*, which contained the following ingredients:

| | |
|---|---|
| Organic and volatile matters, | 0.589 |
| Alumina, oxides of iron and manganese, and phosphates, | .197 |
| Lime, | .104 |
| Magnesia, | .130 |
| Sulphuric acid, | .136 |
| Potash, | .085 |
| Soda, | .163 |
| Silica, | .290 |
| Carbonic acid and loss, | .226 |
| | 1.920 |

Dried at 370° this soil lost 3.54 per cent. of *moisture*, and had the following *composition*, viz:

| | |
|---|---|
| Organic and volatile matters, | 3.670 |
| Alumina, | 2.230 |
| Oxide of iron, | 5.080 |
| Brown oxide of manganese, | .080 |
| Carbonate of lime, | .136 |
| Magnesia, | .633 |
| Phosphoric acid, | .088 |
| Sulphuric acid, | .466 |
| Chlorine, | .003 |
| Potash, | .087 |
| Soda, | .062 |
| Sand and insoluble silicates, | 87.250 |
| Loss, | .215 |
| | 100.000 |

No. 220—Marl. *Labeled "Marl, taken from a bed four feet thick, overlaying a bed of coal eleven inches thick, near the top of a hill, on the land of Francis H. Shouse, Union county, Ky."*

In greenish, slate-colored lumps, containing fragments of encrinal stems, small cyathophilli, pieces of fossil bi-valve shells, and fragments of small coral stems.

One thousand grains, washed with water, with careful trituration in a mortar, left 598 grains of mixed sand and fragments of fossils, of which 309 grains, principally of *fine sand*, passed through fine bolting cloth.

Dried at 400° this marl lost 1.92 per cent. of *moisture*, and had the following *composition*:

| | |
|---|---|
| Organic and volatile matters, | 7 060 |
| Alumina, and oxides of iron and manganese, | 6.700 |
| Carbonate of lime, | 50.850 |
| Magnesia, | .698 |
| Phosphoric acid, | .280 |
| Sulphuric acid, | 1.366 |
| Chlorine, | .062 |
| Potash, | .310 |
| Soda, | .166 |
| Sand and insoluble silicates, | 32.670 |
| | 100.162 |

This might be used as a top-dressing to increase the fertility of poor silicious or exhausted soils, in its neighborhood, but would not pay its carriage for any great distance.

No. 185—Coal. *Labeled "Five feet, or main Mulford coal, Union county, Ky."*

A glossy deep black coal; firm, but not very hard; having a little fibrous coal between the layers, but no marked appearance of pyrites. Over the spirit-lamp it does not decrepitate; softens and swells very much, and agglutinates into a very inflated coke.

| | |
|---|---|
| Specific gravity, | 1.321 |

This coal, of which the *proximate analysis* was given by Dr. Owen in his first report, page 49, has been submitted to *ultimate analysis*, and examined, as to its product of bituminous oils and illuminating gas, by destructive distillation.

*Ultimate Analysis.*

| | |
|---|---|
| Carbon, | 76.200 |
| Hydrogen, | 5.644 |
| Sulphur, | 1.746 |
| Nitrogen, | .552 |
| Oxygen and loss, | 8.258 |
| Ashes, | 7.600 |
| | 100.000 |

The products of the distillation of 1,000 grains of this coal, at a heat gradually increased to redness, in an average of two experiments, were as follows:

136 50 grains of thick dark crude oil;
64.75 grains of ammoniacal water;
684.00 grains of coke;

Leaving 115.75 grains for gas and loss.

The gas collected measured 567.50 cubic inches, on an average, and possessed high illuminating powers.

It will be seen, therefore, that whilst this coal cannot be profitably employed in the manufacture of the bituminous oils, Benzole, and *Paraffin*, it is a very good coal for both gas and coke. Its *ultimate analysis* shows but a small proportion of oxygen and nitrogen. The only drawback to its use is the considerable proportion of sulphur which it is found to contain. This ingredient, however, like the earthy matters which form the ash, is found to vary in its proportion even within the compass of a single lump of the coal.

No. 188—Coal. *Labeled "Ice-house Coal, Mulford's mine, Union county, Ky."*

A not very glossy, but quite dark-colored coal; not very hard, but firm; presenting irised appearances, and some incrustation with sulphate of lime, but no pyritous matter, and little fibrous coal. Over the spirit-lamp it swells up considerably, and agglutinates into a cellular coke.

Specific gravity, - - - - - - - - 1.325

The proximate analysis of this coal was also given by Dr. Owen in first his report, page 51.

*Ultimate Analysis.*

| | |
|---|---|
| Carbon, | 73.419 |
| Hydrogen, | 4.977 |
| Sulphur, | 2.824 |
| Nitrogen, | 1.658 |
| Oxygen and loss, | 10.322 |
| Ashes, | 6.800 |
| | 100.000 |

Submitted to destructive distillation, as above described, the Ice-house Coal gave, from a thousand grains,—

108.00 grains of heavy, thick, dark, *crude oil;*
73.00 grains of dark colored *ammoniacal water,* having the odor of creosote;
714.00 grains of *coke,* (rather dense);
Leaving 105.00 grains for *gas* and loss.

The gas collected measured 465. cubic inches, and did not possess very high illuminating power. It was greatly contaminated with sulphuretted hydrogen, from the large proportion of sulphur contained in the coal.

For comparison with these Kentucky coals, I have appended at the end of this report an ultimate analysis of the Youghiogheny coal, of Pennsylvania, which is generally preferred in this region for the manufacture of illuminating gas.

No. 166—Coal. *Labeled "Coal from Casey's mine, near Caseyville, Union county, Ky."*

This coal, of which the *proximate analysis* was given in the former report, page 361, has been submitted to *ultimate analysis,* with the following results:

| | |
|---|---|
| Carbon, | 74.309 |
| Hydrogen, | 5.244 |
| Sulphur, | .880 |
| Oxygen, nitrogen, and loss, | 11.967 |
| Ashes, | 7.600 |
| | 100.000 |

WARREN COUTNY.

No. 417—Limonite. *Labeled "Hydrated Oxide of Iron, in the ridge above the conglomerate, amongst sandstone; waters of Claylick creek, seven miles above the mouth of Barren river, Warren county, Ky."*

Exterior crust an irregular layer of dense, hard, dark-brown limonite, with a few minute specks of mica; interior friable, yellowish, and reddish ochreous matter.

*Composition,* dried at 212° F.—

| | | |
|---|---|---|
| Oxide of iron, | 67.14 | = 47.02 per cent. of *Iron.* |
| Alumina, | .80 | |
| Phosphoric acid, | .86 | |
| Carbonate of lime, | .27 | |
| Magnesia, | .67 | |

| | |
|---|---|
| Brown oxide of manganese, - | 1.37 |
| Potash, - - - - - | .37 |
| Silex and insoluble silicates, - | 17.95 |
| Combined water, - - - | 11.16 |
| | 100.59 |

The air-dried ore lost 0.70 per cent. of *moisture* at 212°.

WAYNE COUNTY.

No. 450—LIMONITE. *Labeled "Bog Iron Ore, Meadow creek, Wayne county, Ky."*

A friable, dark-brown mineral; adhering to the tongue; presenting many irregular cavities, lined with lighter colored material; powder of a dark-brown color, becoming of a lighter-brown by calcination.

*Composition*, dried at 212° F.—

| | | |
|---|---|---|
| Oxide of iron, - - - | 23.70 | = 16.59 per cent. of *Iron*. |
| Alumina, - - - - | 6.02 | |
| Phosphoric acid, - - - | 1.13 | |
| Lime, a trace. | | |
| Magnesia, - - - - - | .71 | |
| Brown oxide of manganese, - | 6.62 | |
| Potash, - - - - - | .42 | |
| Soda, - - - - - - | .16 | |
| Silex and insoluble silicates, - | 52.35 | |
| Combined water, - - - | 8.91 | |
| | 100.00 | |

The air-dried ore lost as much as 5.80 per cent. of *moisture* at 212° F.

Too poor to be smelted alone, and containing too much phosphoric acid to be desirable for mixture with the richer ores of iron. The proportion of oxide of manganese in this ore is quite considerable.

No. 453—LIMONITE. *Labeled "Iron Ore from the Old Iron Works, Wayne county, Ky."*

A dark, reddish-brown, dense limonite, with numerous irregular cavities. Powder of a rich maroon color.

Specific gravity, - - - - - - - 3.252

*Composition*, dried at 212° F.—

| | | |
|---|---|---|
| Oxide of iron, | 58.30 | = 40.82 per cent. of *Iron*. |
| Alumina, | 1.35 | |
| Phosphoric acid, | .70 | |
| Carbonate of lime, | .45 | |
| Magnesia, | .37 | |
| Potash, | .21 | |
| Soda, | .03 | |
| Silex and insoluble silicates, | 35.35 | |
| Combined water, | 3.99 | |
| | 100.75 | |

The air-dried ore lost 1.90 per cent of *moisture* at 212° F.

A very good silicious iron ore, which requires only the addition of limestone to flux it in the furnace.

No 229—Soil. *Labeled "Average quality of the "Barren" soil of Wayne county, Ky; hickory and black oak land, waters of Meadow creek; based on a reddish ferruginous sub-soil.* (*Sub-carboniferous Limestone Formation, or Stylina Chert.*")

Color of the dried soil dark-brownish-grey. Washed with water, one thousand grains of this soil left four hundred and ninety-eight and a half grains of brown-grey *sand*, generally *very fine*, and containing ninety-one grains of coarser sand, the particles of which, examined with the lens, were hyaline, milky, and yellow quartz, with small rounded fragments of a ferruginous mineral.

One thousand grains of the air-dried soil, digested for a month in water containing carbonic acid, gave up more than two grains and a half of *brown extract*, dried at 212°, which was found to consist of the following ingredients, viz:

| | *Grains.* |
|---|---|
| Organic and volatile matters, | 1.170 |
| Alumina, oxide of iron, and phosphates, | .223 |
| Lime, | .330 |
| Magnesia, | .120 |
| Sulphuric acid, | .151 |
| Potash, | .096 |
| Soda, | .067 |
| Silica, | .140 |
| Oxide of manganese, chlorine, and loss, | .254 |
| | 2.551 |

The air-dried soil lost 3.16 per cent. of *moisture* at 380° F.; dried at which temperature, it was found to have the following *composition*, viz:

| | |
|---|---|
| Organic and volatile matters, | 5.370 |
| Alumina, | 4.326 |
| Oxide of iron, | 2.526 |
| Carbonate of lime, | .256 |
| Magnesia, | .246 |
| Brown oxide of manganese, | .236 |
| Phosphoric acid, | .036 |
| Potash, | .115 |
| Soda, | .136 |
| Sand and insoluble silicates, | 86.066 |
| Sulphuric acid, chlorine, and loss, | .687 |
| | 100.000 |

The only essential ingredient of this soil, which falls far below the average proportion, is the *phosphoric acid.* The application to it of bone-dust, or other *phosphatic* manures, would no doubt be greatly beneficial. Guano, Poudrette, *super-phosphate of lime,* &c., in mixture with ordinary barn-yard manure, would greatly increase its fertility.

No. 234—SOIL. *Labeled "Meadow creek soil, Dougherty farm, Wayne county, Ky." "See Dr. Owen's notes."*

In lumps, like dried clay; nearly black; of the color of onion seed. (*Sub-carboniferous Limestone Formation.*)

Washed with water, one thousand grains of this soil left only 177½ grains of fine black sand, &c., which contain only twenty-two grains of coarser particles, part of which were blackened vegetable remains, which, when removed by burning, left about 16½ grains coarse sand, consisting of rounded particles of milky quartz, carnelian?, and a hard ferruginous mineral.

One thousand grains of the air-dried soil, digested for a month in water containing carbonic acid, gave up more than eight and a half grains of *brown extract,* dried at 212°. The infusion, before evaporation, had a smell like that of stable manure, or rotten straw; and the *extract,* when moistened, had the same odor.

The composition of this watery extracts was as follows:

| | Grains. |
|---|---|
| Organic and volatile matters, | 4.120 |
| Alumina, oxide of iron, and phosphates, | .348 |
| Carbonate of lime, | 3.773 |
| Magnesia, | .023 |
| Potash, | .034 |
| Soda, | .058 |
| Silica, | .178 |
| Oxide of manganese and sulphuric acid, not estimated. | |
| | 8.534 |

It is probable that much of the lime stated as carbonate of lime was, in the extract, united with *organic acids*, which, when burnt out, left it in combination with carbonic acid. This soil contains a considerable proportion of such compounds, and hence the large amount of *extract* taken up by the carbonated water.

Dried at 365° F., the air-dried soil lost 8.28 per cent. of *moisture!* Thus dried its *composition* is as follows:

| | |
|---|---|
| Organic and volatile matters, | 21.560 |
| Alumina, | 10.240 |
| Oxide of iron, | 3.120 |
| Lime, | 1.021 |
| Magnesia, | .922 |
| Brown oxide of manganese, | .078 |
| Phosphoric acid, | .229 |
| Sulphuric acid, not estimated. | |
| Potash, | .351 |
| Soda, | .123 |
| Sand and silicates, | 62.506 |
| | 100.150 |

A remarkable soil, from the very large proportion of organic matters which it contains. Its contents of lime, phosphoric acid, potash, and soda, are also above the average. If properly drained it would prove a very productive soil. Its very dark color would cause it to become very warm under the action of the sun, in consequence of its great power of absorbing heat.

### WHITLEY COUNTY.

No. 231—Soil. *Labeled "Soil, from the Coal Measures of Whitley county, slope of the Clear fork, where the ferruginous shales prevail. Natural growth Beech, White Oak, and Hickory."*

Color of the dried soil yellowish-grey or buff-grey. It contains flat, angular fragments of ferruginous sandstone and iron ore. Washed carefully with water, one thousand grains left 466. grains of dirty buff-grey *sand*, mostly fine enough to pass through the finest bolting cloth, but containing 144 grains of *coarser sand*, the particles of which, examined with the lens, were rounded quartz grains—hyaline, milky, and yellow—with small fragments of a ferruginous mineral, with the angles rounded.

One thousand grains of the air-dried soil, digested for a month in water containing carbonic acid, gave up more than two grains of *brown extract*, dried at 212°, of which the *composition* was—

| | *Grains.* |
|---|---|
| Organic and volatile matters, | 1.160 |
| Alumina, oxide of iron, and phosphates, | .218 |
| Carbonate of lime, | .058 |
| Magnesia, | .023 |
| Sulphuric acid, | .129 |
| Potash, | .054 |
| Soda, | .151 |
| Silica, | .090 |
| Oxide of manganese and loss, | .339 |
| | 2.222 |

Dried at 390° this soil lost 3.28 per cent. of *moisture*, and presented the following *composition:*

| | |
|---|---|
| Organic and volatile matters, | 6.300 |
| Alumina, | 5.260 |
| Oxide of iron, | 5.660 |
| Carbonate of lime, | .076 |
| Magnesia, | .121 |
| Brown oxide of manganese, | .420 |
| Phosphoric acid, | .165 |
| Sulphuric acid, | .322 |
| Potash, | .170 |
| Soda, | .147 |
| Sand and silicates, | 80.786 |
| Loss, | .573 |
| | 100.000 |

No. 447—CARBONATE OF IRON. *Labeled "Carbonate of Iron, the so-called sil:er ore of Swift's mine, Log Mountain, Whitley county, Ky."* (*"White Mineral Hydrated Silicate of Alumina?"*)

A dark-grey nodular carbonate; not adhering to the tongue; exhibiting minute quartz crystals, specks of pyrites, and incrusted, in parts, with quartz and another white mineral, which was found to be the silicate of alumina; powder of a mouse-grey color.

*Composition*, dried at 212° F.—

| | | |
|---|---|---|
| Carbonate of iron, - - - | 78.35 | = 39.20 per cent. of *Iron*. |
| Oxide of iron, - - - | 3.36 | |
| Carbonate of lime, - - | .88 | |
| Carbonate of magnesia, - - | 2.67 | |
| Carbonate of manganese, - | 1.49 | |
| Alumina, - - - - | .58 | |
| Phosphoric acid, - - - | .63 | |
| Sulphur, - - - - | .26 | |
| Potash, - - - - | .29 | |
| Soda, - - - - | .45 | |
| Silex and insoluble silicates, - | 9.88 | |
| Organic matter, trace of copper, and loss, - - - - | 1.16 | |
| | 100.00 | |

The air-dried ore lost 0.20 per cent. of *moisture*, at 212°.

No. 199—CARBONATE OF IRON. *Labeled "Nodular Carbonate of Iron, found in the shale at the Falls of the Cumberland river, Whitley county, Ky. The so-called silver ore of Cumberland Falls."*

Of a dull dark-grey color, with infiltrations of a small quantity of whitish mineral, (silicate of alumina,) in the fissures; scarcely adhering to the tongue; powder of a yellowish-umber color.

*Composition*, dried at 212° F.—

| | | |
|---|---|---|
| Carbonate of iron, - - - | 73.13 | = 38.81 per cent. of *Iron*. |
| Oxide of iron, - - - | 4.94 | |
| Carbonate of lime, - - | 1.15 | |
| Carbonate of magnesia, - - | 1.59 | |
| Carbonate of manganese, - | 3.74 | |
| Alumina, - - - - | .79 | |
| Phosphoric acid, - - - | .16 | |
| Sulphur, - - - - | .09 | |
| Potash, - - - - | .39 | |

| | |
|---|---|
| Soda, - - - - - - | .19 |
| Bituminous matters, - - | 3.25 |
| Silex and insoluble silicates, - | 9.95 |
| Moisture and loss, - - - | .63 |
| | 100.00 |

The air-dried ore lost 0.50 per cent. of *moisture* at 212°.

The above analysis of this somewhat *notorious* ore was made at this laboratory before it was known to me that Dr. Owen had also made a full examination of the same mineral, the results of which are published on page 235 of his first report. Indeed, this ore has been frequently examined, in consequence the wide prevalent belief, a few years ago, that it contained a considerable proportion of silver. Whatever may have been the motives prompting those who originated the statement that the Cumberland Falls Iron Ore was rich in silver, it is certain that a great number of person were deluded into the purchase of shares in a stock company, which was organized for working this new Potosi. The excitement, about the latter end of the year 1850, was so great on this subject that individuals in other states were induced to leave their homes in order to embark in this flattering pursuit; and even now, the writer is informed, a hope still lingers in the minds of some in the neighborhood of the falls that some day a man "well versed in the working of metals" may come along, who, by his metalurgic skill, will change their iron ore into silver—a feat which was for a time played off before the excited stockholders, to the extent of exhibiting five or ten cents worth of silver from his crucibles, by a Cornish miner, who had been employed by the prime movers of the speculation.

The ore is a very good iron ore, approaching the so-called *black-band* ore in its composition, but not containing as much bituminous matter as that variety. It could be quite economically smelted into a good quality of iron.

No. 448—Limonite. *Labeled "Iron Ore, head waters of Mud creek, Whitley county, Ky."*

A dense, compact, limonite, of a dark-brown color; nearly black; exhibiting some lustre; some surfaces covered with red and yellow ochreous mineral; a few irregular cavities throughout the mass; powder of a rich light yellowish-brown.

Specific gravity, - - - - - - - 3.711

*Composition*, dried at 212° F.—

| | | |
|---|---|---|
| Oxide of iron, - - - | 80.50 | = 56.37 per cent. of *Iron*. |
| Alumina, - - - - | 1.88 | |
| Brown oxide of manganese, - | .18 | |
| Phosphoric acid, - - - | .37 | |
| Carbonate of lime, - - - | .18 | |
| Magnesia, - - - - | .80 | |
| Potash, - - - - - | .20 | |
| Soda, - - - - - | .19 | |
| Silex and insoluble silicates, - | 2.48 | |
| Combined water, - - - | 12.66 | |
| Loss, - - - - - | .56 | |
| | 100.00 | |

The air-dried ore lost 1.30 per cent. of *moisture* at 212° F.

A very pure hydrated oxide of iron—so pure that some poorer ore must be mixed with it to smelt it successfully in the high furnace.

No. 449—CARBONATE OF IRON. *Labeled "Carbonate of Iron, well at Mr. Sears', mouth of Poplar creek, Whitley county, Ky."*

A dark grey, fine granular, dense ore; in parts changed into brown and yellowish-brown; adhering to the tongue; powder dark-yellowish-grey.

Specific gravity, - - - - - - - - 3.432

*Composition*, dried at 212° F.

| | | |
|---|---|---|
| Carbonate of iron, - - - | 67.72 | } = 37.60 per cent. of *Iron*. |
| Oxide of iron, - - - | 6.99 | } |
| Carbonate of lime, - - | 3.38 | |
| Carbonate of magnesia, - - | 10.05 | |
| Carbonate of manganese, - | .70 | |
| Alumina, - - - - | 1.58 | |
| Phosphoric acid, - - - | .76 | |
| Potash, - - - - | .30 | |
| Soda, - - - - - | .11 | |
| Silex and insoluble silicates, - | 8.48 | |
| | 100.07 | |

The air-dried ore lost 0.50 per cent. of *moisture* at 212°.

No. 451—LIMONITE. *Labeled "Iron Ore, south part of Pine Mountain, Whitley county, Ky."*

A dark red-brown friable limonite; irregularly fine cellular; powdei of a dull red color.

*Composition*, dried at 212° F.—

| | |
|---|---|
| Oxide of iron, - - - | 63.60 = 44.53 per cent. of *Iron*. |
| Alumina, - - - - | 2.98 |
| Phosphoric acid, - - - | .31 |
| Sulphur, - - - - | .85 |
| Brown oxide of manganese, - | .31 |
| Lime, a trace. | |
| Magnesia, - - - - | .30 |
| Potash, - - - - | .34 |
| Soda, - - - - - | .29 |
| Silex and insoluble silicates, - | 17.25 |
| Combined water, - - - | 13.75 |
| Loss, - - - - - | .02 |
| | 100.00 |

The air-dried ore lost 4.00 per cent. of *moisture* at 212°.

This ore would require no addition but that of limestone to flux it in the furnace.

WOODFORD COUNTY.

No. 547—Limestone. *Labeled "Leptæna Limestone, under the fine Woodford soil, near Versailles, Woodford county, Ky. (Lower Silurian Blue Limestone.")*

Very full of fossil remains, (shells, coral, and crinoid stems;) fresh fracture, of a dark-grey color, glimmering with minute facets of calcareous spar; weathered surfaces dirty-buff, and very irregular from rapid disintegration; powder of a light-buff-grey color.

*Composition*, dried at 212° F.—

| | |
|---|---|
| Carbonate of lime, - - - | 91.33 = 51.25 *Lime*. |
| Carbonate of magnesia, - - | .56 |
| Alumina, and oxides of iron and manganese, - - - | 1.53 |
| Phosphoric acid, - - - | .70 |
| Sulphuric acid, - - - | .33 |
| Potash, - - - - - | .34 |
| Soda, - - - - - | .43 |
| Silex and insoluble silicates, - | 5.18 |
| | 100.40 |

The air-dried rock lost 0.20 per cent. of *moisture* at 212° F.

No. 548—LIMESTONE. *Labeled "Hill at Shryock's ferry, Woodford county, Ky. (Bird's-eye Limestone? of the Lower Silurian Formation.")*

A compact, very fine grained rock, with casts of fucoid stems (?) passing perpendicularly through it, which are filled with pure calcareous spar; of a handsome yellowish-grey color; powder white.

Specific gravity, - - - - - - - - 2.705

*Composition*, dried at 212° F.—

| | | |
|---|---|---|
| Carbonate of lime, - - - | 94.75 | = 53.17 of *Lime*. |
| Carbonate of magnesia, - - | 1.96 | |
| Alumina, and oxide of iron, &c., | .63 | |
| Phosphoric acid, a trace. | | |
| Sulphuric acid, - - - - | .30 | |
| Potash, - - - - - - | .23 | |
| Soda, - - - - - - | .32 | |
| Silica and insoluble silicates, - | 2.18 | |
| | 100.37 | |

The air-dried rock lost 0.20 per cent. of *moisture* at 212°.

This limestone being harder, of less easy disintegration under atmospheric influences, and containing less phosphoric acid and alkalies than the preceding, will not contribute so much mineral fertilizing matter to its super-incumbent soil as that rock, or as the one which immediately follows this.

No. 549—LIMESTONE. *Labeled, "Bellerophon Limestone, ("Niggerhead,") near Versailles, Woodford county, Ky."*

A light-grey, granular limestone, full of fossils, glistening with small facets of calcareous spar, and exhibiting some yellowish-brown infiltrations of oxide of iron.

*Composition*, dried at 212° F.—

| | | |
|---|---|---|
| Lime, - - - - - - | 54.12 | = 96.24 carbonate of Lime. |
| Magnesia, - - - - - | .45 | |
| Carbonic acid, - - - - | 41.90 | |
| Alumina, and oxide of iron, &c., | 1.04 | |
| Phosphoric acid, - - - - | .63 | |
| Sulphuric acid, - - - - | 1.78 | |
| Potash, - - - - - | .48 | |
| Soda, - - - - - - | .39 | |
| Silex and insoluble silicates, - | .78 | |
| | 101.57 | |

The air-dried rock lost 6.20 per cent. of *moisture* at 212° F.

No. 550—Soil. *Labeled "Virgin Soil, from Judge R. C. Graves' farm, water-shed between Greers' creek and Clear creek, near Versailles, Woodford county, Ky. Natural growth—hackberry, ash, walnut, mulberry, box elder, &c. One of the best soils of Ky."*

Color of the dried soil dirty-brown, or light-umber, with a slight tint of reddish. One thousand grains of this soil, carefully washed with water, left about 688. grains of light-umber colored *sand*, of which only about 90 grains was too coarse to go through the finest bolting cloth. This *coarser portion of the sand*, is composed of small rounded grains of soft iron ore, and of harder dark ferruginous mineral, with very few rounded quartzose particles.

One thousand grains of the air-dried soil, digested for two months in water containing carbonic acid, gave up more than six grains of yellowish-brown *extract* of the following *composition*, dried at 212°, viz:

| | *Grains.* |
|---|---|
| Organic and volatile matters, | 0.210 |
| Alumina, oxide of iron, and phosphates, | .888 |
| Brown oxide of manganese, | .498 |
| Carbonate of lime, | 3.377 |
| Magnesia, | .230 |
| Sulphuric acid, | .562 |
| Potash, | .100 |
| Soda, a trace. | |
| Silica, | .149 |
| | 6.014 |

The air-dried soil lost 4.70 per cent. of *moisture* at 400°; dried at which temperature its composition is as follows:

| | |
|---|---|
| Organic and volatile matters, | 7.771 |
| Alumina, and oxides of iron and manganese, | 12 961 |
| Carbonate of lime, | 2.464 |
| Magnesia, | .173 |
| Phosphoric acid, | .319 |
| Sulphuric acid, | .150 |
| Potash, | .394 |
| Soda, | .130 |
| Sand and insoluble silicates, | 75.266 |
| Loss, | .372 |
| | 100.000 |

No. 551—Soil. *Labeled "Same soil as the preceding, from a field in constant cultivation since* 1808, *when a crop of hemp was raised; it has been fourteen years in hemp; average of the last year's* (1855) *crop of corn eighteen to twenty barrels,* (*of five bushels each,*) *to the acre; it has produced thirty-four bushels of wheat to the acre; Judge Graves' farm, near Versailles, Woodford county, Ky.*"

Color of the soil like that of the preceding, but a little lighter. Carefully washed with water one thousand grains of this soil left 490 grains of light-umber colored *sand*, of which fifty-four and a half grains would not pass through fine bolting cloth, and were composed principally of small rounded particles of soft iron ore, and of red and brown ferruginous quartz, and a few irregular fragments of milky quartz.

One thousand grains of the air-dried soil, digested for two months in water containing carbonic acid, gave up more than three and a half grains of *grey-brown extract*, dried at 212°, the composition of which was—

| | *Grains.* |
|---|---|
| Organic and volatile matters, - - - - - - - - | 0.530 |
| Alumina, and oxide of iron and phosphates, - - - - - - | .198 |
| Carbonate of lime, - - - - - - - - - - - | 2.248 |
| Magnesia, - - - - - - - - - - - - - | .163 |
| Sulphuric acid, - - - - - - - - - - - | .223 |
| Potash, - - - - - - - - - - - - | .131 |
| Soda, - - - - - - - - - - - - - | .035 |
| Silica, - - - - - - - - - - - - - | .089 |
| Brown oxide of manganese and loss, - - - - - - - | .103 |
| | 3.720 |

The air-dried soil lost 4.60 per cent. of *moisture* at 400° F.; dried at which temperature its *composition* is as follows:

| | |
|---|---|
| Organic and volatile matters, - - - - - - - - | 5.513 |
| Alumina, and oxides of iron and manganese, - - - - - - | 13.344 |
| Carbonate of lime, - - - - - - - - - - | 2.734 |
| Magnesia, - - - - - - - - - - - - | .333 |
| Phosphoric acid, - - - - - - - - - - - | .306 |
| Sulphuric acid, - - - - - - - - - - - | .037 |
| Potash, - - - - - - - - - - - - | .205 |
| Soda, not estimated. | |
| Sand and insoluble silicates, - - - - - - - - - | 77.594 |
| | 100.066 |

By comparison of this analysis of the soil of the *old field* with that of the *virgin soil* of the same locality, given above, the following instructive facts may be observed, viz: that by cultivation the soil has lost much of its soluble materials, which are dissolved by water containing carbonic acid, as well as of its *organic and volatile matters;* it is therefore lighter colored, and has a somewhat lower power of absorbing heat and moisture, than the virgin soil.

When we examine critically what mineral ingredients have been removed by the long series of cropping, we do not observe that the loss has fallen on the *sand and silicates*, or on the *alumina and oxide of iron, &c.*, but upon those substances which always exist in soils in small relative proportions, and which are essential to all vegetable growth, viz: the potash, soda, lime, phosphoric acid, and sulphuric acid. From some accidental cause the *magnesia*, which is also an element of vegetable tissues, appears to be in larger proportion in the old soil than in the new. Upon the whole, however, there is less loss of these valuable ingredients than might have been expected, probably from the circumstance that in the cultivation of hemp, with which the ground had been occupied for a considerable portion of the time, when the plant is rotted on the ground on which it is grown, and nothing finally removed from it but the lint or fibre, very little is carried off from the soil except lime and potash, and the other ingredients in minor proportion. If the whole hemp plant is removed from the soil, and water-rotted, not even the hemp-herds being restored to it by burning, the deterioration which results is much greater. Had this soil been cultivated wholly in corn, small grain, and such crops as tobacco, potatoes, &c., the chemical analysis would have shown a much greater loss from it of the elements of vegetable nutrition. Probably, also, the corn raised on this ground was habitually fed to hogs and cattle on the spot—a very common practice in Kentucky—so that, finally, nothing was removed from it, of its essential mineral ingredients, but that quantity which entered into the composition of the bones, flesh, and fluids of these animals.

No. 552—Sub-soil. *Labeled "Sub-soil from a field which has been in cultivation ever since* 1808, *farm of Judge R. C. Graves, two miles south of Versailles, Woodford county, Ky."*

Color of the dried sub-soil dark yellowish dirty-brown.

One thousand grains, when washed in water, left 664. grains of brown-grey *sand*, of which only 75. grains were too coarse to pass through the finest bolting cloth, and this was principally rounded particles of soft iron ore, which could be crushed in the fingers, and a few rounded quartzose grains.

One thousand grains of the air-dried soil, digested in water containing carbonic acid for two months, gave up nearly five grains of yellowish-brown *extract*, dried at 212°, which had the following composition:

| | *Grains.* |
|---|---|
| Organic and volatile matters, | 0.850 |
| Alumina, oxide of iron, and phosphates, | .379 |
| Carbonate of lime, | 2.817 |
| Magnesia, | .093 |
| Sulphuric acid, | .419 |
| Potash, | .177 |
| Soda, | .010 |
| Silica, | .129 |
| Oxide of manganese and loss, | .076 |
| | 4.950 |

The air-dried sub-soil lost 4.52 per cent. of *moisture* at 400°. Dried at which temperature its *composition* is as follows:

| | |
|---|---|
| Organic and volatile matters, | 6.450 |
| Alumina, and oxides of iron and manganese, | 13.773 |
| Carbonate of lime, | 3.476 |
| Magnesia, | .354 |
| Phosphoric acid, | .447 |
| Sulphuric acid, | .0[illegible]2 |
| Potash, | .498 |
| Soda, | .095 |
| Sand and insoluble silicates, | 75.434 |
| | 100.607 |

This sub-soil is as rich as the original virgin soil.

No. 553—SUB-SOIL. *Labeled "Red clay, under the sub-soil, from Judge R. C. Graves' farm, near Versailles, Woodford county, Ky."*

Color of the dried sub-soil dirty light-reddish-brown.

One thousand grains left, after careful washing in water, 680 grains of reddish-brown sand, of which 403 grains were too coarse to go through the finest bolting-cloth, and consisted mainly of rounded particles of yellowish-brown and dark-brown iron ore, so soft as to be easily crushed in the fingers, with a very few small quartzose fragments.

One thousand grains of the air-dried sub-soil, digested for two months in water containing carbonic acid, gave up only *one grain* of *brownish extract*, dried at 212°, of which the composition was—

| | *Grain.* |
|---|---|
| Organic and volatile matters, | 0.300 |
| Alumina, and oxides of iron and manganese, and phosphates, | .078 |
| Lime, | .163 |
| Magnesia, | .073 |
| Sulphuric acid, | .185 |
| Potash, | .067 |
| Soda, | .013 |
| Silica, | .099 |
| Carbonic acid and loss, | .022 |
| | 1.000 |

Dried at 400° the air-dried sub-soil lost 5.04 per cent. of *moisture;* thus dried its *composition* is as follows:

| | |
|---|---|
| Organic and volatile matters, | 6.065 |
| Alumina, and oxides of iron and manganese, | 33.377 |
| Carbonate of lime, | .138 |
| Magnesia, | .0[illegible]0 |
| Phosphoric acid, | .383 |
| Sulphuric acid, | .198 |
| Potash, | .234 |
| Soda, | .127 |
| Sand and insoluble silicates, | 59.360 |
| Loss, | .038 |
| | 100.000 |

In view of the large proportion of alumina and oxide of iron, &c., in this 'red clay' it is probable that some of the 6.065 grains, stated above as the *organic and volatile matters,* is simply water.

This clay contains rather more phosphoric and sulphuric acids than the super-incumbent soil, but much less of carbonate of lime; the potash is in about average proportion. Its great peculiarity is the large amount of *alumina and oxide of iron* which it contains; and these, by their strong affinity for organic matters, prevent the solution of much solid matter by the carbonated water.

From the foregoing analyses of the soils and sub-soils of this part of Woodford county it is evident, that whilst deep ploughing into the immediate sub-soil would be quite beneficial to growing crops, the *heavy, red clay* under the sub-soil would not add any thing peculiarly valuable to this rich soil, which already has enough of alumina and oxide of iron in its composition to make it a loam very favorable for cultivation.

ILLINOIS.

No. 554—SOIL. *Labeled "Soil taken from just under the newly upturned original sod of the prairie, opposite to Keokuk, Iowa, a few (about eight) miles back from the Mississippi river, on the Illinois side."*

The dried soil is of a dark mouse-color, almost black; without any appearance of pebbles or gravel; under the microscope showing very fine glimmering grains of sand. This was not submitted to the solvent action of water charged with carbonic acid, to which it would doubtless give up a considerable amount of *solid extract.*

Dried at 300° it lost 3.28 per cent. of *moisture;* and, thus dried, was found to have the following *composition,* viz:

| | |
|---|---|
| Organic and volatile matters, | 9.050 |
| Alumina, | 2.405 |
| Oxide of iron, | 2 350 |
| Carbonate of lime, | .890 |
| Magnesia, | .526 |
| Phosphoric acid, | .175 |
| Sulphuric acid, not estimated. | |
| Potash, | .197 |
| Soda, | .100 |
| Sand and insoluble silicates, | 84.470 |
| | 100.163 |

This analysis of the prairie soil of the north-western part of Illinois was introduced for the purpose of comparison with the soils of Kentucky. The specimen analyzed was collected by the writer himself, in October, 1855.

Notwithstanding the luxuriance of the growth of the first crops on the prairie soil, occasioned partly by the large amount of available nourishing matter afforded by the decay of the thick sod, it is evident, from the above analysis, that, taking into consideration *durability* as well as *immediate fertility,* as ascertained by the chemical analysis of the soil itself, apart from the sod, there are many of our Kentucky soils—which take the second rank when compared with those of the *blue-grass region*—which yet are fully equal to the prairie soil. The reader may turn, for comparison, to the analysis of Mr. Barlow's soil, Barren county; to that of the virgin soil of Mr. O'Bannon's farm, Jefferson county; and to that of the virgin soil, on Benson creek, Franklin county, &c., &c., all in the present volume.

Compared with the *first rate* soil of Kentucky, that of the prairies contains a much smaller proportion of *alumina and oxide of iron,* as well as of *lime, magnesia, phosphoric acid, and alkalies.* It contains a much larger amount of *fine sand,* and doubtless a larger proportion of the *coarser sand,* than our best soils; and, therefore, whilst its large quantity of *organic matters* is held in the soil with a small force of attraction, (because of the large proportion which the *sand and silica* bear to the *alumina and oxide of iron,*) and hence they are readily soluble and immediately available in the production of luxuriant crops, these very circumstances will cause its more speedy exhaustion; and, when this accumulated deposit has been consumed by thriftless husbandry, this soil must sink down to a second-rate position. Yet, from its lightness, it is admirably adapted to garden purposes, sustained, as it should be, by the judicious supply of manures.

PENNSYLVANIA.

No. 555—Coal. *Labeled "Youghiogheny Coal, Pennsylvania."*

A good specimen obtained from the Lexington Gas Works, and analyzed for the purpose of comparison of our Kentucky coals with a coal of well known good qualities.

Specific gravity, - - - - - - - - 1.329

*Proximate Analysis.*

| | | | |
|---|---|---|---|
| Moisture, - - - - | 1.00 | Total volatile matters, - | 36.00 |
| Volatile combustible matters, - | 35.00 | | |
| Carbon in the coke, - - | 58.40 | Light spongey coke, - | 64.00 |
| Ashes, (lilac-grey,) - - | 5.60 | | |
| | 100.00 | | 100.00 |

*Ultimate Analysis.*

| | |
|---|---|
| Carbon, | 78.437 |
| Hydrogen, | 5.689 |
| Nitrogen, | 1.319 |
| Oxygen and loss, | 8.555 |
| Ashes, | .600 |
| Sulphur, not estimated. | |
| | 100.000 |

It will be seen that several of the Kentucky coals compare very favorably with this well known soft bituminous coal, which is much esteemed by the blacksmith, and for *gas* and *coke:* we may refer particularly to Garrard's coal, Clay county, and to several of the coals of Union and Crittenden counties, which are good coking coals.

To make the comparison more extensive this coal was submitted to distillation, at a temperature slowly raised to the red-heat, to ascertain the relative amount of *oils* and *gas* produced. One thousand grains of the air-dried coal gave, of

| | *Grains.* |
|---|---|
| Thick brownish-black crude oil, | 136 |
| Purplish ammoniacal water, | 52 |
| Light cellular coke, | 710 |
| Leaving for gas and loss, | 102 |
| | 1000 |

The gas collected measured 545. cubic inches, and had pretty good illuminating power, but not better than that from Mulford's coal, if as good.

This result does not, of course, represent the relative quantity of illuminating gas which the coal would yield if heated under conditions favorable for the production of gas. When distilled, as this and all the Kentucky coals examined were treated, at the lowest heat which would cause their decomposition, in order to produce as much as possible of the liquid and solid hydrocarbons, the quantity of gas obtained is always very much less than could be produced from the same coal suddenly exposed to a red heat, in the gas retort; but, as all the coals examined were submitted to the same low temperature, it is believed that the relative quantity of gas collected would give a correct idea of their gas producing powers under more favorable conditions.

## TABLE 1. (*A*)

IRON ORES, LIMONITES. *Showing the per centage of each ingredient.*

| Number in the report. | County. | Specific Gravity. | Peroxide of iron. | Carbonate of iron. | Brown oxide of manganese. | Carbonate of lime. | Magnesia. | Alumina. | Phosphoric acid. | Sulphur. | Potash. | Soda. | Combined water. | Silica. | Per centage of iron. |
|---|---|---|---|---|---|---|---|---|---|---|---|---|---|---|---|
| 489 | Bullitt, | 2.984 | 62.01 | – | 0.78 | 0.18 | 1.02 | 0.68 | 0.89 | 0.58 | 0.36 | 0.20 | 12.00 | 21.18 | 43.46 |
| 473 | Carter, | – | 78.42 | – | 3.17 | trace. | .30 | 1.48 | .73 | – | .21 | .18 | 11.94 | 3.77 | 54.93 |
| 412 | Clinton, | 3.503 | 74.30 | – | 1.68 | – | .35 | 1.48 | .18 | trace. | not es | timated. | 12.24 | 9.95 | 52.03 |
| 414 | Edmonson, | – | 60.90 | – | .75 | trace. | 1.15 | .65 | .57 | – | .36 | .32 | 11.15 | 23.68 | 42.64 |
| 415 | Edmonson, | – | 74.70 | – | .35 | trace. | .15 | .45 | .55 | – | not es | timated. | 11.19 | 12.65 | 52.31 |
| 419 | Edmonson, | – | 62.12 | – | .05 | trace. | .29 | 2 45 | .43 | .02 | .38 | .42 | 13.25 | 20.55 | 43.50 |
| 307 | Greenup, | – | 80.30 | – | .35 | – | .40 | not est. | .60 | – | .34 | .01 | 12,12 | 6.55 | 56.23 |
| 474 | Greenup, | – | 68.30 | – | – | .28 | 2.64 | 3.65 | .27 | – | .27 | .22 | 12.09 | 12.28 | 47.83 |
| 476 | Greenup, | – | 72.80 | – | .45 | .18 | 1.19 | 2.17 | – | – | .48 | .02 | 11.20 | 10.57 | 50.98 |
| 316 | Greenup, | – | 76.90 | – | .25 | – | .28 | 1.21 | .64 | – | .23 | .16 | 9.09 | 11.77 | 53.85 |
| 317 | Greenup, | 3.292 | 68.20 | – | .25 | trace. | 1.02 | 2.98 | .99 | – | not es | timated. | 8.57 | 17.17 | 47.76 |
| 318 | Greenup, | – | 61.10 | – | .95 | .45 | 1.09 | .85 | trace. | – | .38 | .10 | 11.67 | 23.85 | 42.78 |
| 478 | Greenup, | 3.083 | 58.30 | – | .65 | .15 | .77 | 1.05 | 1.25 | – | .40 | .08 | 8.31 | 29.77 | 40.82 |
| 289 | Greenup, | – | 24.70 | – | .05 | – | .67 | 3.75 | trace. | – | .32 | .01 | 5.66 | 64.42 | 17.29 |
| 309 | Greenup, | 2.770 | 41.40 | – | .75 | 1.15 | 1.50 | 3.36 | .54 | – | .23 | .01 | 10.54 | 41.47 | 28.99 |
| 293 | Greenup, | 3.026 | 77.50 | – | 1.03 | .76 | .79 | 1.23 | .40 | .50 | .20 | .14 | 9.62 | 7.77 | 54 25 |
| 431 | Greenup, | – | 74.50 | – | 2.43 | .77 | 1.81 | 1.00 | .33 | .57 | .15 | .13 | 3.86 | 14.93 | 52.17 |
| 430 | Greenup, | – | 38.38 | – | 1.23 | trace. | .60 | 3.54 | 1.01 | .05 | .28 | .18 | 8.12 | 46.83 | 26 87 |
| 290 | Greenup, | – | 80.03 | – | 2.03 | .64 | 2.87 | 1.44 | .66 | – | .25 | .16 | 2.01 | 9.93 | 56.02 |
| 291 | Greenup, | 3.018 | 73.90 | – | 1.13 | trace. | .39 | 1.71 | .62 | .09 | .19 | .05 | 11.51 | 10.43 | 51.75 |
| 292 | Greenup, | 3.406 | 81.40 | – | 1.63 | trace. | .35 | .77 | .24 | .07 | .26 | .22 | 6.72 | 8.33 | 57.00 |
| 441 | Greenup, | – | 51.10 | – | 1.83 | trace. | .68 | 1.07 | .76 | .32 | .38 | .10 | 8.13 | 35.93 | 35.78 |
| 442 | Greenup, | – | 83.83 | – | 1.73 | trace. | .32 | .43 | .94 | .21 | .30 | .11 | 11.30 | .83 | 58.70 |
| 444 | Greenup, | – | 53.44 | 24.79 | 1.44 | 0.87 | .62 | .09 | 1.26 | .11 | .34 | .08 | 6.89 | 9.93 | 49.39 |
| 446 | Greenup, | – | 66.76 | – | 1.23 | trace. | .26 | 1.00 | 1.41 | trace. | .34 | trace. | 11.59 | 17.87 | 46.75 |
| 422 | Greenup, | – | 66.03 | – | .55 | trace. | .76 | 4.15 | .67 | .06 | .46 | .11 | .71 | 27.15 | 46.24 |
| 424 | Greenup, | – | 63.60 | – | .55 | trace. | .99 | .25 | .70 | .06 | .25 | .05 | 10.77 | 23.23 | 44.54 |
| 425 | Greenup, | – | 85.91 | – | 2.17 | .17 | .85 | 1.25 | .09 | – | .23 | .18 | 7.90 | 1.25 | 60.16 |

TABLE 1. (*A*)—Continued.

| Number in the report. | County. | Specific gravity. | Peroxide of iron. | Carbonate of iron. | Brown oxide of manganese. | Carbonate of lime. | Magnesia. | Alumina. | Phosphoric acid. | Sulphur. | Potash. | Soda. | Combined water. | Silica. | Per centage of iron. |
|---|---|---|---|---|---|---|---|---|---|---|---|---|---|---|---|
| 428 | Greenup, | - | 84.45 | - | .09 | - | 1.43 | 1.20 | .38 | .06 | .44 | .10 | 2.80 | 9.05 | 59.14 |
| 429 | Greenup, | - | 69.60 | - | .75 | trace. | .35 | .55 | .42 | .07 | .42 | .01 | 12.18 | 15.65 | 48.74 |
| 299 | Greenup, | - | 53.46 | - | .85 | 33.85 | 3.15 | trace. | .87 | .02 | .23 | .07 | 6.45 | 1.05 | 3[illegible].44 |
| 436 | Greenup, | - | 88.51 | - | 1.95 | trace. | .78 | .15 | .19 | .03 | .09 | .17 | 6.00 | 2.23 | 61.98 |
| 437 | Greenup, | - | 80.20 | - | .05 | - | .51 | .05 | .86 | .48 | .02 | - | 11.31 | 6.45 | 56.14 |
| 438 | Greenup, | - | 49.90 | - | .27 | 8.05 | 4.19 | 7.00 | 1.45 | - | .41 | - | 9.61 | 19.15 | 35.06 |
| 439 | Greenup, | - | 73.34 | - | 9.41 | 1.27 | .83 | .27 | .36 | - | .40 | .03 | 9.54 | 4.55 | 51.36 |
| 408 | Grayson, | - | 63.60 | - | .87 | .27 | 1.22 | 2.36 | .89 | - | .25 | - | 12.02 | 19.15 | 44.54 |
| 418 | Monroe, | - | 76.90 | - | .95 | .27 | .73 | .27 | .30 | - | .20 | .08 | 11.79 | 9.35 | 53.85 |
| 405 | Ohio, | - | 39.48 | - | 1.77 | trace. | 1.12 | 1.81 | .60 | - | .34 | .06 | 8.28 | 47.37 | 27.64 |
| 413 | Ohio, | - | 60.18 | - | .27 | trace. | .73 | 4.85 | .60 | - | .40 | 08 | 13.14 | 19.75 | 42.14 |
| 546 | Pulaski, | 2.696 | 27.18 | - | - | - | - | .70 | - | - | - | - | 3 50 | 69.18 | - |
| 420 | Trigg, | - | 56.10 | - | 1.05 | .57 | trace. | .45 | .38 | trace. | .34 | .08 | 10.70 | 30.15 | 39.28 |
| 421 | Trigg, | 3.778 | 79.40 | - | .67 | - | .60 | .45 | .87 | - | .21 | .05 | 11.98 | 5.75 | 55.60 |
| 417 | Warren, | - | 67.14 | - | 1.37 | .27 | .67 | .80 | .86 | - | .37 | - | 11.16 | 17.95 | 47 02 |
| 450 | Wayne, | - | 23.70 | - | 6.62 | trace. | .71 | 6.02 | 1.13 | - | .42 | .16 | 8.91 | 52.35 | 16.59 |
| 453 | Wayne, | 3.252 | 58.30 | - | - | .45 | .37 | 1.35 | .70 | - | .21 | .03 | 3.99 | 35.35 | 40.82 |
| 448 | Whitley, | 3.711 | 80.50 | - | .18 | .18 | .80 | 1.88 | .37 | - | .20 | .19 | 12.66 | 2.48 | 56.37 |
| 451 | Whitley, | - | 63.60 | - | .31 | trace. | .30 | 2.98 | .31 | 0.85 | .34 | .29 | 13.75 | 17.25 | 44.53 |

## TABLE 1. (*B*)

### Iron Ores, Carbonates of Iron.

| Number in report. | County. | Specific gravity. | Carbonate of iron. | Oxide of iron. | Carbonate of lime. | Carbonate of magnesia. | Carbonate of manganese. | Alumina. | Phosphoric acid | Sulphuric acid. | Sulphur. | Potash. | Soda. | Bituminous matter. | Silica. | Water. | Per centage of iron. |
|---|---|---|---|---|---|---|---|---|---|---|---|---|---|---|---|---|---|
| 488 | Bullitt, | 3.446 | 57.59 | 7.77 | 6.28 | 11.76 | 1.32 | 1.55 | 0.71 | – | 0.29 | 0.75 | 0.27 | – | 11.18 | – | 32.62 |
| 493 | Bullitt, | 3.445 | 53.64 | 7.71 | 6.08 | 13.99 | 1.94 | .55 | .10 | 1.37 | .55 | .69 | .20 | – | 11.48 | 2.25 | 31.30 |
| 409 | Butler, | 3.026 | 70.20 | 9.92 | 2.55 | 7.04 | 1.60 | 1.51 | .64 | – | trace. | .42 | .01 | – | 7.65 | – | 39.45 |
| 416 | Edmonson, | 3.507 | 65.15 | 7.98 | 1.95 | 8.45 | 1.83 | .95 | .36 | .67 | – | .57 | .05 | 2.89 | 9.17 | – | 37.04 |
| 482 | Greenup, | – | 70.27 | 10.16 | 2.45 | 5.52 | 1.46 | .15 | .73 | – | – | .40 | .09 | – | 8.15 | – | 40.70 |
| 475 | Greenup, | 3.155 | 28.01 | 14.42 | 29.37 | 5.57 | .18 | 1.38 | .29 | – | – | .42 | .33 | – | 19.98 | – | 23.62 |
| 479 | Greenup, | 3.497 | 67 84 | 5.89 | 3 25 | 4.88 | 1.97 | 1.45 | .60 | – | – | .50 | .09 | – | 13.78 | – | 37.46 |
| 312 | Greenup, | – | 56.92 | 14 14 | 1.25 | 5.28 | 2.04 | 1.05 | .99 | – | – | .61 | .01 | .80 | 16.15 | – | 37.10 |
| 311 | Greenup, | – | 60.49 | 5.25 | 3,15 | 6 52 | .83 | .41 | trace. | – | – | .34 | .29 | – | 21.82 | – | 32.57 |
| 294 | Greenup, | – | 54.42 | 30.24 | .45 | .83 | 1.29 | 1.86 | .43 | – | .35 | .38 | .20 | 2.58 | 6.97 | – | 47.51 |
| 440 | Greenup, | – | 43.90 | 23.06 | 3.87 | 3.28 | .65 | .33 | .23 | – | .18 | .23 | .23 | 2.60 | 22.15 | – | 35.02 |
| 443 | Greenup, | 3.360 | 67.50 | 1.28 | 2.15 | 4.57 | 1.18 | .35 | .36 | – | .17 | .29 | .09 | – | 21.45 | – | 33.12 |
| 445 | Greenup, | 3.567 | 47 84 | – | 3.25 | 3.65 | 6.00 | .55 | trace. | – | 11.51 | .34 | .08 | 1.94 | 4.75 | – | 41.63 |
| 300 | Greenup, | 3.566 | 60.40 | 21.38 | 3.17 | 3.46 | 1.52 | .65 | .63 | – | trace. | .40 | .13 | – | 6.03 | 2 23 | 43.82 |
| 406 | Laurel, | 3 126 | 32 29 | 5.01 | 2.95 | 3.60 | .64 | 1.55 | 1.00 | – | – | .42 | .01 | .98 | 51.55 | – | 19.10 |
| 410 | Laurel, | 3.395 | 68.46 | 3.41 | .75 | 3.73 | 1.31 | 1.43 | .52 | – | – | .34 | .07 | .79 | 19.65 | – | 35.45 |
| 411 | Laurel, | 3.352 | 66.01 | 2 67 | 5.85 | 9.19 | .86 | .35 | .63 | – | – | .34 | .33 | – | 12.68 | – | 33.05 |
| 407 | Lincoln, | 3.334 | 47 97 | 10.66 | 7.25 | 12.13 | 3 03 | 2.99 | .36 | – | .21 | .57 | .24 | – | 13.95 | – | 30.77 |
| 452 | Pulaski, | 3.344 | 53.02 | 20.13 | 5.35 | 7.48 | .71 | 1.95 | 1.13 | – | – | .54 | .08 | .16 | 9.45 | – | 35.60 |
| 447 | Whitley, | – | 78.35 | 3.36 | .88 | 2.67 | 1.49 | .58 | .63 | – | .26 | .29 | .45 | 1.16 | 9.88 | – | 39.20 |
| 199 | Whitley, | – | 73.13 | 4.94 | 1.15 | 1.59 | 3.74 | .79 | .16 | – | .09 | .39 | .19 | 3.25 | 9.95 | – | 38.81 |
| 449 | Whitley, | 3.432 | 67.72 | 6.99 | 3.38 | 10.05 | .70 | 1.58 | .76 | – | – | .30 | .11 | – | 8.48 | – | 37.60 |

## TABLE 2. Coals.

| Number in report. | County. | Specific gravity. | Moisture. | Volatile combustible matters. | Carbon in the Coke. | Ashes. | Total volatile matters. | Coke. | Carbon. | Hydrogen. | Sulphur. | Nitrogen. | Oxygen. | Bituminous oil per 1000 grains. | Designation. |
|---|---|---|---|---|---|---|---|---|---|---|---|---|---|---|---|
| 462 | Christian, | 1.280 | 4 60 | 34.90 | 58.36 | 2.14 | 39.50 | 60.50 | 76.636 | 4.533 | 1.440 | 15.191 | | – | Woolich's. |
| 460 | Clay, | 1.259 | 2.70 | 34.90 | 61 10 | 1.30 | 37.60 | 62.40 | 80.619 | 5.444 | .575 | 1.457 | 10.305 | – | Garrard's. |
| 25 | Crittenden, | – | – | – | – | 3.80 | – | – | 78.500 | 5 333 | 1.040 | 1.344 | 9.983 | – | Sneed's. |
| 189 | Daviess, | 1.275 | – | – | – | 2.00 | – | – | 77.891 | 5.422 | .300 | 1.821 | 12.566 | – | Wolf Hill. |
| 502 | Daviess, | 1.328 | 6.70 | 36.00 | 51.30 | 6.00 | 42.70 | 57.30 | 71.019 | 5.022 | 2.090 | 15.069 | | – | Triplett's. |
| 101 | Greenup, | – | – | – | – | 4.00 | – | – | 79 091 | 5.111 | .734 | 11.604 | | – | Ashland, (main.) |
| 468 | Hancock, | 1.282 | 6.50 | 34.90 | 53.20 | 5.40 | 41.40 | 58.60 | 73.255 | 5.155 | .520 | 15 470 | | – | Hawesville, (1st bed.) |
| 519 | Hancock, | 1.392 | 3.00 | 39.10 | 45.40 | 12.50 | 42.10 | 57.90 | 63.436 | 4.622 | 5.866 | 12.476 | | – | Judge Mayhall's. |
| 520 | Hancock, | 1.266 | 6.30 | 39.80 | 51.40 | 2.50 | 46.10 | 53.90 | 75.328 | 5.600 | .890 | 15 882 | | – | Mr. Pate's. |
| 243 | Hancock, | 1.318 | 1.30 | 54.40 | 32.00 | 12.30 | 55.70 | 44.30 | 68.128 | 6.489 | 2.476 | 2.274 | 5.833 | 318. grs. | Breckinridge. |
| 463 | Hopkins, | 1.277 | 3.20 | 35.40 | 57.80 | 3.60 | 38.60 | 61.40 | 75.491 | 5.088 | 1.520 | 14.101 | | – | Mr. Hall's. |
| 465 | Hopkins, | 1.422 | 5.00 | 28.40 | 53.50 | 13.10 | 33.40 | 66.60 | 66.000 | 4.244 | .820 | 13.436 | | – | Mr. Samuel's. |
| 135 | Hopkins, | – | – | – | – | 2.40 | – | – | 77.400 | 4.999 | 1.060 | 1.620 | 12.521 | – | Wright's Mountain. |
| 456 | Lawrence, | 1.326 | 3.60 | 35.90 | 53.50 | 7.00 | 39.50 | 61.50 | 72.655 | 5.111 | 1.750 | 13.084 | | – | McHenry's. |
| 469 | Lawrence, | 1.358 | 4.10 | 33.70 | 53.00 | 9.20 | 37.80 | 62.20 | 70.200 | 4.777 | 1.470 | 13.953 | | – | Keener's. |
| 240 | Livingston, | 1.356 | – | – | – | 8.60 | – | – | 78.000 | 4.977 | .630 | 0.628 | 7.165 | 148. grs. | Union County. |
| 464 | Muhlenburg, | 1.271 | 3.80 | 41.50 | 53 60 | 1.10 | 45 30 | 54 70 | 79.577 | 5.199 | ·640 | 13.384 | | – | Walker's. |
| 191 | Muhlenburg, | 1.263 | 5 80 | 32.50 | 56.70 | 5.00 | 38.30 | 61.70 | 74.455 | 4.933 | .906 | 1.030 | 13.076 | 102.10 | Roberts', (main.) |
| 156 | Muhlenburg, | – | – | – | – | 3.40 | – | – | 76.091 | 5.222 | 1.350 | 13.937 | | – | Airdrie. |
| 157 | Muhlenburg, | – | – | – | – | 3.40 | – | – | 76.855 | 5.244 | .654 | 13.847 | | – | Eade's. |
| 459 | Ohio, | 1.272 | 5.50 | 41.20 | 48.90 | 4.40 | 46.70 | 53.30 | 71.618 | 5.377 | 1.750 | 16.455 | | – | Pitchner's. |
| 470 | Ohio, | 1.311 | 4.70 | 37.90 | 52.02 | 5.38 | 42.60 | 57.40 | 74.510 | 5.332 | 3.054 | 12.504 | | – | Barrett's. |
| 461 | Ohio, | 1.272 | 5.60 | 38.30 | 53.60 | 2.50 | 43.90 | 56.10 | 75.219 | 5.177 | 1.704 | 14.900 | | – | Jackson's. |
| 160 | Owsley, | – | – | – | – | 3.00 | – | – | 76.791 | 6 177 | .241 | 13.791 | | 248.5 grs. | Haddock's Cannel. |
| 467 | Pulaski, | 1.274 | 2.20 | 38.90 | 57.00 | 1.90 | 41.10 | 58.90 | 78 608 | 5.311 | .380 | 13.451 | | – | Sear's. |
| 471 | Pulaski, | 1.311 | 4.40 | 33.80 | 58.80 | 3.00 | 38.20 | 61.80 | 76.364 | 5.200 | .420 | 14.716 | | – | Cumberland. |

| | | | | | | | | | | | | | | | |
|---|---|---|---|---|---|---|---|---|---|---|---|---|---|---|---|
| 185 | Union, | 1.321 | – | – | – | 7.60 | – | – | 76.200 | 5.644 | 1.746 | .552 | 8.258 | 136.5 grs. | Mulford's five foot. |
| 188 | Union, | 1.325 | – | – | – | 6.80 | – | – | 73.417 | 4.977 | 2.824 | 1 658 | 10.322 | 108. grs. | Ice House. |
| 166 | Union, | – | – | – | – | 7.60 | – | – | 74.309 | 5.244 | .880 | 11.967 | | – | Casey's. |
| 555 | (Pennsylvania.) | 1.329 | 1.00 | 3.500 | 58.40 | 5.60 | 36.00 | 64.00 | 78.437 | 5.6?9 | not est. | 1.319 | 8.555 | – | Youghiougheny. |

## TABLE 3. Iron Furnace Slags.

| Number in report. | County. | Silica. | Alumina. | Lime. | Magnesia. | Protoxide of iron. | Protoxide of manganese. | Potash. | Soda. | Oxygen in the silica. | Oxygen in the bases. | Proportion of O. in the bases to O. in the silica. | Furnace. |
|---|---|---|---|---|---|---|---|---|---|---|---|---|---|
| 491 | Bullitt, | 54.60 | 15.90 | 11.93 | 8.09 | 3.29 | 1.08 | 4.25 | 1.31 | 28.350 | 16.780 | 1 : 1.69 | Bellemont. |
| 492 | Bullitt, | 53.36 | 17.26 | 9.74 | 8.09 | 6.35 | .89 | 4.09 | 1.02 | 27.700 | 16.540 | 1 : 1.67 | Bellemont. |
| 330 | Greenup, | 58.00 | 20 50 | 12.06 | 2.19 | 3.51 | 1.21 | 2.12 | .55 | 28.884 | 15.551 | 1 : 1.78 | Buena Vista. |
| 423 (*a*) | Greenup, | 48.80 | 33.27 | 12.50 | 1.24 | 1.19 | .51 | 1.62 | .18 | 25.338 | 15.499 | 1 : 1.63 | Caroline, (gran'lr.) |
| 423 (*b*) | Greenup, | 48.86 | 33.05 | 12.86 | 2.74 | 1.13 | .51 | 1.54 | .15 | 25.369 | 16.169 | 1 : 1.57 | Caroline, (glassy.) |

## TABLE 4. SANDSTONE, &c.

| Number in report. | County. | Specific gravity. | Sand, &c. | Carbonate of magnesia. | Carbonate of lime. | Potash. | Soda. | Alumina and oxide of iron. | Sulphuric acid. | Phosphoric acid. | |
|---|---|---|---|---|---|---|---|---|---|---|---|
| 496 | Bullitt, | 2.427 | 93.68 | 0.84 | trace. | 0.21 | 0.59 | 3.95 | trace. | - | |
| 497 | Bullitt, | 2.415 | 94.78 | 2.29 | 0.18 | .27 | .14 | 2.85 | trace. | - | |
| 498 | Bullitt, | 2.453 | 94 75 | .70 | .16 | .96 | .10 | 3 48 | trace. | - | |
| 505 | Fayette, | -- | 87.83 | 1.40 | trace | .27 | .14 | 8.65 | .22 | 0.25 | Underlying the Beech Ridge. |
| 506 | Fayette, | -- | 83.45 | 2.30 | 1.79 | .41 | .01 | 10.25 | .92 | .50 | |

## TABLE 5. PIG IRON.

| Number in report. | County. | Specific gravity. | Iron. | Graphite. | Combined carbon. | Total carbon. | Manganese. | Silicon. | Slag. | Alumina. | Calcium. | Magnesium. | Potassium. | Sodium. | Phosphorus. | Sulphur. | Furnace. |
|---|---|---|---|---|---|---|---|---|---|---|---|---|---|---|---|---|---|
| 435 | Greenup, | 7.009 | 90.00 | 1.77 | 0.90 | 2 67 | 0.33 | 4 28 | 1.15 | 0.13 | 0.14 | 0.21 | 0.17 | 0.14 | 0.61 | 0.12 | Laurel. |
| 434 | Greenup, | 6.886 | 89.54 | 1.87 | .16 | 2.03 | .54 | 5.57 | 1.25 | .13 | .19 | .20 | .17 | .11 | .46 | .10 | Laurel. |

## TABLE 6. LIMESTONES.

| Number in report. | County. | Specific gravity. | Carbonate of lime. | Carbonate of magnesia. | Lime. | Magnesia. | Carbonic acid. | Alumina and oxide of iron. | Phosphoric acid. | Sulphuric acid. | Carbonate of iron. | Carbonate of manganese. | Potash. | Soda. | Bituminous matter. | Silica. |
|---|---|---|---|---|---|---|---|---|---|---|---|---|---|---|---|---|
| 484 | Anderson, | – | 96.65 | – | 54.23 | – | – | 1.26 | 0.92 | 0.25 | – | – | 0.57 | 0.39 | – | 0.88 |
| 485 | Anderson, | – | 83.95 | 0.91 | 47.11 | – | – | 2.23 | .25 | .34 | – | – | .38 | .47 | – | 11.28 |
| 486 | Anderson, | 2.653 | 86.45 | 1.57 | 48.52 | – | – | 1.83 | .12 | trace. | – | – | .62 | .11 | – | 9.57 |
| 490 | Bullitt, | 2.766 | 63.13 | 27.76 | 35.43 | – | – | 4.34 | .19 | 3.77 | – | – | .44 | .15 | – | 1.63 |
| 494 | Bullitt, | 2.799 | 63.45 | 29.64 | – | – | – | 3.15 | – | .27 | – | – | .20 | .21 | – | 2.18 |
| 495 | Bullitt, | 2.765 | 50.25 | 31.05 | – | – | – | 5.37 | trace. | 1.46 | – | – | 59 | .20 | – | 10.32 |
| 507 | Fayette, | 2.660 | 92.73 | .63 | 52.03 | – | – | 2.42 | .86 | .34 | Chlorine | 0.05 | .23 | .28 | – | 2.18 |
| 508 | Fayette, | 2.711 | 77.63 | 10.00 | 43.56 | – | – | 3.23 | .70 | 3.12 | – | – | .32 | .15 | – | 4.98 |
| 511 | Fayette, | 2.716 | 51.57 | 29.33 | – | – | – | 3.57 | .37 | .34 | – | – | .71 | .82 | – | 11.58 |
| 512 | Fayette, | 2.703 | 55.54 | 40.80 | 31.16 | – | – | .96 | – | .02 | – | – | .36 | .22 | – | 2.79 |
| 513 | Fayette, | 2.615 | 55.99 | 37.33 | – | – | – | .72 | .25 | .33 | – | – | 2.35 | .25 | – | 3.38 |
| 514 | Franklin, | 2.699 | – | – | 50.19 | 0.66 | 40.15 | 1.24 | .44 | .68 | – | – | .23 | .29 | – | 6.94 |
| 515 | Franklin, | 2.700 | 76.75 | .19 | – | – | – | 2.25 | .09 | .85 | – | – | .48 | .44 | – | 18.86 |
| 516 | Franklin, | – | 92.65 | 1.54 | 51.99 | – | – | 1.19 | .09 | 1.27 | – | – | .30 | .13 | – | 3.68 |
| 481 | Greenup, | 2.691 | 73.90 | 2.08 | – | – | – | 1.19 | .46 | – | – | – | .27 | .05 | – | 21.67 |
| 477 | Greenup, | – | 91.47 | 2.75 | – | – | – | .48 | – | – | oxide 1.82 | .05 | .13 | .10 | – | 3.38 |
| 433 | Greenup, | 2.699 | 97.90 | .74 | 54.93 | – | – | .53 | – | – | – 22.19 | – | .28 | .08 | – | 1.27 |
| 432 | Greenup, | 2.731 | 50.33 | 1.83 | – | – | – | .77 | .77 | .70 | oxide 1.49 | .47 | .38 | .20 | – | 21.43 |
| 426 | Greenup, | – | 65.13 | – | – | 1.41 | – | .13 | .17 | .40 | oxide 25.80 | .17 | .11 | .06 | – | 1.27 |
| 427 | Greenup, | 2.729 | 84.47 | 3.47 | – | – | – | .25 | .62 | – | 7.73 oxide 1.77 | .26 | .32 | .14 | – | .55 |

## TABLEs 6—Continned. LIMESTONE.

| Number in report. | County. | Specific gravity. | Carbonate of lime. | Carbonate of manganese. | Lime. | Magnesia. | Carbonic acid. | Alumina and oxide of iron. | Phosphoric acid | Sulphuric acid. | Carbonate and oxide of iron. | Carbonate of manganese. | Potash. | Soda. | Bituminous matters. | Silica. | |
|---|---|---|---|---|---|---|---|---|---|---|---|---|---|---|---|---|---|
| 456 | Grayson, | 2.651 | 46.83 | 26.84 | 26.28 | – | – | 0.38 | 0.12 | 0.33 | 3.44 | trace. | 0.50 | 0.37 | – | 20.78 | Hydraulic. |
| 521 | Jefferson, | – | 50.43 | 18.67 | 28.29 | – | – | 2.93 | .06 | 1.58 | – | – | .32 | .13 | – | 25.78 | Hydraulic. |
| 528 | Jefferson, | – | 50.76 | 45.00 | 28.49 | – | – | 1.78 | | .04 | – | – | .21 | .35 | – | 2.48 | Magnesian. |
| 530 | Jefferson, | – | 56.36 | 37.07 | 31.62 | – | – | 1.28 | | trace. | – | – | .33 | .35 | – | 5.68 | Magnesian. |
| 325 | Lawrence, | – | 50.95 | 4.53 | – | – | – | 1.91 | .36 | – | 7.63 | – | .57 | .31 | 2.00 | 32.17 | – |
| 455 | Ohio, | 2.721 | – | – | 47.06 | 2.39 | 38.55 | 1.44 | .12 | .80 | – | – | .29 | .24 | – | 9.96 | Hydraulic? |
| 457 | Trigg, | 2.702 | – | – | 43.91 | 7.00 | 40.90 | .36 | .06 | – | – | – | .21 | .09 | – | 8.36 | Hydraulic? |
| 458 | Trigg, | 2.596 | – | – | 28.61 | 14.77 | 38.85 | 1.96 | .92 | .29 | – | – | .27 | .30 | – | 13.68 | Hydraulic? |
| 547 | Woodford, | – | 91.33 | .56 | 51.25 | .26 | 40.38 | 1.53 | .70 | .33 | – | – | .34 | .43 | – | 5 18 | Leptæna. |
| 548 | Woodford, | 2.705 | 94.75 | 1.96 | 53.17 | .93 | 42.61 | .63 | trace. | .30 | – | – | .23 | .32 | – | 2.18 | Bird's-eye? |
| 549 | Woodford, | – | – | – | 54.12 | .45 | 41.90 | 1.04 | .63 | 1.78 | – | – | .48 | .39 | – | .78 | Bellerophon. |

## TABLE 7. Soils, Marls, and Sub-soils.

| Number in the report. | County. | Dissolved from 1000 grains by water containing carb. acid. | Moisture. | Organic and volatile matters. | Alumina. | Oxide of iron. | Oxide of manganese. | Carbonate of lime. | Magnesia and carbonate. | Phosphoric acid. | Sulphuric acid. | Chlorine. | Potash. | Soda. | Sand and silicates. | Formation, &c. |
|---|---|---|---|---|---|---|---|---|---|---|---|---|---|---|---|---|
| | | Grains. | per ct. | per c. | per ct. | per ct. | per ct. | per ct. | per ct. | per ct. | per ct. | per ct. | per ct. | per ct. | per ct. | |
| 233 | Adair, | 2.471 | 2.50 | 4.440 | | 4.841 | | 0.196 | 0.046 | 0.065 | 0.232 | 0.005 | 0.075 | 0.092 | 90.446 | Lower Silurian. |
| 218 | Ballard, | 0.733 | 1.80 | 2.110 | 2.580 | 2.240 | 0.090 | .150 | .860 | .410 | – | – | .120 | .020 | 91.720 | Quaternary. |
| 219 | Ballard, | 1.293 | 2.14 | 2.920 | 2.250 | 3.390 | 0.360 | trace. | .470 | .180 | – | – | .190 | trace. | 90.210 | Quaternary. |
| 225 | Barren, | 3.872 | 2.34 | 5.200 | 3.460 | 2.206 | .234 | .366 | .205 | .159 | .197 | – | .197 | .090 | 87.686 | Sub-carb. Limestone. |
| 227 | Barren, | 0.820 | 3.90 | 4.730 | 10.380 | 6.398 | .256 | .096 | .522 | .075 | .466 | – | .142 | .082 | 77.067 | Sub-carb. Limestone. |
| 312 | Breckinridge, | 1.775 | 6.72 | 7.040 | | 12.170 | | .976 | .413 | .101 | .198 | .002 | .556 | .190 | 78.680 | Do. marly shale. |
| 216 | Christian, | 0.960 | 2.24 | 2.960 | 2.390 | 2.360 | .270 | .130 | .790 | .270 | – | – | .190 | .040 | 90.260 | A sub-soil. |
| 500 | Clarke, | 2.093 | 4.16 | 6.100 | 3.940 | 4.920 | .400 | .470 | .620 | .480 | not est. | – | .320 | .080 | 82.650 | Lower Silurian. |
| 501 | Clarke, | 1.370 | 2.96 | 4.010 | 7.710 | 7.060 | .290 | .990 | 1.040 | .380 | not est. | – | .360 | .030 | 78.030 | Lower Silurian. |
| 222 | Clinton, | 1.481 | 1.96 | 3.970 | 1.766 | 2.466 | .076 | .076 | .131 | .090 | not est. | – | .085 | .099 | 90.720 | Sub carb. Limestone. |
| 232 | Cumberland, | 5.122 | 2.40 | 5.770 | 1.230 | 3.140 | .076 | .336 | .438 | .127 | .734 | .006 | .220 | .029 | 87.110 | Knob Formation. |
| 230 | Daviess, | 3.592 | 1.62 | 3.350 | 2.026 | 2.146 | .126 | .176 | .258 | .088 | not est | – | .096 | .053 | 91.920 | Coal Measures. |
| 504 | Fayette, | 3.520 | 4.12 | 4.881 | | 10.306 | | .276 | .133 | .254 | .109 | – | .139 | .047 | 83.834 | Lower Silurian (beech woods.) |
| 509 | Fayette, | 4.350 | 7.30 | 5.242 | | 19.206 | | 1.196 | .426 | .434 | .054 | – | .308 | .086 | 72.994 | Red sub-soil. |
| 510 | Fayette, | 1.112 | 6.38 | 4.913 | | 20.300 | | .116 | .034 | .383 | .082 | – | .309 | .159 | 73.874 | Red sub-soil. |
| 517 | Franklin, | 3.680 | 5.18 | 9.133 | | 8.100 | | .316 | .517 | .243 | .068 | – | .173 | .049 | 80.754 | Lower Silurian. |
| 518 | Franklin, | 2.637 | 1.98 | 3.790 | | 4.589 | | .196 | .066 | .151 | .054 | – | .135 | .026 | 90.734 | Lower Silurian. |
| 518a | Franklin, | 2.366 | 2.525 | 4.206 | 2.120 | 2.915 | .004 | .173 | .233 | .128 | .043 | – | .130 | .051 | 90.170 | Lower Silurian. |
| 518b | Franklin, | 0.830 | 3.30 | 3.179 | 4.470 | 4.825 | .005 | .082 | .312 | .148 | .033 | – | .282 | .002 | 86.380 | Lower Silurian. |
| 522 | Jefferson, | – | 4.42 | 7.996 | | 7.418 | | .394 | .240 | .205 | .082 | – | .200 | .043 | 83.134 | Upper Silurian. |
| 523 | Jefferson, | – | 2.80 | 4.506 | | 6.240 | | .316 | .200 | .191 | .067 | – | .158 | .070 | 88.318 | Upper Silurian. |
| 524 | Jefferson, | – | 2.98 | 2.844 | | 6.335 | | .256 | .226 | .099 | .082 | – | .181 | .028 | 89.900 | Upper Silurian. |
| 525 | Jefferson, | – | 3.60 | 3.112 | | 17.020 | | .194 | .366 | .497 | .088 | – | .297 | .111 | 77.434 | Upper Silurian. |
| 526 | Jefferson, | – | 3.94 | 4.039 | | 11.840 | | .236 | .216 | .126 | .109 | – | .239 | .043 | 82.694 | Upper Silurian. |

## TABLE 7—Contined. Soils, Marls, and Sub-soils.

| Number in the report. | County. | Dissolved from 1000 grains by water containing carb. acid. | Moisture. | Organic and volatile matters. | Alumina. | Oxide of iron. | Oxide of manganese. | Carbonate of lime. | Magnesia and carbonate. | Phosphoric acid. | Sulphuric acid. | Chlorine. | Potash. | Soda. | Sand and silicates. | Formation, &c. |
|---|---|---|---|---|---|---|---|---|---|---|---|---|---|---|---|---|
| | | grains. | per ct. | per c. | per ct. | per ct. | per ct. | per ct. | per ct. | per ct. | per ct. | per ct. | per ct. | per ct. | per ct. | |
| 527 | Jefferson, | - | 2.700 | - | | 33.900 | 4.280 | .580 | 1.220 | - | - | - | - | - | 50.180 | Upper Silurian (ir. grav. |
| 529 | Jefferson, | - | 3.220 | 3.761 | | 6.952 | | .156 | .240 | .088 | .340 | - | .177 | .031 | 88.294 | Upper Silurian. |
| 224 | Laurel, | 2.404 | 3.600 | 6.190 | | 8.926 | 0.078 | .116 | .280 | .139 | .355 | 0.009 | .239 | .021 | 83.626 | Coal Measures. |
| 217 | Logan, | .585 | 2.800 | 3.140 | 4.770 | 3.660 | .186 | .300 | .400 | .140 | not esti | mated. | .120 | .030 | 89.270 | Sub carb. Limestone. |
| 228 | Monroe, | 2.853 | 1.820 | 4.130 | 2.700 | 2.126 | .116 | .106 | .200 | .075 | not esti | mated. | .119 | .122 | 89.393 | Knob. |
| 223 | Ohio, | 2.330 | 1.74 | 5.080 | | 4.349 | | .176 | .166 | .101 | .413 | .066 | .157 | .015 | 90.166 | Coal Measures. |
| 226 | Russell, | 2 221 | 3.44 | 4.170 | | 4.478 | | .176 | .066 | .088 | .227 | - | .063 | .068 | 90.786 | Knob. |
| 480 | Simpson, | 0.608 | 4.14 | 7.020 | 11.980 | 8.820 | .130 | .210 | .200 | .240 | not esti | mated. | .190 | .060 | 71.130 | Sub cb. lime. red sub soil. |
| 237 | Union, | 2.280 | 2.76 | 4.580 | 2.986 | 2.666 | .056 | .396 | .390 | .115 | not est. | .003 | .139 | .116 | 88.426 | Coal Measures. |
| 235 | Union, | 1.480 | 3.16 | 2.740 | | 9.530 | | .276 | .287 | .147 | .288 | - | .185 | .056 | 86.130 | Coal Measures, sub-soil. |
| 236 | Union, | 1.920 | 3.54 | 3.670 | 2.230 | 5.080 | .080 | .136 | .633 | .088 | .466 | .003 | .087 | .062 | 87.250 | Coal Measures. |
| 220 | Union, | - | 1.92 | 7.060 | | 6.700 | | 50.850 | .698 | .280 | 1.366 | .062 | .310 | .166 | 32.670 | Coal Measures (marl.) |
| 229 | Wayne, | 2.551 | 3.16 | 5.370 | 4.326 | 2.526 | .236 | .256 | .246 | .036 | not esti | mated. | .115 | .137 | 86.066 | Sub-carb. Limestone. |
| 234 | Wayne, | 8:534 | 8.28 | 2.560 | 10.240 | 3.120 | .078 | 1.021 | .922 | .229 | not est. | - | .351 | .123 | 62 506 | Sub-carb. Limestone. |
| 231 | Whitley, | 2.222 | 3.28 | 6.300 | 5.260 | 5.66 | .420 | .076 | .121 | .165 | .322 | - | .170 | .147 | 80.786 | Coal Measures. |
| 550 | Woodford, | 6.014 | 4.70 | 7.771 | | 12.961 | | 2.464 | .173 | .319 | .150 | - | .394 | .130 | 75.266 | Lower Silurian. |
| 551 | Woodford, | 3.720 | 4.60 | 5.513 | | 13.344 | | 2.734 | .333 | .306 | .037 | - | .205 | not es. | 77.594 | Lower Silurian, old field. |
| 552 | Woodford, | 4.950 | 4.52 | 6.450 | | 13.773 | | 3.476 | .354 | .447 | .082 | - | .498 | .095 | 75 434 | Lower Silurian, sub-soil. |
| 553 | Woodford, | 1.000 | 5.04 | 6.065 | | 33.377 | | .138 | .080 | .383 | .198 | - | .234 | .127 | 59.360 | Lower Silurian, red clay. |
| 554 | (Illinois,) | - | 3.28 | 9.050 | 2.405 | 2.350 | | .890 | .526 | .175 | - | - | .197 | .100 | 84.470 | Prairie Soil. |

## TABLE 8. Mineral Waters. (In 1,000 grains of the Water.)

| Number in report. | County. | Specific gravity. | Carbonate of iron. | Carbonate of manganese. | Carbonate of lime. | Carbonate of magnesia. | Sulphate of lime. | Sulphate of magnesia. | Sulphate of potash. | Sulphate of soda. | Chloride of magneisum. | Chloride of sodium. | Silica. | Total saline contents. | Gases. | Name of spring. |
|---|---|---|---|---|---|---|---|---|---|---|---|---|---|---|---|---|
| | | | | | | | | | | | | | | Grains. | | |
| 531 | Lincoln, | – | 0.021 | 0.005 | 0.195 | 0.041 | – | 0.056 | 0.013 | – | – | 0.013 | 0.040 | 0.384 | Carbonic acid. | Grove. |
| 532 | Lincoln, | – | .028 | .005 | .117 | .020 | 0.015 | .112 | .028 | – | – | .018 | .046 | .442 | Carbonic acid. | Brown. |
| 533 | Lincoln, | – | 0.015 | | .139 | .131 | – | .066 | .022 | 0.024 | – | .008 | .041 | .446 | Carbonic acid. | Field. |
| 534 | Lincoln, | 1.00007 | – | – | .013 | .065 | – | .012 | .008 | – | – | .017 | .022 | .164 | Sulphuretted hy. & car. acid. | Howard's. |
| 535 | Lincoln, | 1.0041 | trace. | – | .673 | .116 | .203 | 3.454 | .067 | .774 | – | .081 | .060 | 5.428 | Carbonic acid. | Epsom. |
| 536 | Lincoln, | 1.0068 | trace. | – | .912 | .131 | .185 | 3.520 | .170 | 1.013 | – | .304 | .056 | 6.884 | Carbonic acid. | Foley's. |
| 538 | Lincoln, | 1.0060 | trace. | – | .506 | .375 | 1.566 | 2.989 | .298 | .398 | – | 1.000 | .021 | 7.153 | Carbonic acid. | Sowder's. |
| 539 | Lincoln, | – | .007 | | .118 | .024 | – | .027 | .010 | – | – | .088 | .017 | .291 | Carbonic acid. | Bryant's chaly. |
| 540 | Lincoln, | – | 0.021 | – | .095 | .037 | .010 | .070 | .026 | – | – | .015 | .046 | .320 | Carbonic acid. | Do. Pasture. |
| 541 | Lincoln, | – | trace. | – | .093 | .048 | trace. | .006 | .025 | | 0.042 | .175 | .015 | .404 | Car. acid and sulp. hydrogen. | Valley sulphur. |
| 542 | Lincoln, | – | trace. | – | – | – | .104 | .069 | .016 | .205 | – | .933 | .015 | 1.342 | Do. | Knob sulphur. |
| 543 | Lincoln, | – | .026 | – | .058 | .116 | .012 | .023 | .007 | trace. | – | – | .030 | .272 | Do. | Stone's sulphur |
| 544 | Lincoln, | – | .019 | – | .480 | .013 | .966 | .904 | .066 | .028 | – | .278 | .090 | 2.844 | Carbonic acid. | Well. |

TABLE 9. COMPARATIVE VEGETABLE ASH ANALYSES.

| By whom analyzed— | Merz. | Way and Ogston. | | | | | G. Reich. |
|---|---|---|---|---|---|---|---|
| Kind of vegetable, &c. | Tobacco. | White potatoes. | | Red clover. | Turnip. Green topped white. | | Hemp. |
| | Dried leaves. | Tubers. | Stalks. | T. pratense. | Root. | Leaves. | Whole plant. |
| Potash, | 26.96 | 50.89 | 11.44 | 36.45 | 48.56 | 12.68 | 15.82 |
| Soda, | 2.76 | 2.41 | – | – | – | – | 3.40 |
| Lime, | 39.53 | 2.65 | 37.02 | 22.62 | 6.73 | 28.73 | 35.55 |
| Magnesia, | 9.61 | 4.21 | 6.00 | 4.08 | 2.26 | 2.85 | 7.67 |
| Oxide of iron, | – | 1.06 | 3.78 | .26 | .66 | .80 | 1.08 |
| Sulphuric acid, | 2.78 | 3.19 | 5.12 | 1.85 | 12.86 | 7.83 | 2.76 |
| Hydrochloric acid, | – | – | – | – | – | – | 3.40 |
| Silica, | 4.51 | .91 | 8.22 | .59 | .96 | 2.05 | 7.70 |
| Carbonic acid, | – | 12.14 | 14.09 | 23.47 | 14.82 | 14.64 | 8.38 |
| Phosphoric acid, | – | 17.15 | 2.27 | 6.71 | 7.65 | 3.15 | 14.24 |
| Phosphate of iron, | 4.20 | – | – | – | – | – | – |
| Chloride of potassium, | – | – | – | 2.39 | – | 15.56 | – |
| Chloride of sodium, | 9.65 | 5.38 | 12.06 | 1.53 | 5.44 | 10.67 | – |
| Total, | 100.00 | 99.99 | 100.00 | 99.95 | 99.94 | 99.96 | 100.00 |
| Per centage of ash, | | | | | | | |
| (*a*) in dried vegetable, | 23.33 | 2.98 | 15.00 | 9.56 | 7.40 | 15.20 | *4.60 |
| (*b*) in fresh vegetable, | – | .71 | 2.25 | 1.85 | .59 | 1.82 | – |

* Fresh or dried?

# TOPOGRAPHICAL GEOLOGICAL REPORT

OF THE PROGRESS OF THE

# SURVEY OF KENTUCKY,

THROUGH

Hopkins, Crittenden, Caldwell, Greenup, and Carter Counties,

MADE DURING THE YEARS 1856 AND 1857,

BY

SIDNEY S. LYON,

TOPOGRAPHICAL ASSISTANT.

# INTRODUCTORY LETTER.

To Dr. D. D. Owen, *Principal Geologist:*

Sir: In obedience to your instructions, I herewith submit my report of the progress of the work intrusted to my direction, for the years 1856 and 1857.

The necessary instruments and outfit having been procured for camp No. 2, of the survey, this—the Western—corps was placed under the direction of Mr. Joseph S. Harris, late of the United States Coast Survey, who was dispatched to Hopkins county to resume the work of the late detailed survey at Mr. Watson's, near the line dividing Union and Hopkins counties, on the line of the Caseyville and Providence road.

Having accompanied this party to Hopkins county, a rapid reconnoissance was made of the district in which it had been proposed this corps should operate, meeting Mr. Harris from time to time, and directing his operations.

After having obtained a sufficient knowledge of the country, laid out the work for Mr. Harris during my absence, and left such instructions for the control of camp No. 2 as the requirements of the service seemed to warrant, I proceeded to organize corps No. 3, which was to enter upon the detailed survey of the Eastern Geological District.

For this purpose I repaired to headquarters for the necessary funds. On my return to Louisville I found the sub-assistant, on whom I had relied for the Eastern corps, prostrated by sickness, and unable to take the field. At the time, being unable to procure a proper assistant to supply his place, I was compelled, on this account, to postpone, for a time, the organization of camp No. 3, for the eastern division; meanwhile I concluded to make a reconnoisance of the country lying between Louisville and the margin of the Western Coal Field, in Hancock county, and thence through the country adjacent to the base line, which was to be commenced by the Western corps, No. 2, during the summer. In this examination, having again intersected the line of operation of the Western corps, on Drake's creek, I made all the necessary arrangements with Mr. Harris for commencing the base line, and then proceeded into Crittenden and Livingston counties, to endeavor to determine whether the coal region of Livingston county was an outlier, or an extended peninsula of the Coal Measures connected with the coal field in Union or Hopkins county.

On my return, after consultation with the Principal Geologist, it was decided, as the season was so far advanced, and for the purpose of economizing the funds, to transfer the camp equipage and outfit of corps No. 1 for the use of corps No. 3.

This camp was ordered to Louisville by land, while I proceeded by rail to Cincinnati, for the necessary instruments, and the chronometers which had been sent to the care of Professor Mitchell, of the Astronomical Observatory, who had kindly undertaken to have them rated.

I returned to Louisville by rail, and sent forward, by a special messenger, the instruments for the use of the base line party. Owing to the extreme low stage of the Ohio river, the messenger was detained on the road, and did not join the party for three weeks, and that corps were compelled to begin operations with such outfit as it already had on hand.

The outfit having been completed for corps No. 3, the camp proceeded by land through Paris, Bourbon county, to Greenup county, shipping by the river being out of the question.

After having given the necessary directions for the guidance of corps No 3, to commence operations on Williams' creek, at the mouth of the tunnel of the Lexington and Big Sandy railroad, for carrying out the detailed survey of Greenup county, in the Eastern Coal Field, and having seen that they were making good progress, I then proceeded, in advance of the corps, to make a reconnoissance of the country, and learn the key of these *coal* and *iron* measures, leaving the Eastern corps in successful operation.

On the 12th day of October I left Greenup county and proceeded to join the base line party, in Hancock county.

On the 1st of November, the term of Mr. Harris' engagement having expired, his corps was paid off, and the camp outfit, instruments, &c., were returned to headquarters.

On my return to Louisville I found it necessary to return to Greenup county to settle up the outstanding accounts of corps No. 3.

On my return home I made a rapid reconnoisasnce of the northern part of Greenup and Lewis counties, by the way of Springville and Vanceburg; thence by way of Clarksburg and Mt. Carmel to Flemingsburg, Fleming county; thence by Carlisle, in Nicholas county; thence to Paris, Bourbon county; thence to Georgetown, in Scott county; thence to Frankfort, in Franklin county; thence to Shelbyville, in Shelby county, to Louisville.

Here I engaged the assistance of Mr. Edward Mylotte to aid in making up the office work of the operations, in Greenup and Carter counties, which will be submitted as soon as completed.

I remain, &c.,

SIDNEY S. LYON,

*Assistant Geological Survey of Kentucky.*

# REPORT

OF

# Observations in Hopkins, Crittenden, Livingston, Caldwell, Christian, and Henderson Counties.

In my former report it will be remembered that the out-crop of the lower coal measures, indicating the place of the Bell and Cook coals, which lie at the base of the first thousand feet of coal measures of Union and Crittenden counties, as exhibited on the map accompanying that report, was traced up the line of Tradewater river, until it had been run to the line of Hopkins county, and carried through sections 19 and 20, T. 5 S., R. 2 E. This line requires some modification since it crosses Tradewater river somewhere near the south boundary of section 19, T. 5 S., R. 2 E., and extends thence into Crittenden and Caldwell counties, making, near the corner of these counties, in the Hopkins county line, a long tongue of the coal measures, extending to the south and east of Tradewater. For the position of this tongue see plat of part of Union, Hopkins, Caldwell, and Christian counties, for 1856 and 1857.

The extension of the lower measures of the coal field, into the form above described, as to its outer boundary, has not produced a corresponding change in the line of out crop of the first, second, and third coals of the "Lower Coal Measures," which turn abruptly to the north and even north-westwardly, running in that direction from Providence, in Hopkins county, to the neigbourhood of Steuben's Lick, where the line marking these outcropping beds is deflected more eastwardly, and runs nearly with the line of the Hunting branch of Stuart's creek, to its head, in Wright's ridge, when it takes a bend to the south, and probably crosses the ridge near the Box Mountain Springs, thence down the line of Flat creek, to the Rocky Gap, while the eastwardly boundary of the outcrop of the lower measures, have, by the flattening of the dip, and a succession of waves, faults, and breaks, been spread out on a horizontal surface from one and a half miles to ten or even fifteen miles. The lithological character of the measures has also experienced a change, not less noticeable, viz: the heavy masses of the Finnie Bluff, the Curlew, Ice-house, Little Vein, and the Anvil Rock.

The sandstones are much diminished, and some of them are entirely lost, so much that a section at Wright's ridge, and eastwardly to the outcrop of the lower beds, here known as the Campbell coal, equivalent of the Cook coal, Woldridge, and Terry beds, equivalent to the Bell coal, the associated measures, well developed at the Ohio river, are here very obscure, and though more recognizable at Providence, still it would be hardly possible that the key of these measures could be obtained, either in the line of Wright's ridge or Providence, without first having obtained the clue at the Ohio river, and then having followed the line of outcrop, in all its turnings, to Providence.

Having thus been enabled to identify the equivalent beds at that point, and having obtained a hint of the changes to be expected further to the north, to enable the observer to identify the equivalent beds at Wright's ridge.

At Providence the coals are much thicker and closer together than on the Ohio river, and the associated materials are more calcareous, and the angle of dip seems to be much flatter, since the first thousand feet has been spread out into a belt, ten miles wide, though the spaces between all the coals, where the quantities have been obtained, are less, indicating a positive thinning out of the materials separating the coals, and those materials are of a character indicating a different condition from that controlling the deposition of the equivalent beds, twenty miles to the northwest.

All the sandstones are thinner, and composed of finer grains, than those at the Ohio river; and in their stead we sometimes find limestone, black bituminous shales, and fine micaceous and shaly sandstones.

The same remarks, here made for that part of the coal basin at Providence, and eastwardly to its edge, will apply, with slight modification, to the equivalent measures from the head of the Hunting branch to the edge of the measures, south and east to the margin of the coal-field, near the head of Casselbury and Drake's creeks, in Christian county; these changes are especially noticeable at Mr. Williams', on the Madisonville and Hopkinsville road, on a tributary of Drake's creek, and at the Campbell and Woolridge mines, five miles distant, on the waters of Casselbury creek.

In a line stretching nearly east from Providence, is a range of hills, cut through by various creeks, and which extends to and connects with a range of hills on the south east side of Tradewater river. This range is evidently an axis of elevation, and there are corresponding basins or troughs on the north and south side of this line. That on the south side lies in a line nearly southeast and northwest, beginning in the coal measures, and extending toward the outer edge of the basin, into Caldwell county. This trough is much narrower than that on the north side of the ridge, which covers all the space between its line and the base of Wright's ridge, on its southwest side, being from eighteen to nineteen miles wide in the line of its greatest developement.

This great extent of country, eighteen miles long, with the margin of the outcropping coals, and from ten to twelve miles wide, at right angles with this course, includes a district of country generally level and rich, intersected only by spurs of Wright's ridge, dividing the water courses; many of the valleys are flat and low. These spurs of the ridge may be regarded as the distant, feeble efforts of the mighty power that raised the surrounding margin of millstone grit, and the subcarboniferous limestone, which forms the rim.

Though the prolongation of the Bald hill disturbance is not so conspicuously marked, by high and abrupt ranges in Hopkins as in Union county, still the configuration of the country seems to warrant the opinion that one branch of this disturbance has been extended into Hopkins and Christian counties, and that the same dome-like method observed in Union county has also been exhibited along its course through Hopkins county, and to the margin of the coal field in Christian county.

The detailed surveys necessary to determine this question are not yet sufficiently extended; the subject will be left for further investigation.

The whole energy of the Topographical parties having been engaged in the Topography and Geology of the part of the country which appeared to promise the earliest practical economical results, matters of strictly scientific interest have, for the present, been passed by, and those things only attended to which promised to give results of immediate practical value, except so far only as they were of prime necessity for the proper understanding and investigation necessary to those results.

While awaiting the return of the party who were operating in Muhlenburg county, during the latter part of August, I crossed into Crittenden county, with a determination to find whether there was any continuous connection between the *Union coal* of Livingston county, and those of the Tradewater country. In this excursion I passed by the old site of Bellville, where the counties of Union, Crittenden, and Caldwell corner in the Hopkins county line, and where the line of Hopkins county leaves Tradewater. Passing along the road from Bellville, through Caldwell county, in a southwest direction about three miles, the road then inclines more to the north. Then the intercalated limestones of the millstone grit make their appearance. Two and a half miles further the road makes a southwardly curve, and the Coal Measures re-appear five miles from Tradewater river, as shown in the borings for a well at Shady Grove, which have penetrated the rocks of the Coal Measures, and at a point one mile northeast, where a coal has been opened, said to be four feet thick. From one to one and a half miles from this a coal is to be observed, eighteen inches thick, wedged between heavy sandstones. East of Shady Grove coal has been opened by Mr. J. Land; this coal is said to be four feet thick also. The eighteen-inch coal is again found on the lands of Messrs. Terry and Campbell, and at Mr. Amos Singleton's, three-fourths of a mile east of the grove.

It is highly probable that the intercalated limestone of the millstone grit, before alluded to, near Bellville, has been brought to the surface by a fault.

From Bellville, distant seven and a half miles, in Crittenden county, on the farm of Dr. R. M. Hetherington, coal has been reached in a well; the person boring announced the coal to be one foot thick.

By the line traveled the country is very hilly from Dr. Hetherington's to Piney creek, the hills being caped with from fifty to one hundred feet of the millstone grit, the deep ravines and valleys cutting into the sub-carboniferous limestones.

There are probably one or more faults between Shady Grove and Piney creek, that suddenly bring up the lower rocks, thrusting the Coal Measures forward and to the southwest. Six and a half miles to the northeast of Marion, the county seat of Crittenden county, the upper intercalated limestones of the millstone grit rises to the tops of the hills, being overlaid by a thin capping of from twenty-five to fifty feet, of the debris of the sandstones, which are penetrated by sinking wells, the water being found on top of the limestones. Five miles east of Marion the road crosses a branch of Piney, called Flat creek, which flows in a trough scooped into the masses of the lower interculated limestone.

On passing westwardly from Marion, about five miles, the sub-carboniferous limestone makes its appearance, coming up the dry fork of Livingston creek, here connecting with the same rocks, which are cut into by the waters of the Paroquet fork of Hurricane creek. Where these creeks interlock the sandstones of the millstone grit series are severed, and now all the Coal Measures lying to the west, northwest, and southwest of this point are completely disconnected from the great body of the coal field of western Kentucky.

On the Ohio river the beds of the sub-carboniferous limestone is the surface rock, from a short distance below Crooked creek, in Crittenden county, to the mouth of Deer creek, which enters the Ohio river a short distance above the Union coal mines, in Livingston county. These rocks, as before stated, also form the surface-rock, at the head of Paroquet creek, and from this point extends to the Ohio river; the eastern boundary lies nearly in a north and south direction line. The western limits have not yet been completely traced, it however extends to the north-west from the head of Paroquet creek, for about two miles, forming the beds of the creeks, minor streams, and valleys, the neighboring hills being capped by the lower masses of the millstone grit; then more westwardly, by a great curve, to the mouth of Deer creek, including an area of fifty or sixty square miles of sub-carboniferous limestone country with all the marking characteristics, viz: sink-holes, bold springs, &c.

The belt of millstone grit country lying to the eastward, and between the sub-carboniferous limestone country of Crittenden county, and the productive coal measures on Tradewater river, in the same county, is about twelve miles wide, being very broken from Crooked creek to that river; the dividing ridge between Crooked and Big Hurricane creeks is also capped by the lower masses of the millstone grit and the intercalated limestones, rising rapidly from the Ohio river at the mouth of Crooked creek, into a high table land, with occasional high hills rising above it. The belt of millstone grit, above alluded to, exhibits the evidence of having been much disturbed, the masses having been broken into fissures and cracks, locally much elevated. Places are frequently to be observed where the lower mass of the millstone grit forms the bed of a branch, where it lies in a position nearly horizontal, while the next hills, four or five hundred feet above the level of the stream, have the same rock forming their summits, where it is seen dipping at an angle of ten, fifteen, or even twenty degrees to the southeast, northeast, or northwest, as the case may be, varying with different localities. These remarks are especially applicable to the country north of Piney creek; north of that creek, and eastwardly, to the Caldwell county line, and for some distance into that county, the surface does not present breaks and disturbances on quite so grand a scale. Near the Caldwell county line the measures of the millstone grit at once pass under the rocks of the true coal measures, making the belt of country possesing the remarkable characteristics of the millstone grit country, much narrower in Caldwell and Christian than in Muhlenburg, Butler, and Breckinridge counties, where the same country has been observed. The same remark will apply, with equal force, to Hardin and Pope counties, Illinois, and Perry and Crawford counties, Indiana.

It being established that the coal beds of Livingston county are an outlier, being cut off from the main body of the coal field of which they once formed a part.

It is also worthy of notice, that the upheaving force which has been instrumental in these changes has also brought up the ores of iron, lead, and zinc. It is along the anticlinal axis of this greatest disturbance, which has cut through the millstone grit at the head of and along the line of Paroquet and Big Hurricane creeks, that are to be found the fissures filled with Galena, Fluor spar, and other minerals.

I would, therefore, respectfully suggest that at some convenient time part of the force of the survey be detailed, to investigate the strip of country lying along the axis of this disturbance, extending from the Ohio to the Cumberland river.

Some years since an effort was made to prove the Lead Lodes of Crittenden and Livingston counties; the works were not carried to any considerable extent before they were discontinued, without any profitable result.

The detailed surveys have only been carried to the margin of the coal field bordering on Crittenden, Caldwell, and Christian counties.

The foregoing facts have been obtained, incidentally, in reconnoisance made by myself, for the purpose of obtaining such information as would enable me to direct the operations of the field parties, according to the tenor and spirit of my instructious.

Party No. 2, of the Geological Survey, having begun their operations at the edge of Union county, under favorable circumstances, but the whole party having no previous knowledge of the topography of the country, or its geological features, my operations were restricted mostly to the vicinity of the field-party, thus, by covering but a limited space, I was enabled to make a most critical examination of all the known outcrops of coal, and by pursuing this plan I have, while carrying forward the lineal Survey, discovered many new outcrops of coal, and connected these with my previous observations.

The first line run by party No. 2, this season, was begun at Whitesides' creek, and run northward, and connected with the line dividing the counties of Hopkins and Union, and the work of the previous season. From the point of departure thus obtained, at the termination of the work of Union county, the detailed work of Hopkins county was begun.

After conducting the party a few days, the reconnoisance was carried further. A synopsis of the field notes made, and the facts obtained during these reconnoisances, may aid in arriving at just conclusions as to the structure and value of the Western Coal-field of Kentucky, in Hopkins, Christian, and Caldwell counties.

I shall endeavor to set forth these facts, and the method by which they were obtained, and the impressions they produced on my own mind.

The field-work of the topographical parties not being fully made up, the courses and distances estimated will, for the present, be deduced, from the very imperfect map of Kentucky which I have.

The identification of the different beds of coal, wherever observed, has been made a matter of prime importance, and all coals, spoken of in Hopkins and Christian counties, are referred to their equivalent beds, by the same names by which they are known at the Ohio river.

The first camp pitched on Whitesides' creek, in Hopkins county, was found, on examination, to be on the mass of rocks known at the Ohio river as the rocks covering the three feet or "*Little vein*," and the four-feet coal lying first below it. Neither of these beds, at present, are open here. The "Little vein" has, however, been penetrated, several years since, in digging a well, near the Caseyville and Providence road, within a few yards of the old school house, on Whitesides' creek. About a mile from this the equivalent of the second coal under the Anvil rock has been worked by Mr. Watson, on the southwest side of the ridge, and by Mr. Llewellen, on the northeast side of the same ridge, only a few rods apart.

The following section was taken at the opening into the coal at the "Llewellen bank," on the north side of the ridge, which here is the dividing line between the waters of Tradewater, above the mouth of Crab Orchard creek, of Union county, and Slover creek.

Here the dividing ridge has entirely lost its capping of the "Anvil Rock," there being only about forty feet of materials between the top of the coal and the top of the ridge.

*Section of the Llewellen Coal.*

| Heighth. | Thickness. | |
|---|---|---|
| 39.6 | 30.0 | Sloping ground. |
| 9.6 | 4.0 | Loose pieces of limestone projecting from the surface. |
| 5.6 | .6 | Black bituminous shale varying from 6 inches to 1 foot. |
| 5.1 | .11 | *Coal.* |
| 4.2 | .1 | Parting clay. |
| 4.1 | 2.6 | *Coal.* |
| 1.7 | .2 | Parting clay. |
| 1.5 | 1.5 | *Coal.* |
| | .0 | Top of under-clay, thickness not satisfactorily seen. |

Thickness of the bed is five feet one inch. In this locality there is, therefore, four feet ten inches, in all, of workable coal.

The under-clay was not seen at this place, but the following section of the same coal bed, from the "Watson bank," on the other side of the same ridge, and only a few rods distant, will probably be satisfactory as to the thickness of the under-clay.

*Section of Watson's bank, southwest side of ridge, and equivalent of the second coal under the Anvil Rock.*

| *Heighth.* | *Thickness.* | |
|---|---|---|
| 70.9 | 30.0 | Covered space. |
| 40.9 | 16·0 | Limestone in several beds, much affected by exposure. |
| 34.9 | .6 | Six inches to one foot of black bituminous shale. |
| 34.3 | 1.0 | *Coal.* |
| 33.3 | .3 | Parting clay. |
| 33.0 | 2.6 | *Coal.* |
| 30.6 | .2 | Parting. |
| 30.4 | 1.8 | Coal. |
| 28.8 | 1.0 | Under-clay. |
| 27.8 | 10.0 | Ten to twelve feet of drab colored limestone. |
| 17.8 | 7.8 | Covered space with coal. |
| 10.0 | 10.0 | Sandy shale. |

Thickness of workable coal, in Watson's bed, five feet two inches.

The section heretofore given of the Watson and Llewellen coal, equivalent of the second coal of the lower series, may be further extended by the aid of a partial section obtained about one hundred yards to the east of the opening made on Watson's land.

---

Providence lies in an eastwardly direction from the Llewellen and Watson Coal Banks, about two and a half miles. By an observation of the map of Union county, in the first Geological Report, it will be seen that the outcrop lines of the coals of the Mulford series will be found rnnning eastwardly across section 4, T. 5 S., R. 2 E. After the line has entered section 3 of the same township, it runs south and southwest, to the southeast corner of section 9, when it again curves to the east. The line again curves abruptly to the north, soon after entering section 10. This is probably the centre of a valley of depression in the Coal Measures, which being prelonged extends into Crittenden and Caldwell counties, crossing Tradewater river at or near Bellville. Crossing this valley, as before stated, in section 10, the outcrop line runs northwardly along the eastern edge of the valley to a point near Providence, when the line is again deflected to the south and east, by an elevated fold of the Coal Measures that begins at Providence and runs in a line nearly parallel with the Providence and Princeton road, and on the south side of it, to the outer margin of the Coal Field, near the mouth of Dollison's creek, in Caldwell county. A nameless branch, which rises in section eighteen of the same township, runs northeastwardly into section eight. Out of this section it passes into section seventeen, where it has its bed in the soft materials associated with the fourth coal, under the Anvil Rock, or the "*Four-foot Coal*" of the Lower Coal Measures

| *Heighth.* | *Thickness.* | |
|---|---|---|
| 49.11 | 45.0 | Covered space, with limestones near the coal. |
| 4.11 | 1.00 | Coal. |
| 3.11 | .3 | Parting clay. |
| 3.8 | 2.6 | Coal. |
| 1.2 | .2 | Parting clay. |
| 1.00 | 1.0 | Coal. |

On the north side of the ridge, containing the Llewellen and Watson coal openings, there is evidence of fractures in the masses of the coal measures covering this bed. The drainage having taken the lines of fracture has, by denudation, exposed these beds on the south side of the spurs of the main ridge, and the coal may be entered and worked, on the north of the ridge; with a dip to the north and northwest, carrying these coals under a level country bordering Slover creek.

These beds of the Coal Measures, at Providence, have experienced some slight modifications, the limestones are more ferruginous; there are also beds of chert intercalated in the limestone mass, which is thicker. The dip is greater, and to the northwest. The limestone beds are highly fossiliferous.*

---

If these beds are entered north of the main dividing ridge, the produce of the mines may reach the Ohio river, by a railroad which may be made almost by a naturally graded road bed. By laying the line of the road down the valley of Slover creek, to the Pond Fork; thence up that valley, and on the southwest side of that creek, to the gap between Poplar ridge and Coal hill; thence along the valley of Cypress creek, to a point near the mouth of Pearson's branch; then either up Cypress, and reach the Ohio river by the valley of Hine's creek, or by the Bookham valley pass, through the gap, at the head of the Little vein branch, and thus reach the Ohio; or from Pearson's branch, by the Henry valley pass, through the gap at Winstead's, in section 24, T. 3 S., R. 2 W. For a coal road, with very light grades, I am acquainted with no country where a road, with a better allignment, or lighter grades, could be had, for the same amount of graduation, for a road of this length; while along the line of the road, and at very short distances from it, the best stone for bridges and culverts could, be had; since, near by, along its entire length, the heaviest coal beds in this part of Kentucky find their outcrop, the main trunk road would receive numberless branches from numerous coal mines, that must be opened along the entire length of the line, and only a short distance from it, many of them within from fifty to two thousand yards of the main road. In this connection, I may be permitted to state that a *good* railroad is the only reliable means by which these vast beds of fossil fuel can, with certainty, reach a market. It has been in contemplation to lock and dam Tradewater river for the purpose of forwarding these coals; my opinion is, that Tradewater river, if dammed, and its waters were spread over the surface, as they would be by dams sufficientiy high to obtain the head required, would not, during the dry season of the year, afford a sufficient supply to keep up the pools, much less the water necessary for lockage. The evaporation would probably largely exceed the supply afforded by the river should this be the case, as it most probably is; Tradewater river, therefore, as a means of transporting coal to market, is absolutely useless.

*See specimens "*from Providence.*"

There being no difficulty in the way, the outcropping bench of the Llewellen and Watson coal was easily traced to the equivalent bed opened at Providence. The following section will serve to illustrate, in some degree, the arrangement of the Providence beds. The land holdings here being in very small lots, two banks having different names are included in one section.

These banks are on nearly the same level, and the thickness of the separating material is probably somewhat greater than the work on the ground made them.

*Section of the Lofland Coal Bank.*

| *Heighth.* | *Thickness.* | |
|---|---|---|
| 60.2 | 30.0 | Yellow-grey shales, (place of Anvil Rock?) |
| 30.2 | 3.0 | Black bituminous shale. |
| 27.2 | 1.3 | Slaty coal. |
| 25.11 | .5 | Parting clay. |
| 25.6 | 1.0 | Coal, (in large blocks.) |
| 24.6 | 1.4 | Parting clay. |
| 23.2 | 2.9 | Coal. |
| | 1.9 | Under clay. |
| 20.5 | | |

*Section of Dorris Bank.*

| | | |
|---|---|---|
| 18.8 | 12.0 | Twelve to fifteen feet limestone. |
| 6.8 | 1.3 | Calcareous marly shales, 15 to 20 inches thick. |
| 5.5 | .4 | Black bituminous shale. |
| 5.1 | 4.0 | Coal, fine quality, mining in fine blocks. |
| 1.1 | .1 | Parting clay. |
| 1.0 | 1.0 | Coal to top of underclay. |

The following section is from the Dollison bank, equivalent to Dorris bank:

| *Heighth.* | *Thickness.* | |
|---|---|---|
| 18.2 | 10.0 | Ten to sixteen feet limestone. |
| 8.2 | 2.2 | Grey and black calcareous shale; lower part argillaceous shale. |
| 6.0 | 3.6 | Coal. |
| 2.6 | .1½ | Parting clay. |
| 2.4½ | 1.4½ | Coal. |
| 1.0 | .1 | One to two feet ferruginous limestone. |

A number of sections of these beds could be given, but this will be unnecessary, as they would be a repetition of those already given.

It may be possible that the last section given is of an intercalated coal, between the equivalent of the Mulford bed and the "middle coal;" the

Lofland section is probably the equivalent of the "Anvil Rock," or first coal which has again increased in thickness.

It will be recollected, that at Thompson's mines, in Union county, the Anvil Rock bed had thinned down to fifteen or eighteen inches.

The Lofland, Dorris, and Dollison banks, being three distinct beds of coal, would make the Dollison the third coal under the "Anvil Rock," and equivalent to the Mulford coal, provided there be no new intercalated coal here over the Mulford. This seems to be confirmed by the fact that a coal was opened, partially at my request, on the farm of Mr. Samuel Montgomery, near the eastern edge of section ten, T. 5 S., R. 2 E., which has the characteristic covering and associated shaly materials of the Mulford coal; if this be the equivalent of that coal it will be undoubtedly fall the fourth coal in the series here.

Three-fourths of a mile to the southeast of the last coal alluded to, in the direction across a narrow trough in the measures, a coal is seen in the bank of a small branch, known here as Hunter's bank." The equivalent of this bed is also to be seen on the farm of Mr. Samuel Montgomery, distant about half a mile from the equivalent of the Mulford coal opened on his farm; this last coal is known as the Montgomery coal.

*Section of Montgomery coal.*

| *Heighth.* | *Thickness.* | |
|---|---|---|
| 31.00 | 10.00 | Covered space to top of point. |
| 21.00 | 10. | Thin bedded sandy shale. |
| 11.0 | 8. | Black bituminous shale. |
| 3.0 | 3. | Coal. |

The pit being partially filled with water, neither the bottom of the coal nor clay was seen.

*Section of Hunter's bank.*

| *Heighth.* | *Thickness.* | |
|---|---|---|
| 43.6 | 15.0 | Covered space. |
| 28.6 | 10. | Greyish-yellow sandy shale. |
| 18.6 | 8.0 | Black bituminous shale. |
| 10.6 | 3.9 | Coal. |
| 6.9 | .9 | Parting clay. |
| 6.0 | .7 | Coal. |
| 5·5 | .5 | Under clay. |
| 5.0 | 5.0 | Sandy shales. |
| | .0 | Bed of branch. |

Mr. Alfred Towns had this coal opened, and the parting clay dug through, in doing which the workman discovered the seven inches of coal between the under and the parting clay. When the lowest coal was cut through an abundant spring of water burst forth. When I visited the bank the water had settled, and the spring was flowing in a beautifully clear stream. The day being warm, and being very thirsty, I laid down for a drink, but one mouthful of such water was sufficient—the water being very acid, and largely charged with alum.

In my report for 1855 the map shows the line of the lower coals south of Mr. Imboden's house, section 24, T. 5 S., R. 1 E.

The Winn hill was set down as the equivalent of the "Finnie bluff," or the sandstone mass covering the Bell coal. After the survey was made, in 1855, Mr. Winn dug at the foot of the hill near his house, at the base of the equivalent of the "Finnie bluff," and found a good coal at the place indicated in the report.

Taking the road from Providence to Princeton, the first creek crossed has, at this place, its bed on the soft shale under the Providence coals or over the Mulford coal.

There is evidently a sinking of all the beds for a short distance on the line of this road, when they rise again immediately west of it, while on the right of the same line they lie quite level for two or three miles, as there is only one ridge intervening, made by the members over the Hunter or "Little Vein" coal. On the northwest side of Clear creek the mass of the Curlew hill crosses the road in a low ridge between Clear creek and Mr. Barnhill's house. The mass of the "Icehouse," and its accompanying measures, is probably the northwest bank of the flats of the creek itself, and it is repeated between the forks of the creek. The masses equivalent of the Finnie Bluff extend from the bank of Tradewater river for two or three miles toward Providence, and must be repeated or lie very level.

At two miles beyond Tradewater the millstone grit sets in, and the rocks are raised into high hills. About four miles to the southwest of the road the millstone grit runs up to the river, and the sub-carboniferous limestone forms the hills. The lower masses of the millstone grit still has one, if not both, of the intercalated limestones; the upper mass of this limestone is hard, and broken into polygonal blocks from the size of marbles up to pieces of five hundred pounds weight. The usual buff belt is found less earthy, more solid and compact than at the Ohio, and nearly destitute of fossils. The key obtained by the study of the various members of the coal measures, in Union county, from the millstone grit to the Mobley, has been of the greatest service in identifying the corresponding beds elsewhere; nevertheless, in consequence of some important modifications in the formations in their extension through Hopkins county, it is often necessary, in order to convince ourself positively of identity, to follow out any given bed in all its meanders from its known position in Union county to the locality elsewhere to be established. The changes of lithological character are always accompanied by a corresponding change in the character of of the soil and the growth upon it; this aids greatly in following the outcrops, when the rocks themselves are concealed; indeed, without observing which one would be entirely at a loss.

The first line of country passed over, lying between Providence and Wright's ridge, was through the woods and farms, and by no regular road. Leaving Providence, the route lay to the south of east, crossing Wyer's creek about a mile from its mouth. On the northwest side of the flats of this creek the outcropping sandstones over the "Little Vein" are in sight; beyond this a stretch of bottom land extends for a mile; then a low flat point was crossed, before entering the valley of Clear creek, not far from a pond called "Jenney's hole;" from this point, up the valley of the creek, there is no low water bed for a distance of over two and a half miles, by the path through a flat swampy land, with a succession of small lakes called "holes," with most fanciful names. Pond creek was crossed in the valley of Clear creek, and the route now lay up the southeast side of Clear creek to Lamb's creek, one of its small tributaries. A coal was seen in the hills that bound the south side of Clear creek, near Lamb's creek. From the character of the associated rocks I took it to be either the Ice-house coal, or one of the lower small beds of the Curlew Hill Measures. I am inclined to the opinion that it is the former coal, which has thickened up to four feet four inches, of excellent quality.

It should be mentioned that the ridge on the northwest side of Pond creek, near its mouth, is probably the equivalent of the masses covering the four feet and "Little vein" at the Ohio river. When it was crossed it was about seventy five feet high above the surrounding flat lands. The ridge on the south side of Clear creek, in which the Kirkwood branch takes its rise, is the coal before alluded to, on the south side of the creek; it is opened on the N. W. side of the hill. At the opening the dip is N. 10° W.; the rate of dip is two and a half degrees. In all probability there is a reverse dip somewhere under the flats of Clear creek, or at least a much flatter dip. The rate of dip would bring in the rocks of the "Little Vein" much nearer to this opening than they are, if there were not some decided modification of dip. A limestone is said to exist in the hills over this coal, but it was not seen by myself; there are, however, evidently limestones or else calcareous shales. The sugar-tree, which I consider an invariable sign of the presence of calcareous beds, in the Coal Measures, is quite abundant.

Continuing up the south side of Clear creek, immediately after crossing Lamb's creek, a sharp ridge rises, called Bobb's ridge, probably the equivalent of the ridge at the last coal; this ridge soon receives an additional height, and is capped by another sandstone very like in lithological character to the upper part of the sandstone at Curlew hill. This ridge divides the waters of Richland and Lamb's creeks, and has an elevation of about two hundred and fifty feet above the surrounding flat land. The ridge appears to be the dividing line marking the east side of the great trough extending from Providence to this point, and brings us to the edge of the undulations, marking the character of the measures of Wright's ridge, and the ridge running between Lamb's creek and Richland creek, coming up in the way it does through the direction of the strike line, in a sharp curve to the southeast, then southwardly and southwestwardly to the waters of Cane creek, at least twelve miles south from the mouth of Lamb's creek; so that the productive Coal Measures extend in a tongue twelve to fifteen miles outside of the nearest edge, or smaller diameter, of a regularly shaped basin. The prolongation of the basin, which passes southeast of Bellville, is a depressed fold, while that southeast of Lamb's creek is an axis of elevation, extending on either side of the line of Wright's ridge for several miles; it is, however, most extensive on the westwardly side, also including the body and spurs of the ridge, so that the greatest body of coal, of the Western Coal-field of Kentucky, crops out along this line of elevation. No less than ten beds of coal are here presented to view varying, from three (3) to eight (8) feet in thickness. Five of these are five (5), and two over four (4) feet.

To the eye these coals are not inferior to the best Pittsburgh coal; their true value can however only be determined by analyses.

The road traveled from Richland creek lay up that stream for two or three miles, to the intersection of the Princeton and Madisonville road, near the Sulphur spring; thence up the dividing ridge between Richland and Sugar creeks, crossing Wright's ridge at a point near the head branches of Sugar and Stuart's creeks, where the ridge is nearly severed by a gap lying in a line nearly east and west.

In order to establish the exact geological position of the Coal Measures of Wright's ridge, I determined to trace the measures along the line of the Madisonville and Hopkinsville road, from the thick coal at the Rocky Gap to where the sub-carboniferous limestone is cut through south of the head waters of Drake's creek. From thence, the sub-carboniferous limestone, millstone grit, and the upper intercalated limestone, to the Buttermilk road.

The line of the road from Madisonville to Hopkinsville runs nearly south, crossing the Crab Orchard Fork of Drake's creek, approaching the valley of the latter creek nearly east of the point where the line of Christian and Hopkins counties leaves that creek. Three marked ridges cross this road between the Rocky Gap and the confines of the productive Coal Measures, on the south.

These examinations showed that the eight feet coal, at the Gap opened by Price, Johnson & Co., is the equivalent of the beds on the Hunting branch and Stuart's creek, only the materials separating the two beds, which in some instances amounts to a thickness of several feet, have here diminished to two and a half inches; and for all practical purposes the two beds are here united into one.

On the same horizontal plan, on the west side of the gap, only about five hundred feet distant, there is another bed totally unlike any coal bed which I have had an opportunity of witnessing. The following section will serve to show this bed, which is also on the property of Price, Johnson & Co.

*Section of coal bed at Rocky Gap, north side.*

| *Heighth.* | *Thickness.* | |
|---|---|---|
| 57.03½ | 22.0 | Massive sandstone. |
| 35.03½ | 15.0 | Covered space. |
| 20.03½ | 5.0 | Clay shale, whitish. |
| 15.03½ | 5.0 | Clay shale, yellowish. |
| 10.03½ | 3.0 | Clay shale, grey. |
| 7.03½ | 2.4 | Clay shaie, grey. |
| 4.11½ | .7 | Black bituminous shale. |
| 4.4½ | 2.4 | Black bituminous shale. |
| 2.0½ | 2.0½ | Cannel coal. |

Under clay not measured.

The masses given above are each in distinct beds, with a regular parting between them. The black shale is very hard, black, and rich in bituminous matter. The coal is hard, and in its general appearance much like the finest block mineral at the Breckinridge mines.

The section of the thick coal, on the east side of the gap, is as follows:

*Section of Eight-foot coal at Rocky Gap.*

| *Heighth.* | *Thickness.* | |
|---|---|---|
| 157.11½ | 50.00 | Rounded hill top. |
| 119.11½ | 20.00 | Bench probably sandstone. |
| 89.11½ | 16.00 | Top of slope and foot of bench. |
| 73.11½ | 19.00 | Bench; some sandstone in sight. |
| 54.11½ | 45.00 | Top of covered space. |
| 9.11½ | 1.6 | Blue marly shale. |
| 8.5½ | 3.1 | Coal; soft from exposure. |
| 5.2½ | 0.2½ | Marly shale parting. |
| | | Beds of coal. |
| 5.0 | 5.0 | Coal with 2 small streaks of clay near the center of mass. |

The lower, or five-feet mass of coal appears bright and good. The upper mass will probably be found as good as the lower when the coal has been followed under solid cover.

Starting, as before stated, at these beds at the Rocky Gap, three distinct, different masses of sandstone are passed over, by the line of this road, before the margin of the Coal Field is reached—the distance being about seven miles. The spurs of Wright's ridge, which runs in a line nearly parallel with it, are thrown off at right angles to the road, and are doubtless the sandstones of the "Little Vein," Curlew hill, and Finnie Bluff. There being no repetition of these masses, the inference is that the amount of dip is not greater than two degrees, for a large part of the distance; for suddenly the dip, south of Drake's creek, carries the equivalent of the Cook and Bell coals from the top of the high hills, from two hundred to two hundred and fifty feet high, on the south of the creek, down to the bed of the stream, where the equivalent, probably of the Cook coal, has been mined in the bed of the creek near Mr. Williams.' The covering to this coal being very similar to the covering of the coal mined by Mr. Campbell, on Casselbury creek, five miles to the west, which is certainly the equivalent of the Cook coal of Union and Crittenden counties.

Crossing Drake's creek to the south, and ascending a hill two hundred feet high, which is capped by a sandstone equivalent to the Finnie bluff, a coal outcrop is seen at a place called "Isinglass Glade." On the southwest side of the Glade is the following section:

*Section at Isinglass Glade.*

| *Heighth.* Ft. In. | *Thickness.* Ft. In. | |
|---|---|---|
| 58. | 5.00 | Top of hill at glade. |
| 53. | 2.00 | Black bituminous shale. |
| 51. | 2.6 | Coal. |
| 48.6 | 3.10 | Under-clay. |
| 44.8 | 5.4 | Shales. |
| 39.4 | 5.4 | Belt of carbonate of iron, three inches thick, regular and continuous as far as exposed, in a bed of grey shale. |
| 34.0 | 2.0 | Bed micaceous sandstone. |
| 32.0 | 6.0 | Sandstone. |
| 26.0 | 5.0 | Blue shale. |
| 21.0 | 21.0 | Sandy shale. |
| | .0 | Top of covered space. |

In the under-clay of this coal are a great number of finely formed crystals of gypsum, especially where the coal has disappeared and left the under-clay disturbed; doubtless produced by the abundance of sulphuric acid set free by the wasting pyritiferous coal, which has combined with the lime filtered from the limestone hereafter mentioned; a similar phenomenon was observed at Johnson's, under the equivalent of the Llewellen coal, near Providence, where lime must have been carried from above the coal down to the *acid.*

Above the coal at the Isinglass Glade are two thin beds of carbonate of iron. Between the Glade and the Croft, or Willliams' farm, an additional member comes in above the coal, composed of alternations of sandy shale and beds of flag-stones.

At Petersburg, westwardly from the Glade, the coal measures are flattened, probably by a slip or fault at the south foot of the Glade hill; which has carried down the measures. The hill itself is flattened out to the east and northeast into a high table land. Three miles to the southeast Mr. Lacy has worked a coal bed, which is said to be four feet thick. The workings having been abandoned I did not visit them. About three miles to the west of Petersburg, the dividing ridge between the waters of Tradewater and Pond rivers has its summit, in which interlock the branches of Casselbury creek, the longest branch of Tradewater, and the head branches of McFarland's creek, the longest westwardly branch of Pond river.

On the waters of Drake's creek the coals and associated rocks are much bent and disturbed. One of the lowest of these coals, at Mr. P. W. Cabel's, is either the equivalent of the Cook or the Battery Rock coal. From the character of the rocks associated with this coal I incline to the opinion, that it is the former of these coals. At another opening in this coal, which is seen half a mile further down the branch, the dip was N. 40° E.; the rate of dip was not satisfactorily obtained. The coal is hard and firm around the old pits, although the diggings have not been worked for seven or eight years. The coal is covered with aluminous earth. The top of the hill over this coal is covered by a loose-textured soft sandstone, massive in its character. The sandstone at Chalk-level is probably identical with the sandstone here, and both are probably the equivalent of the sandstone near the base of the Finnie Bluff.

Mr. Felix Bourland, (seventeen miles from Madisonville,) opened the coal last alluded to. Mr. Bourland says the coal is three feet four inches thick. The thickness of the coal is probably over estimated.

At Mr. Brashears' is a coal covered by argillaceous shale; this bed is doubtless the equivalent of the "Cook coal;" and the coal at the Isinglass Glade is the equivalent of the "Bell coal" (?).

The materials between these coals are generally soft and argillaceous, while the mass covering them has decidedly the character of the Finnie Bluff; here, however, it is quite soft in its lower part, while the main sandstone mass is only represented by a few feet* of sandstone,

*About 18 or 20 feet.

flagstones, and sandy shales, the whole mass being about fifty to sixty feet thick.

At Petersburgh* limestone was found at the bottom of a well, which I take to be the equivalent of the limestone afterwards seen, lying above the coal at Mr. Campbell's, and at the Wooldridge "old mine." On the west side of the dividing ridge, this limestone lies at those places thirty-five feet above the coal, equivalent to the Cook coal.

The first mass of the millstone grit seen containing pebbles was near the farm of Mrs. Elizabeth Brashear's, eighteen miles south of Madisonville. This mass forms here the dividing ridge between the waters of Casselbury and Drake's creeks. Limestone, intercalated with sandstone of the millstone grit series, are first met with one and a fourth miles south of Mrs. Brashear's, or nineteen and a fourth miles south of Madisonville.

From information obtained, I infer that there must be a number of openings in the equivalent of the Bell and Cook coals, with perhaps one or two places where the equivalent of the Battery Rock coal is worked, southwest and west of the road.

At Mrs. Brashears' and Mr. Williams', when the intercalated limestones before alluded to were seen, there is much irregularity in the dip, both in quantity and direction, being in most cases conformable to the contour of the hill on which it is observed; the rocks here are generally softer than these equivalent beds of Union and Crittenden counties, and, so far as I have been able to obtain measurements, the rocks are found to be thinner.

At Mr. Williams' I listened to one of the legends of the country, which appears to be fully credited by the people. This story, as related to me, details, with much apparent accuracy, the direction, size, and condition of certain great lodes of lead, not yet worked in this part of the country; also, of certain mines of silver, said to exist near the margin of the coal field. The relator of this information informed me that nothing but his great age and ill health prevented him from opening and operating the mines, whose existence he had communicated to me. Nothing, however, that I was able to observe at these localities, would warrant me in giving any encouragement to these fancies, but rather to discourage any hope of these visions of wealth

*Petersburgh is station No. 1002 of the Nashville and Henderson railroad.

being realized. There may be all that the mineral witches declare there is, of lead and silver, but the Mineralogical and Geological signs do not accompany them here, as they do at localities, where lead and silver are found, elsewhere.

From where the millstone grit and sub-carboniferous limestone cross the road, near Mr. William's house, in Christian county, the margin of the former was traced to the Buttermilk road.

Along the great dividing ridge between the head branches of Pond and Tradewater rivers, its east side is abrupt and precipitous, while on the west side the spurs of the main ridge are thrown off flatter, except on the west side of the head of Tradewater, where the hills are also rough, produced either by lines of drainage ploughed into them, or original lines of abrupt bending, and irregular folding of the uplifted heavy masses of the millstone grit.

On crossing the first hill formed by the masses of the conglomerate, between Mrs. Brazier's and Mr. Williams', you descend into a valley of one of tbe branches of McFarland's creek. On the east side of the road the conglomerate caps the hill from seventy-five to eighty feet thick, with one hundred to one hundred and fifty feet of soft argillaceous shales immediately underlying it.

On the first branch east of Mrs. Brazier's at a short distance north of the road, is to be seen a fine evidence of a fault; the heavy masses of the millstone grit having sunk down. Along a line, on the east side, are the shale beds before alluded to, while on the west is the solid wall of the masses of millstone grit. Near the junction of these measures there bursts from the sandstone side of the fault a bold cold spring of most excellent chalybeate water.

The rocks of the millstone grit, form the bed of Casselbury creek, where it is crossed by the Buttermilk road, near Mr. Alexander Brazier's; on the south side of whose house, and near the *school-house*, is a fine example of the upper intercalated limestone, but no satisfactory section could be obtained. These limestones present the usual "*glady*" appearance. These rocks here dip toward Casselbury creek, i. e., to the north, and sink under the millstone grit. On descending the creek, on a line nearly east and west, the spurs of the hills are crossed at right angles with their length, and are found to be waves formed of solid masses of rocks and shales, from fifty to one hundred feet high from the top of the wave to the bottom of the intervening trough, here represented by the valleys between the spnrs. The band of limestone before alluded to crosses the spurs in a rising line, one wave descending into the succeeding valley or trough, and again mounting and descending. The bending of the strata, by the original force forming these hills, has thrown the limestone up and down with the folds forming these spurs, and on a given line the limestone crosses them like a ribbon resting on their surface. The greatest lie on the axes of the ridges; the least in the valleys between them. On the top of the ridges the dip equals twenty-five to thirty degrees; in the valleys from five to twenty degrees. The buff limestone is here associated with this bed, and is about thirty feet thick; the whole mass of limestone is probably one hundred feet thick. From the base of the intercalated limestone, to the great mass of the sub-carboniferous limestone (?,) the distance is about sixty-five feet, filled with thin bedded sandy shale and flagstones.

In a branch of Casselbury the sub-carboniferous limestone was seen, where it was broken into heavy square and oblong blocks.

The margin of the coal field is deflected very rapidly to the north, from the crossing of Casselbury creek.

I visited the coal bank worked by Mr. Campbell, on the southwest side of the Buttermilk road, where the direction of the dip is north twenty degrees east; the rate two to two and a half degrees; the coal is two feet seven inches thick, the roofing being "grey metal, i. e., grey micaceous shale. This coal rests on a bed of thin under-clay, four to six inches thick; this on hard thin-bedded sandstone. The bank is worked by stripping. The coal is the undoubted equivalent of the Cook coal.

The old Wooldridge bank is an opening into the same bed; it is over this coal that is found the limestone spoken of as being found in the well at Petersburg.

I learned of Mr. Campbell that a bed of thin coal exists on Casselbury creek, one mile to the south of his mine; this is probably the equivalent of the Battery coal; it is said to be one foot thick.

Mr. Wooldridge is now opening another bank, on the east side of a small ridge, about half a mile from the old "*diggings;*" this new opening is in a coal above the limestone, which lies above the coal at the old diggings, but owing to the curving arrangement of the beds I thought *it* useless, with the Lock level, and the time at command, to undertake to determine the thickness of the beds intervening between the two beds of coal.

The coal mined at the new opening is of superior quality; the bed presents a face, where seen, three feet high; it is hard, bright, shining, black.* No sulphur ("*or brass*") was observed in the coals at this opening, which is covered by dark sandy shale ("*grey metal.*") This bed is the equivalent of the Bell coal.

Two and a half miles to the northwest the same bed has been worked by Mr. Patrick Hamby; here the greatest thickness observed was four feet four inches; the coal is covered by argillaceous and sandy shale, probably the counterpart of the shale beds observed at the Wooldridge new mine. The difference in the appearance of the covering materials at the two mines is owing, doubtless, to the wasted condition of those at the Hamby mine. It may not be improper here to state that, all the banks mined here, with one exception, (the Terry mine,) are stripped at the tail of the bed, and no regular pits or slopes are made. The materials associated with these beds are very soft, and sections above and below the coals could not be obtained. The only section as yet seen, north of the millstone grit, on this line, was obtained at the Wooldridge old mine, which gave thirty feet from the top of the Campbell (Cook) coal up to the base of the limestone above it—the space being filled with sandy shale. At Mr. Campbell's the limestone was found to be twelve feet thick and upwards. At both places it contains many fragments of entrochites and spirifer. The shales weather to a yellow-grey color. The Hamby bank is not now worked, the stripping having become very heavy. It is situated one-fourth of a mile from, and on the west side of, the Buttermilk road, and eighteen miles from Madisonville, and about a quarter of a mile from the line dividing Hopkins and Christian counties; it lies in the latter county. In the branch which runs through the bed of coal at the workings, the dip of the beds under the coal was ascertained to be to the northwest; the under-clay here is four inches thick, resting on lumpy irregular bedded sandstone. The bed of the coal is wavy and irregular. This coal has also been worked by Mr George Terry.

About one mile further, to the north, the coal is entered at the north face of a low ridge, where it lies at a much higher level than at the Hamby bank. The same bed is opened in the bed of the branch, one

*See specimens labeled Wooldridge coal, Christian county.

hundred and fifty yards up the stream, from the Terry opening; the coals in the bed of the branch have been stripped, and some coal has been removed; the dip here was north seventy degrees west, at the rate of four or five degrees. On tracing the covering rock down the branch, on the north side of the stream, the coal was traced to a point opposite the Terry opening, on the west side of the road; the coal on this side of the stream having been carried down by the dip thirty feet below the bottom of that exposed in the Terry opening. On the south side of the branch the coal may be traced to the Terry bed, which is entered by the usual form of slope entry, which enters with an ascending dip rising to the south or with the direction of the entry, and the coals are at the entry dipping to the north. A little below the opening on the north side of the branch, sugar-trees, the usual accompanyment of limestone in these measures, are to be found in great abundance. From the data here obtained it is inferred that the limestone is twenty-five feet below the coal, and that the distance between the coals, equivalent to the Cook and Bell coals, is fifty five feet. Hence there appears to be a thinning of the member between these coals, with the insertion of a limestone mass in some localities, as at Mr. Campbell's, from ten to twelve feet thick. The Terry coal mine is on a branch of Buffalo creek. On the left of the road, near Terry's mine, Buffalo creek receives a small branch on the north side. One fourth of a mile up this branch from its mouth is to be seen the following section, which is illustrative of the measures of Wright's ridge. At the locality of this section the end of a ridge is worn off by the action of the waters of the branch, producing an escarpment one hundred feet long and thirty feet high, rising above the surface of the pool at its foot. The face of the escarpment stands in a line nearly north and south, and presents a remarkable stratification, dipping from the centre of the bed, to the north and south, and burying the top of the coal under the water at both ends of the pool.

*Section on branch of Buffalo creek, near the Buttermilk road.*

| *Heighth.* Ft. In. | *Thickness.* Ft. In. | |
|---|---|---|
| 31.5 | 12. | Light colored yellow-grey sandstone. |
| 19.5 | 11. | Grey metal. |
| 8.5 | 4. | Coal. |
| 4.5 | .9 | Under-clay. |
| 3.6 | 3.6 | Grey metal. |
| | .0 | Surface of pool. |

This coal is the equivalent of the Terry coal.

Ascending the branch, toward the north, several waves are apparent in the measures, gradually rising with the ascending valley. Near the head of the valley, the measures of the section are lost to view under the hill; at the base of which, on the west side, Mr. Croft made an opening into, and has taken coal from, the bed equivalent to the Terry bed, which here lies at a much greater elevation than at that mine. The opening is near the summit of one of the many waves into which the measures are thrown in this part of the coal field. This bed rising and falling with the measures until it disappears under the south side of the first great hill south of Caney creek, and is finally lost to view on the line of this road.

The top of the hill south of Caney creek is evidently capped with the equivalent of the Ice house, and the lower part of the Curlew hill measures. This hill is separated from the southern prolongation of Wright's ridge by a deep valley, through which a trial line has been run for the Nashville and Henderson railroad, and it was stated to me that a road, with a maximum grade of forty feet to the mile, can be made through this gap, by a cut of sixteen feet at the deepest part, and a fill of twenty-five feet for the valley of Caney creek.

Near the mouth of Cane run, on Caney creek, two miles west of the Buttermilk road, there is an outcrop of coal under black bituminous shales, and over these eighteen to twenty feet of grey metal, which is covered by loose blocks of heavy sandstones, none of which were seen in place; this is probably the equivalent of the Bell coal.

In a northeast direction from the mouth of Cane Run, new and superior members begin to appear.

It now became apparent that the line traveled had been nearly parallel with the strike line; also, that the dip conformed largely to the external surface of the country, and that few of the hills are raised to a sufficient heighth, above the equivalent of the Bell coal, to contain the next superior bed; until I had reached the first hill south of Caney creek, around whose sides might be traced the denuded edges of the superior measures.

Overlying the coal seen at the mouth of Cane run, near Mr. Joseph Woodruff's, on the Princeton and Greenville road, are to be seen the measures lying first above that coal. The line of the following section lies from Cane run, along the road toward Mr. Woodruff's house.

*Section near Mr. Woodruff's, in the valley of Cane run.*

| *Heighth.* Ft. In. | *Thickness.* Ft. In. | |
|---|---|---|
| 45.00 | 15.00 | Thin bedded fine sharp grit sandstone. |
| 32. | 3.2 | Sandy shale. |
| 28.10 | 2.6 | Coal. |
| 26.4 | 1.2 | Under-clay. |
| 25.2 | 1.6 | Clay filled with limonite ore. |
| 32.8 | .8 | Black bituminous shale. |
| 23.0 | 2.4 | Coal. |
| 20.8 | .8 | Under-clay. |
| 20.0 | 20.0 | Covered space. |
| .0 | 0. | Bed of Cane run. |

There are many blocks of limestone associated with the above section; their bed could not, however, be certainly traced in place; these blocks of limestone have been used to repair the public road in the vicinity of the coal outcrop, as they can be found in the washes and water-worn gullies cut into the soft part of the mass, and are buried in the debris of the coal and clay, both above and below both beds of coal. A more extensive opening on this bed would easily determine the place of the limestone.

This bed, at this point, dips to the northeast at the rate of ten to fifteen degrees.

On the north side of Mr. Woodruff's house, four hundred yards distant from the point of the foregoing section, black bituminous shales are seen in the bed of a branch; these beds are much bent and curved, having been disturbed by a great number of waves, but they have much less dip than the measures of the preceding section. They are covered by the soil and clay of the valley of Cane run, but are no doubt connected with, and are part of a bed of shales seen on Caney creek, at the mouth of Cane run, about three-fourths of a mile distant, to the southeast.

Some segregations of bituminous carbonate of iron are seen here. No decided dip could be obtained.

Half a mile eastwardly lies the foot of "Dozier" hill; here is presented the next member succeeding the section last given. The Princeton and Greenville road has laid bare a bed of sandy micaceous shale, near the "Christian Privilege" meeting-house. The bed of shale forms the body of the point on which the meeting-house stands, rising about fifty-five feet above the ravine, on the northwest side of it.

In this ravine, and probably at the base of the shale beds above, are to be seen several blocks of ferruginous limestone, which could not be traced to any regular bed. Irregularly diseminated in the shale bed, above the limestone, are a number of thin broken bands of clay ironstone.

Having now arrived at the base of the great hills, which are here known under many specific names, they will, for the sake of clearness, be treated of under the general appelation of Wright's ridge.

The ridge is prolonged towards the west, in a high range on the north side of Caney creek, presenting, on its south face, a bold front deeply indented by narrow and nearly parallel ravines, from which flow, during the wet season of the year, Cane, Buck, Fox, and Pigeon runs, besides a number of nameless drains, all entering Caney creek on the north side of that stream—the longest of these runs being about five miles, by a direct course from the head to the mouth of the stream, all nearly south— some of the steepest of the steep hill sides presenting, along their length, the outcrop of four different beds of coal—the same beds being repeated on all of these branches with more or less modiification.

Flowing towards the west, from the ridge, are first: Richland creek, with a number of small tributaries, still further north; Sugar creek, and its branches, on the north slope of the range; and from its folds and wrinkles, Stuart's creek, and its branches, descend nearly due north.

The equivalent of the first, second, third, and frequently the fourth coals of Union county, under the Anvil Rock, appear in natural outcrop at numerous places on those creeks, from Caney creek to the Hunting branch. What is here said of the west side of the ridge is true of the east side, also, with certain modifications. Flat creek and Pleasant run rise in the ridge and flow toward the east. Between these streams the ridge is prolonged nearly to the mouth of Drake's creek. In many places, however, nearly surrounded by deep valleys, called gaps, which cross its line.

Ten miles north of Caney creek the Hunting branch and Stuart's creek join their waters, and form Clear creek—the Hunting branch flowing from the east, and joining Stuart's creek, which flows from the south, forming nearly a right angle with Clear creek, which flows to the southwest.

A territory here, with its southern boundary on Caney creek, with a breadth six of miles from the mouth of Cane run, extending to the north ten miles, covering about fifty-four thousand acres, presents probably as many natural outcroppings of six different beds of fine workable coal, as are to be found in a district of like size in any coal field, if not more.

The facts here exhibited may be explained on the hypothesis that the axis of Wright's ridge, which has a geneıal course north and south, is crossed nearly at right angles by a series of waves, elevating and depressing the measures composing the ridge.

These waves carrying the same measure, from the anticlinal to the synclinal axes, upward and downward, from fifty to one hundred and fifty feet, with a distance from north to south, of from a half to three-fourths of a mile, from the summit of the wave to the summit of the succeeding one, gradually sinking deeper and deeper below the horizon of the last preceding wave. At the same time there is an axis of elevation rising from the flat land on the east side of the ridge attaining the greatest elevation, near the longitudinal axis of the ridge; then again dipping toward the flat lands of Richland and Clear creeks. The amount of this elevation and depression crossing the ridge varying, in different places, from seventy to two hundred and twenty feet.

On the south face of the spur of the ridge lying parallel with Caney creek, and near the Christian Privilege meeting-house before alluded to, is an out-crop of coal, known as the Charles Woodruff bank. The coal has been stripped, and a small amount taken from the tail of the bed, which is four feet seven inches high. The dip here is south, or a little west of south, at the rate of two or three degrees. There is some doubt as to the precise direction of the dip, but the coal certainly dips from the axis of the hill. This is the first bed on the south side of the basin, in this district, that exhibits calcareous spar in the fractures of the coal. The coal is hard and firm, and not unfrequently breaks with a choncoidal fracture. The roof of the coal is very black bituminous shale, from five to six feet thick.

There is no heavy sandstone to be seen here, but about one hundred feet above the coal, and one-fourth of a mile to the west, is a bluff of solid sandstone, weathering into rock houses, resting on fifteen feet of blue argillaceous shale. The sandstone, measured in a favorable place, is nineteen feet thick; its upper exposed surface is eighty-five feet below the top of the hill, mostly of soft materials. About forty-one feet above the rock, is a belt of calcareous material, some fifteen or twenty feet wide, which seems to result from the waste of a limestone; its out-crop does not show on the surface at this place, but was found afterwards in place fifteen to sixteen feet thick.

Further west the equivalent of the Charles Woodruff coal outcrops in the bed of a small branch, at about the same horizontal level as that coal. Descending this branch to Cane run I examined the several out-crops on the land of Mr. John Davis.

The following section, taken from one of these out-crops of the bed here exposed to view, will give its general character, although it is variously modified at every different locality.

*Section on Cane run, from one of the so-called beds of black band ore.*

| *Heighth.* Ft. In. | *Thickness.* Ft. In. | |
|---|---|---|
| 17.4 | 10.0 | Black shales, top of exposure. |
| 7.4 | .6 | Pyritiferous shales and coal. |
| 6.10 | .4 | Hard black shale. |
| 6.6 | 2.0 | Pyritiferous shales, eighteen to twenty-four inches. |
| 4.6 | .6 | Irregular sandstone. |
| 4.00 | 2.6 | Pyritiferous argillaceous shale. |
| 1.6 | 1.6 | Sandy shale. |
| | .0 | Coal in bed of creek from three inches to two feet thick. |

At one of the points on Cane run, where the foregoing section was made, a six-inch coal rests on a bed of sandy micaceous shales, full of fine specimens of stigmaria, in a good state of preservation. They are generally flattened, or partially crushed, and sometimes eleven feet in length, but too delicate to preserve entire* with the means then at my disposal.

By a section obtained near here of strata, exposed forty-five feet above the last section, no openings having been made, I am not able to say whether there is a workable coal or not. I am rather led to believe it is only a thin band of black bituminous shales, covered by bluish argillaceous shale, and thin micaceous sandy shale. It is highly probable that the former bed has been thrown up, and that this out-crop is only a part of the upper members of the last cited section.

All the materials in this particular locality are very soft. The hills

*See specimens collected, marked Cane run, Hopkins county.

and points are round and smooth; the ravines are deep, and cut through the clay down to the soft argillaceous or sandy shales.

At all the points observed on Cane run, the dip was in either the south, southeast, or southwest direction. No rocks were observed dipping at the northward, except at Mr. Joseph Woodruff's.

On one of the head branches of Cane run, which runs from the east toward the west on the northside of the spur of Wright's ridge, called "*Dozier*," is to be observed the equivalent of the Charles Woodruff coal. This last out-crop is on the land of Mrs. Nancy Morgan, and the bed is here known as the "Nancy Morgan coal."

The ravine in which the branch has its course being one of denudation, the out-cropping coal bed is seen on both sides of it. The dip is from one to two degrees to the southwest.

Crossing a spur of "*Dozier*" to the head of Fox river, in an eastwardly direction, I visited, on the run, two exposures of one of the most distinctly marked beds of coal in this region. One of these exposures is near the water level of the run; the other lies about one hundred yards further to the east, i. e., down the run.

This last exposure gives the following section:

*Section of Fox Run coal, on the lands of the "Hopkins Mastodon coal company."*

| *Heighth.* Ft. In. | *Thickness.* Ft. In. | |
|---|---|---|
| 44.0 | 15.0 | Covered space. |
| 29.0 | 8.0 | Bituminous shale. |
| 21.0 | 4.0 | Coal. |
| 17.0 | 1.6 | Under-clay. |
| 15·6 | 3.3 | Sandy shale. |
| 12.3 | 2.3 | Rough ferruginous limestone. |
| 10.0 | 10.0 | Micaceous and sandy shale. |
| | .0 | Bed of Fox run. |

This is bright and hard, exhibiting very little sulphur. A small quantity having been mined three years before my visit, was found lying near the outcrop; it was very bright, and is evidently a good coal to resist the destroying effects of exposure.

At this point the measures dip eastwardly; nearly south the bed has fallen to a lower level by a dip in that direction, and at forty feet from the place of the section, on the north side of it, the measures are falling toward the north. Here is probably the summit of one of the waves before alluded to, as crossing Wright's ridge.

On a branch of Flat creek, on the east side of Wright's ridge, four miles north of Caney creek, and near the "*Mitchell Old Field*," is an out-crop of the equivalent of the bed of the John Davis section on Cane run.

From the locality, at the *Mitchell Old Field*, the bench covering the next coal above this horizon, may be traced up a drain coming into Flat creek from the south side, to the place of the section which will be hereafter given. From which place, the heavy mass before alluded to, as covering the coal, may be traced down Flat creek to the crossing of the Madisonville and Hopkinsville road, where the coal also out-crops; at seventy-five feet above the equivalent bed, at the *Mitchell Old Field* locality; it is also seen out-cropping in the bed of the drain leading from the tunnel of the Nashville and Henderson rail-road to Flat creek, showing the upper part of the "Black Band Bed."

The same coal bed may also be traced, by the covering mass of sandstone, up Flat creek to the Box Mountain springs. Up the right hand, or north branch, the Black Band may be traced through the gap across the ridge, to the headwaters of Stuart's creek, while up the south, or left hand branch, the coal above it is easily traced, by the sandstone before alluded to, to the very source of the stream, there being many places, on both sides of the branch, when the coal is itself exposed in outcrop. Some of these exposures have been slightly opened.

There is great difficulty in obtaining the thickness of the rocks, or the interval between the coal beds, which arises from the uncertainty as to the direction of the dip for any considerable distance, and from the fact that the dip so generally conforming, in some considerable degree, to the sloping hill sides, the thickness is almost certain to be made too great by a quantity equal to the amount of the dip, and the data for the correction of this error cannot be obtained with any degree of certainty.

All the localities of the so-called Black Band ore, on Cane run, Stuart's, Richland, and Flat creeks, are no doubt the out-crop of the same bed, and although there is great difference in the character of the bed, in different localities, but from the relation this bed bears to the next succeeding measure in an ascending order, (which is so distinctly marked as to be unmistakeable,) I have no hesitation in placing all the Black Band localities in the same Geological horizon.

The differences to be found in sections taken of this bed, at a distance of three or four miles asunder, are not greater than of those at the same out-crop, within a few feet of each other. The last remark will apply with great force to the localities on Cane run.

The following section taken across one of the spurs of Wright's ridge, called "*Barney's ridge*," is probably the most reliable section—reaching from one coal bed to another—obtained in the ridge country. The line of the section being very short, and measured nearly with the direction of the strike line, while the vertical distance differs from the horizontal dent about as three to one.

*Section at Barney's Ridge, half a mile south of the Mitchell's old fiield.*

| *Heighth.* Ft. In. | *Thickness.* Ft. In. | |
|---|---|---|
| 166. | 20.0 | Covered space to top of ridge. |
| 146. | 14. 0 | Sandstone? |
| | | Coal?* |
| 132.2 | 37.4 | Steep bank, partly covered and partly sandy shale. |
| 95.0 | 26.0 | Steep face of bluff, principally thick bedded sandstones. |
| 69.0 | 12.00 | Hard mass of sandstone. |
| 57.0 | 16.0 | Steep bank, covered with loose sandstones. |
| 41.0 | 4.0 | Black bituminous shale. This mass probably extends higher. |
| 37.0 | 5.0 | Coal. |
| | | No under-clay. |
| 32.0 | 2.6 | Pyritiferous sandy shale. |
| 29.6 | 4.0 | Top of limestone, loose blocks. |
| | 25.0 | Base of rough blocks limestone; covered space mostly sandy. |
| 7.0 | | Shales, seven feet in sight, at |
| | .0 | Bed of branch. |

At the head of the south fork of Flat creek, one and a half miles from the Box mountain springs, the out-crop of the coal, placed at one hundred and thirty-two feet in the last section, is to be seen in out-crop, in a notch in the hills at the heads of that and Richland creek, at the very summit of the ridge dividing these two streams. The coal, at the base of the section before given, having been traced the entire length of the valley. Here the ridge is much wasted by denudation, which has cut out the upper coal immediately at the notch, on either side of which the hills are much higher than at the notch itself.

*This bed was afterwards seen in place; it is the equivalent of the Bear Wallow coal, &c.

The following section, taken on the southeast side of the notch, includes the upper part of last section:

*Section ot Bear Wallow, head of Richland and Flat creeks.*

| *Heighth.* Ft. In. | *Thickness.* Ft. In. | |
|---|---|---|
| 69. | 18.0 | Covered space above sandstone, composed of waste of sandy shale. |
| 51.6 | 10.0 | Heavy sandstone. |
| 41.6 | 1.6 | Black bituminous shale. |
| 40.00 | 2.0 | Coal. |
| 38.0 | 1.0 | Under-clay. |
| 37.0 | 37.0 | Sandy and argillaceous shale, near the base of which are several thin, broken bands of carbonate of iron, from half an inch to two inches thick. |
| .0 | 0.0 | Base of shales at covered space. |

The bed of shale, and the carbonates at the base of the above section, strikingly resemble the covering mass at the "*Gordon coal*" bank, on the south east side of the ridge.

The coal at the Bear Wallow is seen in out-crop on both sides of the notch, descending by a rapid dip towards it; conforming, in a great degree, to the surface of the ridge. On the south side of the notch the dip is to the north; while on the north side the dip is south. The Bear Wallow is evidently situated upon, or near a line of fracture which runs up the south branch of Flat creek, crossing the ridge at the Bear Wallow. It may also be traced for some distance down the valley of Richland creek; a similar line of fracture also crosses the ridge by the line of the north branch of Flat creek, and is finally lost to sight in the valley of Stuart's creek.

From the top of the ridge, near the Bear Wallow, an extensive prospect is opened to the south and west, extending from the ridge across the flats of Richland, Clear creek, and Tradewater river, to the margin of the coal field; around which extends a barrier of hills, formed by the upturned edges of the beds of millstone grit and sub-carboniferous limestone; taking in at one view the line of hills from the headwaters of Casselbury creek, the longest branch of Tradewater river, to the hills at the mouth of Piney creek. Thus exhibiting, at one view, the line of the margin of the coal field extending through Christian and Caldwell into Crittenden county, a line seventy-five or eighty miles in length, varying in distance, from the observer, from twelve to thirty miles.

On the south side of Clear creek, near the Madisonville and Princeton road, is to be seen an opening into a coal, here known as the *Barret bank.* This opening is situated on the north face of the hill, about seventy-five feet from the bed of the creek.

This bank gave the following section:

*Section of the Barrett coal bank.*

| *Heighth.* Ft. In. | *Thickness.* Ft. In. | |
|---|---|---|
| 185.0½ | 100.0 | Covered space to top of hill. |
| 83.0½ | 2.4 | Marly shales, grey and dove colored, with segregations of iron stones. |
| 82.8½ | .4 | Black marly shale. |
| 82.4½ | .8 | Black marly shale with segregations of limestone. |
| 81.8½ | .4 | Black bituminous shales. |
| 81.4½ | 4.5 | Coal. |
| 76.11½ | .2 | Parting clay. |
| 76.9½ | 1.9½ | Coal. |
| 75.0 | 75.0 | Covered space. |
| | 0.0 | Bed of creek. |

The same bed of coal was visited, where it had been worked by Mr. P. M. Robinson, on the south side of Clear creek. The coal lies near the level of the creek, and gives the following section:

*Section of P. M. Robinson's coal bank.*

| *Heighth.* Ft. In. | *Thickness.* Ft. In. | |
|---|---|---|
| 89.2 | 70.0 | Space covered by soft materials. |
| 19.2 | 1.0 | Shale in sight, at foot of covered space. |
| 18.2 | 3.4 | Coal. |
| 14.10 | .4 | Coal rust. |
| 14.6 | 1.2 | Parting clay. |
| 13.4 | 1.5 | Sandy shale. |
| 11.11 | 3.0 | Limestone. |
| 8.11 | .8 | Marly shales. |
| 8.3 | 1.7 | Black bituminous shale. |
| 6.8 | 4.2 | Coal. |
| 2.6 | .2½ | Parting clay. |
| 2.3½ | 2.3½ | Coal. |
| | .0 | Top of under-clay, the thickness not seen. |

The coal is soft, iridescent, much marked with pyritiferous matter. The beds, at the opening, are dipping to the northeast at one or one and a half degrees. The beds again come to the light near the bed of the creek, about half a mile above the Robinson bank. No good section could be obtained here. The bed has the same limestone covering, which lies much nearer to the lower coal here than at the Robinson bank.

The dip observed at this bank was six degrees, in a direction north fifteen degrees west. Still further up the creek, an opening has been made into a coal called the *Marston* Hall bank. A section of all the materials, at the openings, could not be obtained, I am, therefore, unable to identify this as the equivalent of any of the beds heretofore seen. The distance from this to the last *bank* is about half a mile. About fifty yards to the northwest the bed of the creek is a pyritiferous limestone?

A section from the bed of the creek up to the coal is as follows:

*Section on Clear creek which includes the Marston Hall coal.*

| *Heighth.* Ft. In. | *Thickness.* Ft. In. | |
|---|---|---|
| 98.0 | 20.0 | Covered space to top of point. |
| 78.0 | 3.0 | Black bituminous shale. |
| 75.0 | 4.6 | Coal. The whole bed was not seen. |
| 70.6 | 55.0 | Covered space up to bottom of exposed coal. |
| 15.6 | 2.6 | Sandy shale and flagstones. |
| 13.0 | 12.0 | Sandy shale with ironstones. |
| 1.0 | 1.0 | Limestone? |
| .0 | .0 | Bed of Clear creek. |

Dip to northwest; rate one degree.

The two last localities observed are separated by the valley of a branch entering Clear creek from the south. On the line of this valley is to be observed great irregularity, both in the amount and direction of the dip. Near this last branch, called Stuart's creek, and high in the hills, an examination was made of a coal opened by Mr. Hiram Oldham. The following arrangement was presented here:

*Section of the Hiram Oldham coal bank.*

| *Heighth.* Ft. In. | *Thickness.* Ft. In. | |
|---|---|---|
| 9.10 | 2.6 | Black bituminous shale. |
| 7.4 | 0.5 | Marly shales with segregations of limestone. |
| 6.11 | 1.3 | Rotten or partially decayed coal. |
| 5.8 | 3.2 | Solid coal; good quality. |
| 2.6 | .2 | Parting clay. |
| 2.4 | 2.4 | Coal. |

The under-clay is not exposed, except at its upper surface.

The belt of calcareous shales presented at this opening a lenticular mass, fifteen feet in length, and one foot thick in its thickest part, gradually thinning away to a line at either end.

It is probable the limestone at the Barrett and Robinson banks are here represented by this mass of marly material.

No bed of coal was discovered in the ravine below the Oldham coal. This coal lies from one hundred and fifty to two hundred and fifty feet above the bed of Clear creek, about three-fourths of a mile ot the southeast of the Robinson bank.

Before closing my report on the Wright's ridge country I will add a few remarks, to those already made, on the subject of the so-called bed of Black Band. All the known exposures of this bed in the country having been visited, the following section on the headwaters of Stuart's creek, will be added.

*Section of so-called Black Band* bed at head of Stuart's creek.*

| *Heighth.* Ft. In. | *Thickness.* Ft. In. | |
|---|---|---|
| 1.9 | .6 | Black bituminous shale. |
| 1.3 | .1 | Black *band*, (productive?) |
| 1.2 | .1 | Black shale. |
| 1.1 | .1 | Black *band*, (productive?) |
| 1.0 | .3 | Black shale. |
| .9 | .1 | Black band, (productive?) |
| .8 | .3 | Black shale. |
| .5 | .1 | Black band, (productive?) |
| .4 | .2 | Black shale. |
| .2 | .2 | Black band, (productive?) |
| | .6 | Shale. |
| | 1.0 | One to two feet bluish clay. |

Segregations of limestone, probably part of a bed of limestone.

It may be also added that no two localties ever furnish the same section of ore or separating masses. The average of many localities may be set down at eight inches, and that all the out-crops of the Black Band are in the same Geological horizon. It will also be necessary that special examination be made of the ores of each locality to determine their value.

Northwardly of the head of Flat creek, and on the east branches of Stuart's creek, several out-croppings of the coals appear near the

* See analysis of specimen, No. 132, page 337, report 1854–55.

road from Madisonville to Hopkinsville—one known as the Arnold bank.

An entry has been driven into the coal about sixty yards, in a direction south fifteen degrees east, which direction lies across a spur of a hill.

The coal descends from the mouth of the entry for a short distance, with a slight dip; then it runs level for a short distance, descending again, with step like grades, at the rate of about six inches in twenty-five feet.

The section of the Arnold bank, east fork of Stuart's creek, is as follows:

| *Heighth.* | *Thickness.* | |
|---|---|---|
| Ft. In. | Ft. In. | |
| 22.8½ | 14.00 | Covered space, including yellow micaceous shale. |
| 8.8½ | 1.6 | Black shale. |
| 7.2½ | 4.9 | Coal. |
| 2.5½ | .2 | Parting clay. |
| 2.3½ | 2.3½ | Coal. |
| | .0 | Top of under-clay. |

Down the same branch, northwestwardly from the Arnold bank, is to be seen the Bart. Sisk bank; here the direction of the dip is north fifty degrees east, and the rate two and a half to three degrees, being greatest at the most westwardly exposure. This bank gives the following section:

*Section of the Bart. Sisk bank.*

| *Heighth.* | *Thickness.* | |
|---|---|---|
| Ft. In. | Ft. In. | |
| 19.4 | 4.0 | Black bituminous shale. |
| 15.4 | 2.0 | Coal. |
| 13.4 | 1.0 | Under-clay. |
| 12.4 | 1.6 | Marly shale. |
| 10.10 | 4.6 | Shale with segregations of bituminous limestone. |
| 6.4 | 4.3 | Coal. |
| 2.1 | .2 | Parting-clay. |
| 1.11 | 1.11 | Coal. |
| | .0 | Top of under-clay, thickness not ascertained. |

On the opposite side, a little further down the same branch, to the west and southwest, the same bed has been opened. On the north side of the branch the coal soon disappears, under the bed of the branch as you descend it, or to the west, where the upper bed of coal

44

in the foregoing section is covered by a massive sandstone, twelve to fifteen feet thick.

Northwardly, for two and a half miles, no out-crop is seen, until immediately in the vicinity of Madisonville, where a limestone is exposed resting on a bed of argillaceous shale. Under this limestone is an abundance of water; springs break forth wherever it is worn through. From Madisonville the limestone dips westwardly for some distance, at the rate of about sixty feet to the mile.

Continuing northwardly from Madisonville and Ashbysburg road, I passed over a country mostly composed of soft materials. Thin bedded micaceous sandy shales were observed in the road two miles north of the town. The country is gently undulating, with high lands three and a half to four miles to the northwest. At the deep cut the line of the Henderson and Nashville railroad was intersected, where is a bed of loose textured sandstone, forty feet thick; it rests on a mass of blue pyritiferous shale; the dip is northeast, and the cut exposes many lines of fracture in the mass. The line of the railroad was followed until it crosses the Asbysburg road, at a distance of three miles from the deep cut. As the road lies in the valley of a branch there are no good Geological sections here. The boundary of the valley consists of low hills and undulating land.

The Ashbysburg road leads off to the west and northwest, crossing the head branches of Deer creek, and intersecting the Madisonville and Henderson road near Mr. Orr's. There are said to be some coal openings on the branches of Deer creek, which I was unable to visit.

Near the intersection of the Madisonville and Henderson road with the road from Morganfield to Madisonville, the country begins to exhibit protrusions of the millstone grit, extending from the disturbances; at which are situated, two and a half miles distant to the northwest, bold and constant springs of limestone water (?) The wasted and worn condition of the farms along the road mark the area of the underlying millstone grit, brought up by the extension of the Bald Hill fault.

The transition on the east side of the road here is probably from the millstone grit to the Coal Measures, equivalent to the Jerusalem school-house range, of Union county, which lie high in the series between the "Blue" (Bonharbour?) and the Newburgh beds. It would be exceedingly interesting to settle these opinions, by an examination of the country here in detail. This being on the northeast side of the fault (?) it is probable that a different state of case may exist on the southwest side of it; it is probable that the *lower coals* may be found out-cropping on the slopes of this great ridge on the south and southwest side, duplicating the coals of the Coal Measures, as at Chalybeate and Bald hill. Having, as yet, made only three hurried journeys through this district, I am not prepared to give any precise information of the country lying between the surveyed part of Union and Hopkins counties.

# CHAPTER II.

OBSERVATIONS IN

*Greenup and Carter, and incidentally in Bourbon, Bath, Fleming, and Lewis counties.*

Before proceeding to report on the progress of the surveys in Greenup and Carter counties I shall make a few remarks on observations made in a few other counties along the line of travel into these counties.

All the rocks observed westwardly of Owingsville, in Bath county, belong to the Lower Silurian division, as indicated by the contained fossils, viz: *Leptœna alternata, Orthis occidentalis, Orthis* (*spirifer*) *lyra, Murchisonia bicincta?*

The disintegrated rocks around Paris produce a reddish-brown soil, increasing in depth of color, with local exceptions, to within four miles of Mt. Sterling, where it gradually becomes less red as you go northeastwardly toward Owingsville, beyond this the soil is of a dirty lemon-yellow color, with locally strips of whitish clay (crawfish) lands. Near Mr. Jackson's, four miles from the Licking river, towards Owingsville, the surface gives some indications of iron ore.

The top of the hill east of Owingsville is capped with a yellow earthy calcareous stone, containing a few entrochites. This yellow member was first noticed at this locality; from observations made on this formation in December I should estimate its thickness to be from seven hundred to one thousand feet. From Slate creek to near the mouth of Triplet's creek the road runs over this member, which is the principal rock of the first range of Knobs, first seen in the distance toward the southeast, after leaving Mt. Sterling.

A short distance, after crossing the main fork of Licking river, the Devonian Black Slate was first observed on the line of this road. Near the mouth of Triplet's creek, on the south side of the road, the Black Devonian Slate presents bold bluffs, about fifty feet high, with a slope of an equal heighth above the bluff, evidently the wasted materials of the same formation, which is at least as thick, if not thicker, here than it is in the vicinity of the falls of the Ohio.

On either side of the valley of Triplet's creek, along which the road runs, the hills are based on the slate, and are capped with the same knob sandstones which form the Salt river hills in Bullitt county.

Triplet's creek has its course along a fracture or valley of this formation, nearly parallel with the strike line, and its waters flow upon it for twenty miles. The dividing ridge at the head of Triplet's creek, and between it and the waters of Tygert's creek, rises from four to five hundred feet high, above the waters of the mouth of Triplet's creek, and are composed of the knob-stone.

Three miles from the mouth of Triplet's creek the knob-stone is surmounted by masses of chert breccia. I am not certain whether these masses belong to the upper or lower division of the sub-carboniferous group, as I have not yet observed any fossils in it.

The general dip of the rocks on Triplet creek is with the line of the drainage, and towards the stream itself, which flows in an original sloping valley, whose contour is not produced, as one might at first suppose, altogether by denudation.

After descending six miles from the dividing ridge at the head of Triplet's creek, down a branch of Tygert's creek, the road then leaves the branch, and turning to the left crosses a heavy hill two hundred and fifty to three hundred and fifty feet high, above the valley on either side of it. Towards the top a bed of limestone ten or twelve feet thick, is in place above the road, which, on the northeast side, continues to traverse the knob-stone until within a short distance of Olive Hill P. O., where the limestones of the sub-carboniferous limestone set in. One mile further northeast the road crosses Tygert's creek, which here has its bed in a gorge cut into the sub-carboniferous limestone; and the rocks are dipping to the northeast.

One mile from the crossing of Tygert's creek the millstone grit is first seen. The belt of the sub-carboniferous limestone is about two and a half miles wide along the line of the road, and is probably three hundred to three hundred and seventy five feet thick, including the intercalated grey shale beds. For about one mile the road runs northeast and southwest, on the top of the masses of the millstone grit, which is here nearly horizontal; then the road takes a bend more to the east, and crosses the hills, which are capped with the shale beds at the base of the Coal Measures, consisting of soft shaly sandstones and grey shales, without any beds of hard sandstones. After passing this hill, and descending on the opposite side, a limestone is cut by the road. The sandstone of the millstone grit thins out rapidly as we go northeastwardly.

From the top of the hill to the southwest of Grayson, to the Little Sandy river, the rocks dip toward the river rather more rapidly than the surface of the country, which carries down the rocks forming the hill tops before alluded to, and at the ford of the river forms the bed of the stream.

The following approximate section extends from the rocks at Little Sandy river, to the foot of the hill, southwest of Owingsville:

| Total thickness. Feet. | Thickness of each member. Feet. | |
|---|---|---|
| 2520 | 100 | Soft beds at the base of the Coal Measures, in Carter county, with locally a bed of limestone, twenty inches thick, intercalated. This member varies in thickness in different localities. |
| 2420 | 75 | Seventy-five to one hundred feet millstone grit. This member, as well as the sub-carboniferous limestone, thins out toward the Ohio river near the mouth of Tigert's creek, where this member forms a mass fourteen feet thick, and the sub-carboniferous limestone is only twelve feet thick. |
| 2345 | 100 | Calcareous muddy shale, with a few thin beds of limestone. |
| 2245 | 350 | Sub-carboniferous limestone, thinning rapidly in the direction of the Ohio river. |
| 1895 | 20 | Twenty to seventy-five feet grindstone grit, (upper part of Knob formation?) |
| 1875 | 725 | Knobstone, Waverly sandstone of Ohio. |
| 1150 | 120 | Black (Devonian) slate, 100 to 150 feet thick. |
| 1025 | 700 | Buff porous limestone of Lewis, Fleming, and Bath counties. |
| 325 | 75 | Limestone producing red earth by disintegration. |
| 250 | 100 | Slaty mudstone, thin bedded. |
| 150 | 150 | Lower silurian or blue limestone, forming the base of the Owingsville hill. |

In reporting upon the progress of the work in Greenup and Carter counties, it will be necessary to premise that the facts embodyed in the report have been mostly derived from a reconnoisance, in advance of the detailed work of the topographical party.

The eastern coal field was entered in a favorable direction for studying its members from the base upwards, which I consider the best method, since the millstone grit is always a good base of departure.

At the crossing of the Little Sandy river this stream here flows in a bed worn into a mass of dark sandy micaceous shales, immediately above which rests the little, or eight inch, coal of Stinson's creek, which I consider the lowest coal in these measures. An opening has been made into this coal, about one hundred yards to the left of the road, at the foot of the first hill beyond Stinson's creek, but no satisfactory view of the coal and the associated measures could be obtained, as the drift had fallen in. The Stinson cannel coal shows itself in a number of places in the hills at the right of the road adjacent to Stinson's creek. This bed is worked here by Mr. James Clark; the bed at this place lies about forty feet below the top of the hill.

The following approximate section, here given, is the best I am at present able to furnish; most of the quantities are from actual measurement:

*Section of the Stinson creek connel coal, and the associated beds, descending to the millstone grit.*

| *Thickness.* Ft. In. | |
|---|---|
| 18.00 | Covered space, consisting principally of sandy shales. |
| 2.0 | Black bituminous shale, partly wasted. |
| .8 | Bituminous coal. |
| .4 | Clay streak |
| 1.9 | Cannel coal. |
| 1.0 | Under-clay. |
| 40.0 | Sandy shales. |
| 18.0 | Heavy sandstone. |
| 32.0 | Shaly sandstone, soft and generally thin bedded. |
| .8 | Coal, (lowest coal here.) |
| 1.2 | Under-clay. |
| 50.0 | Sandy shales, equivalent of shales at cropping of Little Sandy. |
| 1.8 | Limestone. |
| 4.4 | Soft shales and shaly sandstone. |
| .0 | Upper part of millstone grit.* |

All the workings I have seen of the cannel coal have been made by stripping the coal bed. The softness of the covering materials renders a perfect section, from natural outcrop, nearly impossible.

* This is a continuation of section heretofore given, pages 153, 54, and 55.

At three and a fourth miles from the crossing of Stinson's creek the heavy sandstone at Star Furnace caps the hill, dividing the waters of Stinson's and William's creek.

Under this heavy sandstone are two beds of argillaceous shales, separated by seventy feet of sandy shales; the upper bed is only a few inches thick, while the lower is several feet. The lower bed has been worked for iron ore, and furnished some good patches of ore, but no regular bed has been discovered in any of the shale beds at this locality.

Northeast of these ore diggings, and beyond the hill, is the Star Furnace Branch of Williams' creek. The Star Furnace is founded upon the rock which caps the hill to the southwest, and over which rests the Star Furnace Coal; this sandstone also underlies the Catlettsburgh, Reiley, Barrett, Cushing, and Williams' coal; also, the lower coal at the Williams' creek tunnel, and the twin coal near Ashland.

The following section applied above the Stinson creek cannel coal, will bring that section up to the Star Furnace coal:

*Section at the Star Furnace, Carter county.*

| *Thickness.* Ft. In. | |
|---|---|
| 25.0 | Covered space. |
| 4.0 | Sandy shale. |
| 2.0 | Coal. |
| .4 | Parting-clay. |
| 2.6 | Coal. |
| 1.6 | Under-clay. |
| 20.0 | Alternations of sandstone and shale. |
| | Ore bed, Kidney ores in shale. |
| 35.0 | Alternations of shales and sandstones. |
| .5 | Black clay streak. |
| 26.0 | Star Furnace sandstone. |
| 80.0 | Dark-grey sandy shales, |
| | Sandy and argillaceous alternating. |
| | Place of Stinson creek cannel coal. |

Thirty-seven feet below the twin coal of the foregoing section, is the position of the "Blue Ore" beds, which are estimated at the furnace at about three feet thick.

Descending the branch from the furnace, toward Williams' creek, the rocks are found dipping, in an eastwardly direction, a little more rapidly than the slope of the country. A mile and a fourth from the furnace the road makes a sudden bend down Williams' creek, which runs more to the north, (about N. 20° E.;) here the rocks are seen along the side of the road nearly on the strike line, and are generally found to conform, in a very considerable degree, to the contour of the spurs of the hills that come down to the rocks, rising the points, and descending to the valleys between them, so that every point is an anti-clinal, and every ravine a synclinal axis—the rise being about equal to the fall. This state of case continues, with slight modifications, along the line of the road from the mouth of the Star Furnace Branch to the mouth of Catletts' creek; sometimes the rocks are found rising as high as three hundred and fifty feet.

From the mouth of the Star Furnace Branch, by the line of the road, to the Williams' creek tunnel, no new measures come in. The rocks rise and descend with the road line, or the measures rise and de-scend again in waves, the greatest elevation of which is probably fifty feet.

The materials, as well as the thickness of the individual members, are very much modified as they are traced from the Star Furnace Branch to the tunnel on Williams' creek; and were they not followed step by step, and never lost by the eye, they would hardly be recog-nized as the measures of the Star Furnace, yet they are their equiva-lent.

A starting point for the detailed surveys of Greenup and Carter counties was made at the south end of the Williams' creek tunnel, on the Lexington and Big Sandy railroad. The first line was carried from this point to Catlettsburgh, and from thence extended over the county. The lines of this work have been plotted, and reduced to the scale of $\frac{1}{50.000}$. There being large tracts of country not reached by these lines, the streams and roads only have been laid down; when the work has been sufficiently completed, and the lines brought sufficiently near each other, it is proposed to lay down the hills, and cross the map, in seve-ral directions, with geological sections, which will be done during the present season.

Passing over the hill at the tunnel, which is the dividing ridge be-tween Williams' creek and the east fork of Little Sandy river, the wav-ing arrangement of the rocks is such that the upper member of the coals at the tunnel rises nearly to the top of the highest ridge, and again fall toward the valley of the east fork of the Little Sandy, thus bringing the lowest coal at the tunnel, i. e., the equivalent of the Star Furnace coal, down to the bed of the stream.

On the east side of the stream, near the water line of the creek, (now nearly dry,) is to be seen the equivalent of the lower coal at William's creek tunnel, which I shall hereafter distinguish as the Twin coal.

*Section of Twin coal near bridge, east fork of Little Sandy.*

| *Heighth.* Ft. In. | *Thickness.* Ft. In. | |
|---|---|---|
| 30.8 | 10.0 | Grey shale, with Kidney ore and small bands of carbonate of iron ore. |
| 20.8 | 2.4 | Blue shale, (or grey metal.) |
| 18.4 | 2.4 | Coal. |
| 16.0 | 6·0 | Sandstone with oblique lines of deposition, equal to 45° from the horizon. |
| 10. | 1. | Coal. |
| 9. | 1. | Under-clay. |
| 8. | 2. | Fine grained sandstone. |
| 6. | 2. | Dark sandy shales. |
| 4. | 4. | Covered space probably shales. |
| | 0. | Pool above bridge. |

The reader may contrast this with that at the north end of the tunnel.

*Section at north end of Williams' creek tunnel.*

| *Heighth.* Ft. In. | *Thickness.* Ft. In. | | |
|---|---|---|---|
| 95.11 | 35. | Covered space to top of hill, mostly shales. | |
| 60.11 | 1. | Sandy shale. | |
| 59.10 | 1.1 | Argillaceous shale. | |
| 58.9 | 1.2 | Argillaceous shale, dark. | |
| 57.7 | 1.6 | Grey sandy shale. | |
| 56.1 | 1.5 | Black bituminous shale. | |
| 54.8 | 1.0 | Slaty coal. | |
| 53.8 | 1.0 | Yellowish-grey argillaceous shale. | |
| 52.8 | 1.6 | Bituminous coal in blocks. | |
| 51.2 | .7 | Under-clay from five to eight inches. | |
| 50.7 | 5.10 | Indurated *fine clay.* | |
| 44.9 | .7 | Coal. | |
| 44.2 | 9.4 | Yellow and bluish shale. | |
| 34.10 | 31.0 | Sandstone, heavy bedded. | |
| 3.10 | 1.6 | Fire clay. | Twin coal. One member lost. |
| 2.4 | 1.8 | Coal, 1 foot to 2 feet. | |
| .8 | .8 | Under clay. | |
| | .0 | Bottom of tunnel. | |

*Section at south end of Tunnel, about 800 feet distant from place of last section.*

| *Heighth.* Ft. In. | *Thickness.* Ft. In. | |
|---|---|---|
| 80.4 | 29.0 | Sandstone. |
| 51.4 | 1. | Coal. |
| 50.4 | 2. | Under-clay. |
| 48.4 | 6.0 | Shaly sandstone and shales. |
| 42.4 | ·6 | Coal six inches to two feet. |
| 41.10 | .2 | Carbonate of iron. |
| 41.8 | 5.0 | Argillaceous shale. |
| 36.8 | 1.8 | Pyritiferous ore? |
| 35.0 | 35.0 | Upper part of sandstone, equivalent of the Star Furnace sandstone. |

With the exception of the six feet sandstone, in the last section, there is a great similarity between this coal and the lowest coal of the previous section; this sandstone also thins out and disappears in the short distance between the ends of the tunnel.

There is associated with the sandstone and shales below the Star Furnace coal a ledge of rock of a dark-grey color, about four feet thick, in many places exfoliating in thin scales of a ferruginous character, to which I would call your attention. (See specimen, No. 1001.)

The timber of the country in the northeast part of Carter county is generally white oak, beech, poplar, and some maple in the valleys. The soil is of a light yellow drab color, and thin.

Half a mile from the "east fork" is the village of Cannonsburgh; from this point the road, which runs up a tributary of the east fork, towards the mouth of Big Sandy, is nearly in the direction of the strike line; the conformity of the rocks to the shape of the hills is not so apparent immediately on the line of the road as it is on either side of it, at a short distance from it.

The sandstone above the upper coal at the tunnel generally rises to the tops of the hills—sometimes in an unbroken bluff on both sides of the road in the valleys, above which it rises sometimes about seventy five feet.

Descending the hill forming the western limit of the water shed of the Big Sandy this sandstone presents a remarkable appearance, which appears to have originated in a sliding motion of the beds from the S. W. toward the N. E. Here, also, certain bands of different colored earths were first observed, which subsequently materially aided me in defining the Coal Measures elsewhere, both in Greenup and Carter counties, where horizons of sandstone and limestone were absent, as is the case in about two hundred feet of these measures lying above this sandstone, with a few exceptions, and these are quite local, or of very limited area; while these bands of colored earths are of great extent, and may be found in roadways and water-worn ravines, on almost every hill side composed of this geological equivalent. These bands will consequently be referred to frequently, as they form a distinguishing feature in many of the succeeding sections.

The sandstones of equivalent beds are more massive at the mouth of Catlett's creek than they are at the head of the creek, or on the head of Key's creek. Two of the distinguishing masses of sandstone, i. e., the mass over the Star Furnace coal, and the Star Furnace sandstone, which are separated by many feet of shales, coal, &c., come together and form one mass, on the farm of Mr. Gartrell, on the Ohio river; also, the bluff between Ashland and Hood's creek presents a similarly constituted mass; where these beds are thus brought together the coal equivalent to the Star Furnace, or twin coal, is absent.

From the mouth of Catlett's creek, (the line of which is a fault where it emerges from the hills,) the rocks rise as you go down the Ohio river; they also rise as you ascend the valley of the creek, equal to the ascent of the valley nearly to its head. The rocks also rise rapidly as you ascend the right hand branches of the creek, or to the northwest and south.

On the Horse branch, one mile below the mouth of Catlett's creek, the twin coal is worked by Dr. Cushing, on the east side of the branch —the coal having been again brought down to the level of high water of the Ohio river by a curving dip, conforming partially to the shape of the ridge dividing the Horse branch from Catlett's creek. A short distance from the mine of Dr. Cushing, Mr. Williams has opened the equivalent bed on the west side of the branch. The coal here gives the following section, varied at different openings:

*Section Dr. Cushing's bank.*

| *Heighth.* | *Thickness.* | |
|---|---|---|
| 3.7 | 2.0 | Coal. |
| 1.7 | .7 | Parting-clay. |
| 1.0 | 1.0 | Coal. |
| | .0 | Top of under-clay. |

The equivalent bed of coal near the mouth of Catlett's creek, is as follows:

*Section of coal, mouth of Catlett's creek.*

| *Heighth.* | *Thickness.* | |
|---|---|---|
| 3.2½ | 1.5 | Coal. |
| 1.9½ | .9½ | Parting-clay. |
| 1.0 | 1.0 | Coal. |
| 0.0 | .0 | Top of under-clay. |

This opening is estimated at sixty-five feet above low water of the Ohio river; it is covered at high water.

On Catlett's creek, half a mile above the place of section last given, the following section, with the associated materials, was obtained.

*Section of (Twin coal) at Reiley's mill.*

| *Heighth.* | *Thickness.* | |
|---|---|---|
| Ft. In. | Ft. In. | |
| 182.3 | 100.0 | Covered space. |
| 82.3 | 21.0 | Heavy sandstone. |
| 61.3 | 8.0 | { Place of coal? |
| 53.3 | 18.0 | { Covered space. |
| 35.3 | 12.0 | Sandstone. |
| 22.3 | 1.1 | Coal. |
| 21.2 | .8 | Parting-clay. |
| 20.6 | 1.0 | Coal. |
| 19.6 | 3.0 | Under-clay. |
| 16.6 | 10.0 | Drab argillaceous shale. |
| 6.6 | .6 | Irregular belt of carb. of iron, varying from 8 to 3 inches. |
| 6.0 | 6.0 | Yellow-grey shales. |
| .0 | .0 | Sandstone bed of creek. |

The bluff on the Gartrell farm, before alluded to, is one hundred and twenty-two feet eight inches in heighth, with a soft space near the center dividing it into two hard masses; the soft space is mostly soft sandstone, with a few hard thin bedded flagstones; it varies, where best seen, from two to eight feet. The rocks forming the lower end of this bluff dip to the southwest, at the head of a short ravine. At this point are seen a few thin and broken sheets of coal, wedged into and between the flagstones of the centre, or soft mass of the bluff; also, many impressions of fossil plants. The bluff is carried by the dip, about three-fourths of a mile from the Ohio river, to the rear of the city of Ashland; where it is composed of the mass forming the upper part of the bluff above the city and the mass, over the Clinton Furnace coal, which is worked in several places near the city. Below the city the bluff, equivalent to the Gartrell bluff, again approaches the river, the two masses of sandstone still in close contact. This bluff continues unbroken to the bank of Hood's creek, where the great sandstones are again separated, and receive between them the equivalent of the *twin coal* and a small amount of shales. Here the coal is quite thin; near the centre of the bluff it is locally capped with limestone, which extends from the railroad, near Ashland, to Hood's creek, one and a half miles below. This is only about a mile from Bellefont Furnace. In this short space a most noticeable change has taken place, which is the introduction of a bed of argillaceous shales between the base of the limestones capping the bluff above Hood's creek and the sandstones on which they rest at that point.

The following sections of the equivalent measures at Hood's creek and Bellefont Furnace, will illustrate the character and extent of this change:

*Section at Hood's creek.*

| *Heighth.* Ft. In. | *Thickness.* Ft. In. | |
|---|---|---|
| 70.1 | 8.0 | Earth stripping above limestone. |
| 62.1 | 3.4 | Shales, argillaceous. |
| 58.9 | .3 | Limestone ore, three to ten inches. |
| 58.6 | 4.0 | Limestone, four and a half to five feet. |
| 54.6 | 1.0 | Shales. |
| 53.6 | 21.0 | Coarse sandstone, frequently full of waterworn pebbles. |
| 32.6 | 2.6 | Streaks of coal and shale. |
| 31.0 | 31.0 | Sandstone. |

Under the sandstone at Hood's creek ford, twenty-eight feet below base of section, sandy shale.

*Section at Bellefont Furnace.*

| *Heighth.* Ft. In. | *Thickness.* Ft. In. | |
|---|---|---|
| 104.10 | 15.0 | Clay stripped to raise ore and limestone. |
| 89.10 | .10 | Limestone ore, eight to twelve inches. |
| 89.0 | 3.00 | Limestone, one to four feet. |
| 86.0 | 2.00 | Fire clay. |
| 84.0 | 18.6 | Bluish-grey shale. |
| 65.6 | 15.0 | Sandstone; no pebbles observed. |
| 50.6 | 14.0 | Black and blue-grey argillaceous shales. |
| 36.6 | 3.6 | Coal. Thin clay parting. |
| 33.0 | 2.0 | Under-clay. |
| 31.0 | 31.0 | Sandstone. |

At three-fourths of a mile from the Ohio river the road from Ashland to the Clinton Furnace intersects the line of bluffs, and ascends then by a gentle ascent. The coal mined near the road here is the equivalent of the coal mined near the Clinton Furnace. In the vicinity of these openings in the coal is to be seen a sandstone, remarkable for the coarseness of its texture, and for the softness and want of cohesion of the particles composing the upper part of the mass. This mass of sandstone overlies the coal, and in some districts serves as a a Geological horizon. The rocks dip towards the valleys, viz: On the east side of the road towards the valley of Key's creek, and on west and southwest towards the valley of Hood's creek. Half a mile from where the hills* commence the land is nearly level, or gently roling; further from the Ohio the measures dip to the southwest towards the small branches of Little Hood's creek, while to the left, or east side of the road, the dip is to the northeast, or toward the small branches of Key's creek. Two miles from the Ohio river a synclinal axis crosses the road at right angles with it; beyond this the rocks again dip to the northeast and north, towards this depression of the measures, after this on to the ridge dividing Key's and Little Hood's creeks.

The soft sandstone alluded to, as being above the coal on this road, may be traced, around the hills, either up Key's or Hood's creeks, to the neighbourhood of Clinton Furnace. The upper part of this sandstone is excessively soft in many localities. Above this soft sandstone the same variagated earths were observed, elsewhere alluded to. The indellible tints of these earths, derived from the weathering of the strata overlying this sandstone, afforded an excellent clue to the identification of the equivalent measures of Greenup and Carter counties, where the rocks themselves were not seen in place—since these colored earths may be seen in the cuts of the roads, the gullies, and slopes of the hill sides, when no rocks are in view.

The following section, taken in the cut of the Clinton Furnace road, is an illustration in point, as are several of the succeeding sections:

*Boulders and gravel of the drift period, have been deposited upon these hills, and on the flat country north of the main dividing ridges. These hills are from one hundred and fifty to one hundred and seventy-five feet high above the banks of the Ohio, at Ashland. No drift was observed at a greater elevation than these hills, nor in the valleys very slightly elevated above the banks of the river, nor on any of the streams behind the range of hills separating the Ohio river from the small streams running parallel to its course.

*Section in road from Ashland to Clinton Furnace.*

| *Heighth.* Ft. In. | *Thickness.* Ft. In. | |
|---|---|---|
| 186.0 | 8.0 | Sandstone. |
| 148.0 | 3.0 | Yellow shale. |
| 145.0 | 2.0 | Black clay streak. |
| 143.0 | 5.0 | Yellow shales. |
| 128.0 | 11.4 | Top of red streak. |
| 106.8 | 10.8 | Top of sandstone, and foot of red streak. |
| 96.0 | 11.0 | Sandstone. |
| 85.0 | 5.0 | Top of red streak. |
| 80.0 | 16.0 | Bottom of red streak, above sandy shales. |
| 64.0 | 5.0 | Sandstone, fifteen inches thick. |
| 59.0 | 32.4 | Yellow-grey shales. |
| 26.8 | 16.0 | Foot of yellow streak. |
| 10.8 | 10.8 | Small gravel ore in road. |
| 0. | .0 | Top of soft sandstone above Ashland coal, equivalent of the Clinton Furnace coal. |

The following section, taken near the Clinton Furnace, starts from the top of the sandstone under the Clinton Furnace coal, (it is also the equivalent of the sandstone over the Williams' creek tunnel coal, Star Furnace, Catlett's creek, and Horse Branch coal, and is distinguished here as the sandstone between the coal in the well and the Clinton Furnace coal,) will serve to show the relation of the ore beds here to the quantities between them:

*Section at point of hill northeast of Clinton Furnace.*

| *Heighth.* Ft. In. | *Thickness.* Ft. In. | |
|---|---|---|
| 117.8 | 38.0 | Top of the bench, and bottom of a five-foot red streak. |
| 79.8 | 42.4 | Ore diggings, Kidney ore. |
| 37.4 | 37.4 | Ore diggings in sandy shale. |
| 0. | 0. | Top of sandstone over Well coal, and under Clinton Furnace coal. |

The following section taken of the hill from Mr. Burwell's house, near the Clinton Furnace, will further connect the ore beds, and will also serve to show their relation to the bands of colored earths and associated materials:

*Section at Mr. Burwell's house.*

| *Heighth.* Ft. In. | *Thickness.* Ft. In. | |
|---|---|---|
| 216.4 | 8.00 | Top of sandstone covered with red earth. |
| 218.4 | 12.0 | Sandstone in two beds, locally underlaid by fourteen inch bed of ore. |
| 206.4 | 14.0 | Yellow sandy shales, much disturbed by slipping. |
| 191.4 | 5.4 | Top of sandstone in solid ledges five feet. |
| 186.0 | 5.4 | Foot of sandstone. |
| 180.8 | 10.0 | Fossilliferous sandstone, fine grained, probably united by a calcareous cement; the fossils are calcareous; locally this is the place of the "top hill Bastard limestone ore." The fossiliferous bed is eight inches thick, and lies between thin beds of sandy shales. |
| 178.8 | 24.0 | Sandy shales fifteen feet thick. |
| 154.8 | 1.0 | Top of black streak 10 to 15 inches thick. |
| 153.8 | 3.0 | Brown-red fire clay. |
| 150.8 | 3.0 | Brown-red fire clay. |
| 147.8 | 10.8 | Block ore beds, (from 147.8 to 156. are the beds producing the fifteen feet red streak or band.) "Red block" of the furnace men, ten to twelve inches thick. |
| 137.0 | 21.0 | Sandy and clay shales alternating. |
| 116.0 | 8.0 | Top of five-foot red streak. |
| 108.0 | 2.8 | Whitish shales, argillaceous above and sandy below. |
| 105.4 | 1.4 | Top of sandstone, four ledges. |
| 104.0 | 3.0 | Sandstone. |
| 101.0 | 16.0 | Clay shales. |
| 85.0 | 16.0 | Sandstone in shale beds eighteen inches thick. |
| 69.0 | 27.0 | Sandstone one foot thick in sandy shales. |
| 42.10 | 10.0 | Sandstone eighteen inches thick in sandy shales. |
| 32.0 | 11.8 | Thin sheet carbonate of iron, two to three inches thick. |
| 20.4 | 4.4 | Top of Clinton Furnace coal. |
| 16.0 | 16.0 | Mouth of entry ditch, fire-clay bottom. |
| .0 | | Sandy shales and shaly sandstone. |
| 5. | | Hard sandstone over cistern or well coal. |

The following section was taken from the equivalent measures of Carter county, about one and a half miles from Mount Savage Furnace. This section will illustrate the remarkable increase in thickness of the strata, as well as the great change in the materials composing them:

*Section at Mount Savage, Carter county. This section starts at a point equivalent to thirty-two feet in the preceding section.*

| *Heighth.* Ft. In. | *Thickness.* Ft. In. | |
|---|---|---|
| 378.0 | 6.0 | Top of hill near "Iron road." |
| 372.0 | 20.10 | Top of heavy sandstone. |
| 351.2 | 10.6 | Foot of exposed part of heavy sandstone. |
| 340.8 | 48.0 | Heavy sandstone, top of steep slope. |
| 292.8 | 16.0 | Argillaceous shales, highest point in road. |
| 276.8 | 1.0 | Yellow streak. |
| 275.8 | 15.0 | Red streak, large amount of small surface ore. |
| 260.8 | 10.8 | Top of sandy shales. |
| 250.0 | 5.4 | Soft sandy shales. |
| 244.8 | 16.0 | Top of red streak, and foot of sandy shales. |
| 228.8 | 4.4 | Top of yellow streak, and bottom of red sandy shales. |
| 223.4 | 5.4 | Bottom of "Rough Block Ore;" three feet thick at this point. |
| 218.0 | 5.5 | Foot of sandstone. |
| 212.8 | 16.0 | Place of Kidney ore diggings in road. |
| 196.8 | 5.4 | Place of limestone ore, as seen on next hill to the eastward, where the bed is extensively worked, both by stripping and drifting. |
| 191.4 | 16.0 | Top of sandy shale. |
| 175.4 | 5.4 | Top of black clay streak, between shales four and a half feet thick. (Place of coal?) |
| 170.0 | 32.0 | Whitish clay. |
| 138.0 | 3.0 | Sandy shales. |
| 135.0 | 14.0 | Top of sandstone, twenty inches thick. |
| 121.0 | 10.8 | Ore diggings, "Grey Kidney Ore," in whitish argillaceous shale. Immediately above this ore bed is the five feet "*Red Streak.*" |
| 110.4 | 10.8 | Top of sandstone, twenty inches thick. |
| 99.8 | 5.4 | Three black streaks, whitish clay between. |
| 94.4 | 10.8 | Top of sandstone, fourteen inches thick. |
| 83.8 | 21.4 | Top of black streak, one foot thick, under yellow shales three to four feet thick, (sandy.) |
| 62.4 | 5.4 | Yellow streak, and top of whitish earth gravel ore. |
| 57.0 | 4.8 | Top of sandstone in two beds, twenty inches thick, sandy shale between the bed, four inches thick. |
| 52.4 | 10.8 | Top of sandstone, fifteen inches thick. |
| 41.8 | 26.8 | Top of slope. |
| 15.0 | 15.0 | "Iron road." |
| 0.0 | | Bed of branch, sixteen feet above Gum branch coal, equivalent to Clinton Furnace coal. |

Passing round the head of the Gum branch the different beds of ore, equivalent to the measures of the preceding section, have been opened and extensively worked, on the slopes of the hills, at a greater or less elevation, depending on the displacement of the rocks. On the north side of the hill, from which the section was taken, a bed of block ore of excellent quality has been worked, about seventy-five feet below the base of heavy sandstone capping the hill. This bed was not seen on the south side of the hill.

Near Mount Savage Furnace two beds of coal are seen. The bed out-cropping near Straight creek, below the Furnace, is doubtless the equivalent of the Star Furnace (twin coal.) The bed opened above the furnace, and on the Gum branch, is the undoubted equivalent of the Clinton Furnace coal.*

The following section is from the equivalent of the Clinton Furnace coal, worked near the Furnace, Mount Savage Iron Works:

| *Heighth.* Ft. In. | *Thickness.* Ft. In. | |
|---|---|---|
| 27.6 | 20.0 | Sandstone. |
| 7.6 | 3.0 | Grey shales. |
| 4.6 | 2.0 | Bituminous shale. |
| 2.6 | 2.6 | Coal. |
| .0 | .0 | Top of under-clay. |

*Section of same bed, three-fourths of a mile to northeast, on Gum branch.*

| *Heighth.* Ft. It. | *Thickness.* Ft. In. | |
|---|---|---|
| 22.0 | 15.0 | Slope, partially exposing sandstone. |
| 7.8 | 3.0 | Three to ten feet grey argillaceous shale. |
| 4.8 | 1.2 | Black rash with clay streaks. |
| 3.6 | 1.0 | Bituminous shale. |
| 2.6 | 2.6 | Coal 1.6 to 2.6 inches. |
| .0 | .0 | Top of under-clay. |

This out-crop faces the N. E., and the covering rock is much softer than at the first section near the Furnace, which faces the S. W. It is quite common to find the same bed of sandstone softer where the exposure faces the north than where it faces the south. These sandstones are much less affected, by the action of water and frost, on the southern than upon northern exposures.

The last section has a horizontal elevation of about seventy-five feet above the first. The rocks dip down the Gum branch toward Straight creek.

*See section at Reiley's Catlett's creek coal.

To the north, one-fourth of a mile, the same coal is seen near the road side, from one hundred and fifty to one hundred and sixty feet higher than the Gum branch section, no doubt by displacement occasioned by a fault, which has let down the Gum branch coal and its measures, or has lifted up the measures, on the north, or hill side. The line of this fault runs up the left hand branch of Gum branch by a course to the northwest. It is probable that it crosses the ridge at the head of the ravine. Its extent and character has not been particularly examined. Passing down Straight creek, a short distance below the Furnace, the equivalent of the twin coal of the Star Furnace, Catlett's creek, &c., is seen with the associated materials, on the north side of the creek. It is probable that a continuous section of all the beds may be obtained on this creek, from the Stinson creek cannel coal to the top of the section at Mt Savage. The materials under the twin coal, were seen in place here. A fine grained sandstone, lying in beds, from six to eighteen inches in thickness, forms the first member below the under clay of the coal, and is about twelve feet thick. The under clay here is from two and a half to three feet thick. Under the sandstone bed, above alluded to, lies a bed of ash colored shales, which is no doubt the equivalent of the shale beds seen at the foot of the ridge between Greenup and Buena Vista Furnaces. The same mass of shale is well presented in the coaling grounds of the Star Furnace on Cane creek, about two and a half miles south of Greenup Furnace, where a thin bed of coal may be seen in the shales twelve feet below the sandstone. This coal begins in a thread-like line on the north, and increases toward the south, in half a mile to fourteen inches thick. It has no under-clay. From the mouth of Straight creek, in the direction of the road to Grayson, the measures, rise but they have not yet been traced by the detailed survey. The sandstone above the lowest coal, in this part of the basin, is seen in the bed of the little fork of Little Sandy, at the first crossing below the mouth of Straight creek, also three fourths of a mile lower down the little fork, fifty feet above the level of the stream.

At the salt wells on Little Sandy, the brine is found in the upper part of the millstone grit. It may be possible that the bottom of the wells do not penetrate the millstone grit, but reach only the fractures through which the brine rises. The shale beds resting on the millstone grit here are of no great thickness. Passing down Little Sandy, about a mile below Grayson, the millstone grit is seen on Barrett's creek, and the sub-carboniferous limestone, about one and a half miles above the intersection of the road with the creek, where the same order is observed as on the turnpike, which fully confirms the previous remarks made in reference to the sub-carboniferous limestone, on the N. E. side of the dividing ridge between Tygert's creek and Little Sandy. The measures are decidedly thinner on Barrett's creek, than where the turnpike crosses Tygert's creek, but no detailed measurements have been made here.

The Kenton salt well is situated in the bed of Tygert's creek, which is here worn down into the knob-sandstone. The well is on the farm of Mr. Jacobs, about six miles northwest of Grayson.

On the west side of the creek the following section was obtained:

*Section near Mr. Jacobs' house.*

| *Heighth.* Ft. In. | *Thickness.* Ft. In. | |
|---|---|---|
| 164. | 32.4 | Elevation of hill at Mr. Jacobs' house. |
| 131.8 | 31.4 | Top of millstone grit. |
| 100.4 | 13.4 | Sub-carboniferous limestone in place. |
| 90. | 57.0 | Drab shales in place. |
| 42. | 42.0 | Ledge of Knob sandstone in place. |
| 0. | 0. | Bed of Tygert's creek on Knob sandstone. |

In the bed of the creek, at the base of the above section, are seen numerous lines of fracture, crossing the creek by a course N. 30° E. From one of these fractures rises the brine, which was manufactured into salt on the first settlement of the country by Simon Kenton.

On the east side of the creek one-fourth of a mile from the Kenton well are several caves which are formed in a local bed of coarse grindstone grit lying immediately under the sub-carboniferous limestone. The grains of sand forming the grindstone grit are probably cemented by a calcareous cement and the rock is the cavernous member of Carter county. In some localities the bedding faces of this rock is thickly studded with angular fragments of Hornstone or flint. Extensive diggings may be observed in many places in this neighborhood, at the outcrop of the sub-carboniferous limestone, often covering half an acre or more of ground. The excavations are shallow, only about six or seven feet deep. The people of the country have various opinions as to the origin and object of these works. Some of the old workings having been recently opened, a good opportunity was thus afforded to make an examination. The materials thrown from the recent opening were a reddish friable earth, fragments of limestone, fragments of flint, and kidney shaped seggregations of flint, weathered out of the adjacent limestone. These are quite symmetrical, about five inches long, three inches wide, and from one and a half to two and a half inches thick. They split at right angles with the long diameter, and perpendicular to the thickness, thus producing long oval pieces with parallel faces. The largest produced are four inches long, from one and a half to two and a half inches wide, and from a quarter to three eighths of an inch thick. Bushels of the ends of these nodules were discovered near some of the workings, with some fragments of the oval pieces, few or none in an entire state, hence I infer that these diggings were made by the aborigines of the country for the purpose of procuring the material from which they made their arrow-heads.

Where the grindstone grit was observed, on the first branch above Mr. Jacobs' farm, it separates into slabs from three to six inches thick, in the line of the deposition of the sand, which forms an angle of 30° with the superior and inferior faces of the regularly stratified layers. These rocks will afford good grindstones, both for neighborhood use and shipment. The bed is, as we have said, local, and does not cover a very great extent of country. For the present I refer them to the upper members or last beds of the knob-stone division of the sub-carboniferous rocks. Its thickness varies from a few inches to twenty-five feet. My examinations were next directed across the drainage of the country, nearly north from the small creek entering Tygert's creek at the narrows, to Grassy creek, and thence over the hills to the North fork; thence to a large creek called Three Prong; (not laid down on Millne & Bruder's map;) thence down Three Prong, three-fourths of a mile; thence across the hills to Leatherwood creek; thence down that creek two miles; thence across the hills to the head branches of the south fork of White Oak creek, and Kenton Furnace; thence down White Oak by the road to Greenupsburg; then with the line of the Ohio river to Springville.

All the creeks crossed by this route to Greenupsburg have thin beds of this grit on the knob-stone, which sometimes rises sufficiently high to compose the entire dividing ridge; at other localities a thin capping of sub-carboniferous limestone and millstone grit; at others, in addition, a few feet of the lower member of the Coal Measures, are to be found on the top of the highest ridges. Along the entire line from Tygert's creek there is manifestly a thinning of the sub-carboniferous limestone, millstone grit, and the Coal Measures, evidently marking the margin of the coal basin towards the northwest. It is highly probable that many of the beds found in considerable force east of Little Sandy river will be found, on examination, to have entirely thinned out, even before they reach that river; other beds lying three or four hundred feet higher in the series east of Little Sandy, may, on White Oak creek, be found resting on the millstone grit, which has thinned out to twelve feet, and here rests on the knobstone—the sub-carboniferous limestone having entirely disappeared. It is highly probable that the line of the margin of the basin was frequently changed, from the time of the deposition of the knob-stone to the end of the coal period.

The coal basin being shallower near the margin, the measures between the ore beds are much thinner here, and the ore beds much closer together here than east of the Little Sandy. There is reason to believe, from the evidence seen on my route from Tygert's creek to the mouth of White Oak creek, that the line of least disturbance during the coal period, was not far from this line; where the route crossed Grassy creek the line of least disturbance was three or four miles to the northwest. It is worthy of particular note that the ferruginous depositions have here extended to the edge of the basin, and ferruginous materials have been infiltered, or have even run over the denuded sub-carboniferous limestone, as may be well observed at the limestone quarry of the Kenton Furnace.

Beds of reddish marly clay are frequently seen resting on the millstone grit, from five to ten feet in thickness. At every step, from Tygert to White Oak creek, are convincing proofs of the thinning out of the sub-carboniferous limestone and millstone grit; at some points the sub-carboniferous limestone either was never deposited, or has been swept away by denudation before the deposition of the millstone grit, which in some localities is seen resting on the knob-stone series: at other localities constituting a mass from twelve to twenty-five feet thick. The underlying member—the knob-stone—has, on the other hand, experienced an enormous expansion. At one locality it was found six hundred and thirty feet thick, without including a portion of the base of the formation under the drainage.

The lowest ore bed observed locally rests on a bed of chert; the same bed, when the chert is absent, rests on the sub-carboniferous limestone and frequently fills fisures in that rock, extending down to the top of the knob-stone. The ores found in this geological horizon belong to the Hematitic class.

On Grassy creek a new furnace is in process of erection. The stack is built upon and into the masses of the knob-stone, the lower part being excavated into the solid mass of this member in place, the material for the walls and inner lining are obtained from the same formation. I think it extremely doubtful whether this stone will be found to stand fire well.

The ores on which the main dependence is placed to supply the furnace lie on the top of the hills, associated with the chert bed of the sub-carboniferous limestone. I visited some of the localities where the ores are now being tested, at one, lying one and a quarter miles from the furnace, the ores are found between chert beds, and are of variable thickness; from one inch to ten or fifteen. The bed on which the ores rest slopes toward the valley at an angle of about fifteen degrees, and is of a very uneven surface, full of irregular shaped cavities, of unequal size and depth, the margin of these cavities touching the margin of all surrounding ores. It is into these holes that the ore has infiltrated, filling some and partially filling others, leaving a surface of ore and a few points of the bottom to form a surface somewhat less rough than the bed on which the ore rests. Upon this last surface rests a bed of chert and the debris of the millstone grit above it. The opinion expressed at the furnace is that the best ores lie at the head of Grassy, or to the west. In my opinion the best and most abundant ores are to be sought down Grassy or to the east.

The following section was taken at one of the ore beds of Grassy Furnace:

| *Heighth.* | *Thickness.* | |
|---|---|---|
| Ft. In. | Ft. In. | |
| 48. | 23.0 | Top of hill; masses of millstone grit. |
| 25. | 25.0 | Foot of millstone grit; covered space. |
| 0. | | Ore bed from one inch to fifteen inches thick, resting on a bed of chert. |

Some few of the highest ridges north of Grassy, and south of Three Prong, are capped by a few feet of the materials forming the base of the Coal Measures; and the ore beds, equivalent to those worked at Kenton Furnace, may be found in some of them. The valleys are from one hundred and fifty to three hundred feet deep, and as before stated, are in part or entirely sunk into the masses of the knob-stone. The sides of the valleys are in many places perpendicular walls.

In traveling through the country, from one valley to another, the small branches are usually followed to the top of the dividing ridge; the ridge is then followed until another suitable branch offers for the descent into the next valley. Good roads, except by the lines of the valleys and branches, are an impossibility. Practicable roads may be had along the valleys.

The remarks, in reference to the dip of the rocks, and their conformability to the contour of the hills made of the measures upon Triplet's creek, are equally applicable here, and of the whole line traveled from the narrows of Tygert's creek to Greenupsburgh.

*Section on White Oak, near Kenton Furnace.*

| *Heighth.* Ft. In. | *Thickness.* Ft. In. | |
|---|---|---|
| 272.10 | 36.2 | Top of hill. |
| 236.8 | 20.8 | Red streak, shown by road wash. |
| 216.0 | 15.6 | Loose chert fn road. |
| 200.6 | 185.0 | Top of second bench of knob-stone. |
| 15.6 | 15.6 | Foot of abrupt part of hill. |
| .0 | 0.0 | Bed of white oak creek. |

On the north side of this hill ores are now being dug. A section was carried from these openings, and the place of the ores located in the foregoing section.

*Section on north side of hill.*

| *Heighth.* Ft. In. | *Thickness.* Ft. In. | |
|---|---|---|
| 272. | 15.6 | Top of hill, and top of the foregoing section. |
| 256.6 | 15.6 | Top of thin masses millstone grit? and locally, place of clay ironstone. |
| 241. | 10.4 | Base of sandstone; millstone grit? |
| 228. | 0. | Top of block ore, six to eighteen inches thick, under ten feet four inch clay shales. |

There is a thin bed of shales under the ore bed, which rest on the knob-stone. It may be possible that the sandstone quarried as millstone grit in the above section, may be one of the sandstones of the Coal Measures. If this be the case the sub-carboniferous limestone and millstone grit are both absent at the place of the section.

The bed of ore heretofore worked at the Kenton Furnace, lying under and in the fissures of the sub-carboniferous limestone, is probably of the same geological period as the ore beds under the chert at some, and upon the beds of limestone in other, localities. The ores in the fissures of the limestone at Kenton Furnace are peculiar to this locality, so far as I am at present advised.

From the mouth of Little Sandy, river and for some distance up that stream, the knob-stone forms the mass of the hills, which have a capping of the superior measures of from fifty to seventy-five feet. The remark is equally applicable to the river hills of the Ohio. From the mouth of Sandy to Springville along the line of the Ohio river hills, in a few places, are found beds of limestone, probably belonging to the sub-carboniferous period.

The following section, taken near Springville, does not include the entire thickness of the knob-stone division—part of the mass lies below the bed of the Ohio river:

| *Heighth.* Ft. In. | *Thickness.* Ft. In. | |
|---|---|---|
| 677.6 | 69 6 | Top of hill. |
| 608. | 22.0 | Top of grindstone grit. Member of knob-stone? |
| 586. | 218.0 | Covered space, principally shales. |
| 3 8. | 19.4 | Top of ledges of rock, and foot of shales. |
| 348.8 | 17.6 | Thick masses of drab-grey rocks. |
| 331.2 | 5.2 | Thick masses of iron stained rocks. |
| 326. | 5.2 | Thick masses of blue-drab rocks. |
| 320.10 | 32.0 | Ledge flesh-grey rocks. |
| 298.10 | 21.0 | Ash-grey shales. |
| 277.10 | 28. | Ash-grey shales with flagstones intercalated. |
| 149. | 16. | Slope covered by land slide. |
| 133. | 91. | River road. |
| 42. | 42. | Sawmill. |
| 0. | | Low water Ohio, October 1856. |

At a short distance from Greenup Furnace is a thin coal, lying under the bed of Cane creek. The openings into this coal, which is mined for blacksmith's use, were fallen in, and no measurements could be made of the coal.

The following section was taken of the hill immediately above the coal:

| *Heighth.* Ft. In. | *Thickness.* Ft. In. | |
|---|---|---|
| 154.10 | 21.8 | Top of hill at point. |
| 133.2 | 47.6 | Second bluff. |
| 86.6 | 40.0 | Top of first bluff, upper part of shales. |
| 26.6 | 16.6 | Foot of sandstone bluff. |
| 10.0 | 10.0 | Blue shales. |
| 0. | | Top of coal. |

This coal is slaty, and said to be eighteen inches thick.

The following section is from the equivalent beds of the section last given, immediately opposite Greenup Furnace, where the materials composing the hill are better exposed:

*Section at Greenup Furnace.*

| *Heighth.* Ft. In. | *Thickness.* Ft. In. | |
|---|---|---|
| 144·8 | 32.0 | Top of hill, sandy shales. |
| 112.8 | 32.0 | Foot of red streak. Some ore has been taken out here. |
| 80.8 | 24.8 | Top of black bituminous shale, eighteen inches thick. |
| 56. | 6.0 | Top of thick sandstone, and foot of thin sandy shales. |
| 50.8 | 13.0 | Foot of thick bedded sandstone. |
| 39.0 | 17.8 | Thick bedded sandstone, thin bedded ledges 7 feet thick. |
| 21.4 | 21.4 | Foot of thick masses sandstone, containing many casts of fossil shells, mostly spirifer. |
| .0 | .0 | Top of shales above coal 16.6 inches thick. |
| 16.6 | 16.6 | Top of coal, under bed of branch. |

The following section is from the point of the ridge between Greenup and Pennsylvania Furnaces, ascending, from the valley of Cane creek, by the line of the road from Pennsylvania Furnace to the quarry of limestone. The section begins at the top of the sandstone, at thirty-nine feet in last section:

| *Heighth.* Ft. In. | *Thickness.* Ft. In. | |
|---|---|---|
| 244.8 | 16.0 | Top of hill, loose rough sandstone. |
| 228.8 | 11.8 | Rough block ore bed. |
| 217.0 | 44.4 | Top of covered slope. |
| 162.8 | 92.8 | Soft sandstones on top of sloping covered space, mostly beds of sandy shale. |
| 70. | 11.0 | Foot of steep sloping covered space. |
| 69. | 32.4 | Small ore in road. |
| 37.8 | 10.8 | Bottom of bed of argillaceous shale. |
| 27. | 7. | Sandy shales in place. |
| 20. | 20. | Top of steep rounded slope. |
| 0. | | Top of sandstone. |

About a mile to the northeast of the terminating point of the foregoing section, and on the same ridge, is the quarry from which the limestones used at Greenup Furnace are procured.

The following section taken at this quarry will exhibit the arrangement here:

| *Hetghth.* Ft. In. | *Thickness.* Ft. In. | |
|---|---|---|
| 42.10 | 26.6 | Top of hill, covered space, loose sandstone on top. |
| 16.4 | 5.0 | Surface exposed by cut at quarry. |
| 11.4 | 3.0 | White argillaceous shale. |
| 8.4 | 1.0 | Black clay streak. |
| 7.4 | 3.3 | Silicious argillaceous clay. |
| 4.1 | .6 | Limestone ore bed. |
| 3.7 | 1.4 | Blue-grey sandy shales containing numerous silicious segregations, hard and compact. |
| 2.3 | .6 | Ledge of limestone. |
| 1.9 | 1.0 | Ledge of limestone. |
| .9 | .9 | Ledge of limestone. |
| 0.0 | .0 | Blackish sandy shales, with carbonaceous partings. |

It was very desirable, that this limestone should be traced to other localities. Its geological place could be traced, but owing to the softness of the covering masses no outcrop could be found. The thirty feet of materials immediately above are very soft, and the chances of the limestone being found in outcrop, unless exposed by a land slide, are rare indeed.

About one hundred and fifty yards from the quarry, and on the east side of the same ridge, and nearly on the same horizontal level, is an ore bank which has recently been opened in argillaceous and fine sandy shale, producing beautiful, fine-textured, clear-bright red ore, in part, and part grey—"*see specimen* 91, *Greenup county.*" This ore, in its external character, resembles the *eisenkalkstein* of Fallenburgh, which you kindly furnished me for comparison with the Greenup county ores. The German ore possesses more evenness of fracture.

About one fourth of a mile to the southeast of the ore bank last mentioned the hill is a few feet higher. An ore bed was formerly worked on the top of the sandstone, supposed to be the equivalent of the sandstone at the top of the section at the quarry.

The following section, taken from a sag in the top of the ridge, will give a few feet of these measures:

| *Heighth.* | *Thickness.* | |
|---|---|---|
| 41. | 4.0 | Top of hill capped by four feet clay shales. |
| 37. | 21.0 | Sandy shale. |
| 16. | 8.6 | Ledge of sandstone, one foot thick, in sandy shales. |
| 7.6 | 7.6 | Bottom of red streak. |
| .0 | .0 | Sag above limestone bed. |

In the vicinity of Greenup Furnace a thin bed of cannel coal is seen, lying high in the hills; it is eleven inches thick at the place of exposure. The coal lies ninety-eight feet below the top of the hill, and about one hundred and forty-six feet above the level of Cane run, at the furnace. It is probably the equivalent of the *black* streak, 18 inches thick at 80 feet 8 inches, in the section opposite the Furnace.

# CHAPTER III.

*Report of the progress of the work on the base line.*

Uniontown, Union county, having been fixed as the initial point of the base line, Mr. Joseph S. Harris, with his corps, No. 2, were detailed from the survey of Hopkins, Christian, and Muhlenburg counties, where that corps had been operating during the summer, to conduct this important work.

An approximate determination of the latitude of the initial point was first obtained, at Uniontown, which gave the latitude of this place 37° 46′ 4″.

A monument of stone, three feet long, was placed in the inclosure of Dr. John T. Berry, (marked, on the top, as "initial point of Base line, Kentucky Geological Survey.") The theodolite used for this work was six inches, No. 48, of Windeman make, Washington City, D. C. Reading to 02′. The chain was of steel, thirty-three feet, and adjusted by Christuman's patent two pole steel tape.

Two sets of observations were made to determine the magnetic declination. It had been contemplated to make observations for declination at distances not greater than six miles apart, along the entire line, but it was found to consume too much time, and the weather being frequently unfavorable for such observations, those at No. 1,192 and at No. 3,326 were the only ones found to be practicable. The first gave 5° 58′ 4″ east; the last 6° 22′ 3″ east.* The direction of the line was checked, from time to time, by observations for azimuth, and such corrections made as were necessary.

The distances made by the line are kept in feet. The base line was carried from Uniontown, Union county, to Wm. Smith's farm, near the Hawesville and Hartford road, Hancock county, 322,975 feet, or sixty-one miles eight hundred and eighty-five feet. The line was intended to run due east from the initial point to the Virginia state line, near the corner of Pike county, Kentucky.

*This should be repeated; it is undoubtedly too great, unless part of the effect be attributed to local attraction.

At 187,369 feet, crossing of Green river.

At 233,300 feet, Center of third street, Owensboro'.

At 269,280 feet, Panther creek.

At 296,318 feet, Knottsville.

The plan of the base line will be plotted and ready for the report of the operatives of this summer, 1857.

The map of Hopkins, part of Christian, and part of Muhlenburg, and the map of Greenup, part of Carter counties, &c., are also in a state of forwardness, and will also be ready for the engraver in July or August.

# INDEX.

www.ingramcontent.com/pod-product-compliance
Lightning Source LLC
LaVergne TN
LVHW020126110826
845151LV00001B/285

* 9 7 8 1 4 2 5 5 4 2 3 5 1 *